Springer Proceedings in Mathematics & Statistics

Volume 285

Springer Proceedings in Mathematics & Statistics

This book series features volumes composed of selected contributions from workshops and conferences in all areas of current research in mathematics and statistics, including operation research and optimization. In addition to an overall evaluation of the interest, scientific quality, and timeliness of each proposal at the hands of the publisher, individual contributions are all refereed to the high quality standards of leading journals in the field. Thus, this series provides the research community with well-edited, authoritative reports on developments in the most exciting areas of mathematical and statistical research today.

More information about this series at http://www.springer.com/series/10533

Maria José Pacifico · Pablo Guarino
Editors

New Trends in One-Dimensional Dynamics

In Honour of Welington de Melo
on the Occasion of His 70th Birthday
IMPA 2016, Rio da Janeiro, Brazil,
November 14–17

 Springer

Editors
Maria José Pacifico
Federal University of Rio de Janeiro
IM-UFRJ, Institute of Mathematics
Rio de Janeiro, Brazil

Pablo Guarino
Institute of Mathematics and Statistics
Fluminense Federal University
Niterói, Rio de Janeiro, Brazil

ISSN 2194-1009   ISSN 2194-1017   (electronic)
Springer Proceedings in Mathematics & Statistics
ISBN 978-3-030-16835-3   ISBN 978-3-030-16833-9   (eBook)
https://doi.org/10.1007/978-3-030-16833-9

Mathematics Subject Classification (2010): 37E05, 37E10, 37E20, 37F10, 37F45, 37C05, 37D25, 37A35, 37A25

Preface

This volume presents the proceedings of the meeting New Trends in One-dimensional Dynamics, celebrating the 70th anniversary of Welington de Melo, which was held at IMPA, Rio de Janeiro, November 14–18, 2016. The occasion was particularly gratifying for us because of the active participation of a number of experts in this field, many of Welington de Melo's co-authors and all his former doctoral students. Collecting the articles for this volume was initially intended as an opportunity to celebrate the success of the meeting and Welington de Melo's joy at the excellent scientific level of the lectures and the friendly atmosphere that week. Unfortunately, Welington de Melo passed away a month after the meeting. He is sorely missed by the Brazilian mathematical community, and the publication of this volume became more a tribute to Welington de Melo. His role in the development of mathematics is indisputable, especially in the area of low-level dynamics, and his legacy includes, in addition to numerous articles with fundamental contributions, books that are mandatory references for beginners in this area. Welington had only seven formal Ph.D. students, and we were both honored of being two of them. It is worth mentioning that Artur Avila, the 2014 Fields Medal winner, was also Welington's Ph.D. student.

We are including in this volume an obituary by Sebastian Van Strien and Edson de Faria, two dear friends and co-authors; an article by Jacob Palis and Fernando Lenarduzzi describing how Welington came to know his Ph.D. Advisor Jacob Palis at IMPA, how he began his successful scientific career, and comments on the outcomes of his thesis; the first of his published articles; and photos of Welington taken throughout his life.

We are indebted to our colleagues and friends for their contributions, which made the meeting possible. We are also indebted to Leticia Ribas and Renata Maiato, from the Department of Events at IMPA, who helped with the meeting's logistics and secretarial work. We would like to express our deepest gratitude to all of them.

Niterói, Brazil
Rio de Janeiro, Brazil

Pablo Guarino
Maria José Pacifico

Contents

Welington de Melo and Jacob Palis: Their First Meeting, Some of Their Work on Structural Stability and a Lifetime of Friendship

Jacob Palis and Fernando Lenarduzzi

Abstract The present publication is a symbol of the appreciation of those who attended to the conference in memory of Welington's important contribution to the theory of Dynamical Systems. They have sailed in Angra, proved theorems and wrote a book together, but the most valuable thing for them was the friendship they shared. Throughout the text, we will tell the story of how they became friends and scientific partners, telling a bit about Welington's thesis and its relation to the Stability Conjecture.

Keywords Stability Conjecture · Structural Stanility and Axiom A

1 Brave and Bold

How did Jacob first meet him? Welington was his first Ph.D. student, an excellent one, whose participation in IMPA's Dynamical Systems Seminars was always precious. The seminars started upon Palis' return from the University of California Berkeley, in 1969, where he got his Ph.D. under the guidance of Steve Smale that had recently been awarded with the Fields Medal.

The Seminar that Jacob organized in that occasion took place in week days, including Saturday mornings. In one of those Saturdays, Welington knocked at Jacob's door at IMPA and presented himself as a student accepted by IMPA to obtain a Master Degree. He informed Palis that he had attended Elon Lima's course in a recent Brazilian Mathematical Colloquium in Poços de Caldas. He added that he would very much like to study Dynamical Systems to obtain a Ph.D. in this area. At first Jacob had some doubts but the resemblance of his own case at the University of California in Seminars removed those doubts.

J. Palis (✉) · F. Lenarduzzi
Instituto de Matemática Pura e Aplicada-IMPA, Rio de Janeiro, Brazil
e-mail: jpalis@impa.br

F. Lenarduzzi
e-mail: fernl@impa.br

© Springer Nature Switzerland AG 2019
M. J. Pacifico and P. Guarino (eds.), *New Trends in One-Dimensional Dynamics*, Springer Proceedings in Mathematics & Statistics 285,
https://doi.org/10.1007/978-3-030-16833-9_1

Curiously, one Saturday, when returning home from one of those seminars, in the neighborhood of Laranjeiras, Palis unexpectedly met José Pelúcio Ferreira, who was working at FINEP, a Funding Authority for Studies and Projects. He told Jacob that he was very surprised to learn of IMPA's Seminars on Saturday mornings and that he would consider the case of some financial support for such promising scientific activity. Palis was, of course, very happy with that possible support and so, as soon as he got home, he phoned Maurício Peixoto, Elon Lima and the Director of IMPA, Lindolpho de Carvalho Dias.

Jacob also looked for Welington's colleagues at the University of Minas Gerais to obtain further information about his performance as a student there. That was when he was informed that de Melo was an excellent student and very determined to learn new relevant topics in mathematics. With all the positive reactions Palis have received, he asked Welington to present one of the important Seminar topics. It was a success! With that, Jacob have formalized his request to the Direction of IMPA to accept de Melo as a Ph.D. student, under his guidance. A couple of weeks after that, Welington came to Jacob's office to say that he had discovered a serious gap in one of his mathematical papers. Happily, it turned out it was not a serious gap.

In two years Welington de Melo concluded his Ph.D. with a beautiful thesis denominated "Structure Stability in 2 Dimensional Manifolds", that was considered excellent by a number of mathematicians. It was published in the high level journal Inventiones Mathematicae [9].

Right after the presentation and overwhelmingly approval of Welington's thesis, Palis proposed him to be hired by IMPA as a researcher. Actually he had, indeed, a brilliant performance at IMPA since the beginning. After some time, he focused his studies especially in unidimensional dynamical systems, becoming a high level mathematician guiding excellent younger researchers. A very exceptional example is that of Artur Avila that some years later received the famous Fields Medal.

Welington's scientific production was indeed continuous and profound. Jacob always says that it is a pleasure to point out the good international repercussion of the book "Introdução aos Sistemas Dinâmicos" that they wrote together. Welington's text with Sebastien Van Strien on unidimensional dynamics have also become a basic reference in this topic.

With all these successful achievements, they realized that they could indeed organize an excellent Symposium on Dynamical Systems, which took place in Salvador in 1971. It represented a very special moment to the consolidation of IMPA's important place in the Brazilian intellectual structure. In fact it had the participation of worldwide outstanding mathematicians such as Jurgen Moser, Maurício Peixoto, Rene Thom, Christopher Zeeman, Sheldon Newhouse, Floris Takens, John Mather, as well as young talented mathematicians like, Ricardo Mañé, Jorge Sotomayor, César Camacho, Welington de Melo himself, among others. The Proceedings of this important Congress was edited by Maurício Peixoto and published in 1973 by Academic Press, which has one preview of Welington's thesis. This Congress launched the development of several other areas of research in mathematics at IMPA.

In 2006 another congress took place in Salvador, celebrating Welington 60th anniversary, under the organization of his first Ph.D. student, Maria José Pacifico.

It is important to stress how meaningful the presence of his wife Gilza de Melo guided all aspects of Welington's life, including the scientific one.

2 The Stability Conjecture and Welington

Here we want to say something about Welington's early work and to say some words about how the subject he worked with still evolving. We want to link what was his first contribution to research and first steps to become the successful mathematician he would come to be. The key words to his thesis are *structural stability*.

Just to be clear, we would like to recall some basic concepts to make this text fully comprehensible and establish some notation. We say that a diffeomorphism $f : M \to M$ is C^r structurally stable if there exists a C^r-neighborhood of f such that every element g of it is topologically conjugated to f, that is, there exists an homeomorphism h of M such that $f \circ h = h \circ g$. We will denote by $Per(f)$ and $\Omega(f)$ the set of periodic points of f and the set non-wondering points f, respectively.

One of the first works that dealt with this subject we can cite is the one Peixoto proved in [5] the stability for flows in two manifolds, however the proof depends highly on the geometry of two dimensions of the manifolds.

We can also say something about Anosov's work [1], in which he proves the structural stability for diffeomorphism and vector fields when the whole manifold M has a hyperbolic structure, the Anosov systems.

Then we had the work developed by Palis and Smale They proved in [4] the structural stability of diffeomorphisms and flows such that the non-wandering set $\Omega(f)$ is a finite number of orbits and the *strong transversality condition* is satisfied, that is, for all $x, y \in \Omega(f)$, we have that stable manifold $W^s(x)$ and the unstable manifold $W^u(y)$ have a transversal intersection. The precise statement is the following

Teorema A (Palis-Smale) *Let the set of non-wondering points $\Omega(f)$ to be finite and hyperbolic. Also, assume that it has the strong transversality condition. If $f \in Diff^r(M)$ then f is structurally stable.*

The idea consists into using the simple structure of the non-wondering set find fundamental domains and unstable disks, to get a foliation in a neighborhood around each one of the orbits in $\Omega(f)$. The way these foliations are constructed is made compatible to the iterates, the so called *compatible tubular families*.

Definition An unstable tubular family for Λ is a continuous foliation $\mathcal{F}^u$ in a neighborhood V of Λ that satisfies the following properties

(i) the leaves are C^r manifolds immersed in M and transversal to the stable manifolds to the points in Λ;
(ii) the leaf that contains a point $x \in \Lambda$ is the connected component of $W^u(x) \cap V$ containing x;
(iii) the tangent spaces of each one of the leaves vary continuously;

(iv) the foliation is f invariant: $f^{-1}(F^u(x)) \subset F^u(f^{-1}(x))$, where $F^u(x)$ is the leaf that contains x.

The proof lies into using the tubular families as a system of coordinates to do the conjugacy for two diffeomorphisms.

However, the construction of the neighborhood lies in the finitude of the non-wondering set and that is one of the fundamental steps Welington was able to take in his thesis. He was able to mimic the construction of the compatible tubular families into a more general setting: two-dimensional manifold and a diffeomorphism with the Axiom A property, that is, $\Omega(f)$ is hyperbolic and $\overline{Per(f)} = \Omega(f)$.

Teorema B (de Melo) *If M is two-dimensional and $f \in Diff^1(M)$ is Axiom A, satisfying the strong transversality condition, then f is structurally stable.*

The proof here is based in the geometry and the dimension 2 and it is not clear how to take the same ideas to higher dimensions. Welington's remarkable thesis also had two other stability results about the stability of proper hyperbolic sets.

Teorema C *Let M be a n-dimensional manifold, $f \in Diff^1(M)$ Axiom A and Λ be an attractor of f whose stable manifolds have co-dimension 1. Then f is locally stable with respect to Λ.*

Teorema D *Let M be a two-dimensional manifold, $f \in Diff^1(M)$ Axiom A and Λ be a basic set of f, then f is locally stable with respect to Λ.*

Using the same ideas of construction the compatibility, in 1971 Robbin proved that if f is C^2 then hyperbolicity is a sufficient condition for structural stability [7], and in 1976 Robinson reduced the C^2 hypothesis to C^1 [8].

Recall the conjecture stated by Palis and Smale in their paper in 1968:

Conjecture (The Stability Conjecture) *$f \in Diff^r(M)$ is structurally stable if and only if f satisfies*

(a) *Axiom A: $\Omega(f)$ is hyperbolic and the set of periodic points is dense in $\Omega(f)$;*
(b) *Strong Transversality Condition: for all $x, y \in \Omega(f)$, we have that $W^s(x)$ and $W^u(y)$ have a transversal intersection.*

As we already mentioned, some beautiful and meaningful mathematics was made trying to prove this conjecture. So far, we have only talked about the implication that tries to prove that "Axiom A implies Stability" and one the most remarkable work that explores the converse was written by Mañé in [3].

Teorema E (Mañé) *Every C^1 structurally stable diffeomorphism of a closed manifold of any dimension satisfies Axiom A.*

The C^r structural stability conjecture for $r \geq 2$ remains wide open. Every technique known today is restricted to the C^1 topology. The C^1 perturbations with controlled dynamical properties is an obstacle that holds progress in the $r \geq 2$ direction:

the closing lemma, connecting lemma and Frank's lemma are either unknown or they are false in higher topologies.

The conjecture itself can still inspire some interesting and relevant mathematics. Instead of trying to adapt the C^1 techniques to higher dimensions, one could try to attack the conjecture into different approaches, surfaces or using new types of systems that emerged, as proposed by Pujals [6] and [2] for example.

References

1. D. Anosov, *Geodesic Flows on Closed Riemannian Manifolds of Negative Curvature*, (PYOC. Steklow Inst. 90, 1967), Amer. Math. Sot. Transl., 1969
2. S. Crovisier, E. Pujals, *Strongly Dissipative Durface Diffeomorphisms*, arXiv:1608.05999
3. R. Mañé, *A Proof of the C^1 Stability Conjecture*, Inst. Hautes Études Sci. Publ. Math. 66 (1988), pp. 161–210
4. J. Palis, S. Smale, *Structural Stability Theorems*, 1970 Global Analysis, vol. XIV (Proc. Sympos. Pure Math., Berkeley, Calif., 1968), Amer. Math. Soc., Providence, R.I, pp. 223–231
5. M. Peixoto, Structural stability on two dimensional manifolds. Topology **1**, 101–120 (1962)
6. E. Pujals, Some simple questions related to the C^r stability conjecture. Nonlinearity **21**, T233–T237 (2008)
7. J. Robbin, A structural stability theorem. Ann. Math. **94**, 447–493 (1971)
8. C. Robinson, Structural stability of C^1 diffeomorphisms. J. Diff. Equ. **22**, 28–73 (1976)
9. W. de Melo, Structural stability of diffeomorphisms on two-manifolds. Invent. Math. **21**, 233–246 (1973)

Welington de Melo (1946–2016)

Edson de Faria and Sebastian van Strien

Abstract In memorian: Welington de Melo

Keywords One-dimensional dynamics

Classifications 01A70

Welington de Melo passed away on December 21st, 2016, shortly after a conference held at IMPA celebrating his 70th birthday. The present volume was originally meant to be a tribute in life to Welington's accomplishments as a mathematician, but sadly it has turned instead into a memorial volume. The world has lost a great mathematician, a leader in the area of Dynamical Systems, and we have lost a great colleague, a mentor, and a dear friend.

Welington was born on November 17th, 1946, in the city of Guapé, Minas Gerais. He became an electrical engineer in 1969, but found mathematics more interesting and completed his Ph.D. under the supervision of Jacob Palis at IMPA in 1972. His thesis dealt with the structural stability, via geometric methods, of Axiom A maps on two-dimensional manifolds. After his Ph.D. at IMPA, he held a postdoc position for two years with Steve Smale in Berkeley, stayed for a few months in Warwick, and for the remainder of his mathematical career worked at IMPA.

Throughout his career, Welington wrote 39 papers and 5 books (including an unpublished book on differential topology). Welington believed that a mathematical paper should present results that are as complete as possible, and disliked accordingly

E. de Faria (✉)
University of São Paulo-USP, São Paulo, Brazil
e-mail: edson@ime.usp.br

S. van Strien
Imperial College London, London, UK
e-mail: s.vanstrien@imperial.ac.uk

© Springer Nature Switzerland AG 2019
M. J. Pacifico and P. Guarino (eds.), *New Trends in One-Dimensional Dynamics*, Springer Proceedings in Mathematics & Statistics 285,
https://doi.org/10.1007/978-3-030-16833-9_2

the idea of publishing mere partial results. Hence his papers tended to be few in number but quite substantial, in content as well as in length.

His first book was written in collaboration with J. Palis, originally in Portuguese [1] and later translated into English [2], Russian [3] and Chinese. It became a very influential book in the area of Dynamical Systems, and in particular a standard textbook reference on structural stability. His second book [4], written in collaboration with one of us (SvS), represented at the time of publication the state of the art in One-dimensional Dynamics, and it is still regarded as the most complete and authoritative reference on the subject. Welington also wrote two other books with one of us (EdF) on two completely different subjects [5–8].

Those of us who knew Welington well have always been impressed by his intellectual integrity, his frankness, and his overall honesty. Dennis Sullivan adds the adjective *relentless*. In pursuing the details of the proof of period doubling rigidity, Welington was relentless about understanding and confirming every point. Indeed, as Dennis recalls, the last detail was actually completed by Welington himself at the Orsay pool with his observation "but that map is injective so by Koebe Distortion …" at which point Dennis said "Yes, bingo!". For Welington, the most important thing was always the pursuit of truth and excellence, in Mathematics and elsewhere. He will be sorely missed.

1   His Mathematical Contributions

What follows is a brief account of Welington's major accomplishments as a mathematician.

1.1   Structural Stability: Thesis

One of the main problems in dynamical systems is to describe the systems that are structurally stable. In the early 1970s this led to:

Theorem 1 (Palis, Palis-Smale, Robin, de Melo, Robinson) C^1 *diffeomorphisms on a compact manifold that satisfy Axiom A (a hyperbolicity assumption) and the strong transversality condition are structurally stable within the set of C^1 diffeomorphisms.*

This goes back to Jacob Palis' 1968 thesis in which he proved this for C^1 Morse-Smale systems in dimension two, and a year or so later this was extended by Jacob Palis and Steve Smale to arbitrary dimension. This work introduced the notions of tubular families (foliations in neighbourhoods of the stable and unstable manifolds).

About the same time, Joel Robin in the C^2 topology and somewhat later Clark Robinson in the C^1 topology proved the above theorem using a functional analytic approach (in the spirit of Anosov, Moser and others).

The functional analytic approach is appealing but is less flexible than the geometric approach. This was the inspiration for *Welington's thesis* (which predates the functional analytic proof in the C^1 setting). In his beautiful thesis [9], which was published in Inventiones Mathematicae, Welington was able to show that the geometric approach works for C^1 Axiom A maps in the two-dimensional setting (or rather in the case where stable foliations are codimension one).

1.2 Economics (1974–1979)

Welington was very much influenced by Steve Smale who in the early 1970s was quite interested in economics. Inspired by Steve's work, Welington studied the problem of simultaneously optimizing several functions. Forever the pure mathematician, he wrote that this 'is a natural question to consider, even in some basic models in economics'.

More formally, let $f = (f_1, \ldots, f_d): M^n \to \mathbb{R}^d$ be a C^∞ mapping on the compact n-dimensional manifold-without-boundary M. A point p in M is a *local Pareto optimum (LPO)* of f if there is a neighborhood U of p in M such that

$$q \in U, \ f_i(q) \geq f_i(p) \ \forall i \implies f(q) = f(p).$$

A smooth curve $a: [0, 1] \to M$ on which each f_i is strictly increasing is called an *admissible curve* for f.

One result of Welington in this direction is [12]:

Theorem 2 (1976, Welington de Melo) *For a generic f, there exist an admissible curve from any point q in M to some LPO provided* dim $M > \max\{d - 1, 2d - 4\}$.

The proof uses a combination of the singularity theory of maps and ideas from dynamical systems.

1.3 Geometric Theory of Dynamical Systems (1978)

After returning to IMPA, Welington started to write lecture notes on dynamical systems together with Jacob Palis [1]. No doubt the resulting text book *Geometric Theory of Dynamical Systems* was among the most influential text books in dynamical systems. It was translated into English [2], Russian [3] and Chinese. Without this book, dynamical systems would have become a much more disjointed discipline, because unlike many textbooks it was also showing a research programme.

1.4   Moduli of Stability (1979–1987)

What do we know about maps which are not structurally stable? Can we classify conjugacy classes of such maps? Jacob Palis had initiated this research programme in the mid 70s. Many people, in particular Welington but also Floris Takens (the thesis advisor of SvS) had started working on this.

More precisely, a diffeomorphism $f \in \text{Diff}^\infty(M)$ is said to have *modality k* if this is the minimal integer for which there exists a small neighbourhood U which contains a countable number of k-parameter C^1 families of diffeomorphisms such that each diffeomorphism in U is conjugate to at least one diffeomorphism from these families. In particular, if f is structurally stable or if there are at most a countable number of different conjugacy classes in a neighbourhood U of f then the number of moduli is defined to be *zero*.

An example where one needs a one-parameter of diffeomorphisms to parametrize conjugacy classes is when one has a two-dimensional diffeomorphism f with saddle-points p and q such that

$$W^u(p) \text{ and } W^s(q) \text{ have a quadratic tangency.}$$

Jacob Palis showed that such a map has modality ≥ 1: if another nearby diffeomorphism $\tilde{f}$ also has a quadratic tangency then

$$\frac{\log|a|}{\log|b|} = \frac{\log|\tilde{a}|}{\log|\tilde{b}|}$$

where $a, \tilde{a}$ and $b, \tilde{b}$ are the eigenvalues corresponding to $W^s(p)$, $W^s(\tilde{p})$ resp. $W^u(q)$, $W^u(\tilde{q})$.

Let $\mathcal{A}$ be the space of Axiom A diffeomorphisms on compact surfaces. In [21], SvS and Welington managed to describe the subspace of $\mathcal{A}$ of diffeomorphisms with finite modality.

Theorem 3 (1987, Welington de Melo and SvS) *If $f \in \text{Diff}^\infty(M^2)$ is in $\mathcal{A}$, then f has finite modality if and only if $f \in \mathcal{M}$ where $\mathcal{M}$ is defined below.*

Here $f \in \mathcal{M} \subset \mathcal{A}$ whenever f has no cycles and moreover

- if $x, y \in \Omega(f)$ are such that $W^u(x)$ is not transverse to $W^s(y)$, then the basic sets containing x and y consist of periodic orbits;
- there is only a finite number of nontransversal intersections between stable and unstable manifolds and the contact between these manifolds along each of these orbits is of finite order;
- if $p, q \in \text{Per}(f)$ are such that $W^u(p)$ has an orbit of nontransversal intersection with $W^s(q)$ then the number of orbits in $W^s(p)$ [resp. in $W^u(q)$] belonging to some unstable [resp. stable] manifolds of a periodic saddle point of f is finite;

- if x is a point of nontransversal intersection of $W^u(p)$ and $W^s(q)$ then there exists an arc Σ transversal to $W^u(p)$ at x such that no connected component of $\Sigma - \{x\}$ contains points of both stable and unstable manifolds of saddles;
- if $W^u(p)$ has a point of nontransversal intersection with $W^s(q)$, and $W^u(q)$ has a point of nontransversal intersection with $W^s(r)$, then there is no saddle point of f whose unstable manifold [resp. stable manifold] intersects $W^s(p)$ [resp. $W^u(r)$].

Since every diffeomorphism in $\mathcal{A}$ satisfying the transversality condition is structurally stable, Theorem 3 generalizes the structural stability result for two-dimensional manifolds.

The proof of this theorem is interesting and builds on the geometric construction that Welington knew so well from his Ph.D. The flexibility of the 'geometric approach' in constructing topological conjugacies was crucial in being able to tackle the types of tangencies of invariant foliations that occur.

1.5   One Dimensional Dynamics (1987–2016)

Inspired by the following results, Welington realised in the late 1980s that it was time to concentrate on interval dynamics:

- May's paper on chaos had made the study of iterations of $x \mapsto ax(1-x)$ popular, but the number of results was still small;
- Milnor-Thurston's combinatorics paper;
- the introduction of Schwarzian derivative;
- the Feigenbaum-Coullet-Tresser conjectures on renormalization had been proved in special cases;
- Sullivan had introduced the notion of quasiconformal maps into iterations of holomorphic maps, and was working hard to prove his theorem on renormalization.

Welington's intuition was right because since that time a rich, beautiful and remarkably complete theory has emerged:

- a complete combinatorial description (Milnor-Thurston theory);
- a complete topological description (Denjoy-Fatou-Sullivan theory in the real case);
- a measure-theoretic description (number of ergodic components, existence and non-existence of absolutely continuous invariant measures);
- geometric rigidity (real bounds, quasiconformal rigidity, density of hyperbolicity);
- a fairly complete renormalization theory;
- results towards the Palis conjecture in dimension 1: prevalence of systems which are hyperbolic or admit absolutely continuous invariant measures.

In many of these developments Welington played a major role.

Let us describe Welington's first result in this area. Consider maps $f: M \to M$ where $M = S^1$ or $M = [0, 1]$. What can you say about their dynamics? One of the most basic results is the absence of wandering intervals, as was proved by Denjoy for circle diffeomorphisms. The analogue of this result for smooth unimodal interval maps was proved in [24]:

Theorem 4 (1987, de Melo and SvS) *If $f : M \to M$ is unimodal, C^3 and satisfies a non-flatness condition then it has no wandering intervals.*

Under the assumption of negative Schwarzian, this result was proved previously by Guckenheimer. The main contribution of the above paper was the cross-ratio (inspired by a paper of Yoccoz who had previously used a particular cross-ratio for his study of smooth circle maps with critical points). The cross-ratio came to be a standard tool in one-dimensional dynamics, because it can be used to control the non-linearity of high iterates of a map.

Encouraged by the above result, and the work of Sullivan, Welington and SvS continued working in this direction.

In 1918 Julia gave the following description of the dynamics of rational maps on the Riemann sphere: there are periodic domains attracting an open set of points, and outside this set there is a closed set (the Julia set), and may be some open sets which 'wander' and which are attracted to this Julia set. The remaining problems were:

- Can a rational map have infinitely many periodic domains?
- Do such wandering domains exist?

The answer in both cases is no (Fatou, 1919) and (Sullivan, 1985).

In the circle diffeomorphism case the corresponding results were obtained by Poincaré and Denjoy. For general one-dimensional (piecewise monotone) maps the combinatorial study was done by Milnor and Thurston (and others). Building on work of Denjoy, Guckenheimer, de Melo and SvS, Blokh and Lyubich, Lyubich, Welington, together with SvS and his Ph.D. student Marco Martens were able to prove the following [26]:

Theorem 5 (1992, Martens, de Melo and SvS) *If $f : M \to M$ is C^2 and satisfies a non-flat condition then*

- *it has no wandering intervals*
- *the period of periodic attractors is bounded.*

Subsequently, a completely different proof of this result was given by Edson Vargas, a former Ph.D. student of Welington, and SvS. This approach also provides 'real' bounds, that one can get rid of the Schwarzian derivative condition in the most general setting.

1.6 One-Dimensional Dynamics: The 1993 Monograph

Welington always felt that one should spend time to obtain optimal results and proofs, and that the success of a research area depended on well-written text books which also described a research programme. He convinced SvS to join in this endeavour and together we spend a couple of years to write the research monograph 'One-dimensional dynamics. We tried very hard to make the material up-to-date. Of course one of the hardest, and most interesting parts of working on this book was to extract from Dennis Sullivan his proof on renormalization. Welington (and also SvS) visited Dennis many times in New York at CUNY, and in Paris at IHES. Typically, Dennis would first explain something to Welington who then would try it out on SvS.

To our delight, the resulting monograph [4] was well received. Over the last decade, Welington and SvS often discussed a new edition of the monograph which was to include the exciting developments since the early 1990s.

1.7 Renormalization and Rigidity: Circle Maps (1994–2016)

The phenomenon of *rigidity* was first observed in Geometry, through the work of G. D. Mostow, and was later seen to be present in Dynamics as well. In the latter context, rigidity means that a small number of dynamical invariants completely determines the fine scale structure of orbits. Thus, roughly speaking, the rigidity paradigm states that, under a minimum number of preset conditions, maps that are topologically conjugate are in fact more smoothly conjugate. In low-dimensional dynamics, especially in dimension one, rigidity is oftentimes achieved through the study of an underlying *renormalization operator*. Renormalization of a dynamical system means a suitably rescaled first return map to a neighborhood of a special point in phase space—usually a critical point. The starting point—in the late 1980s—for the modern study of rigidity and renormalization (in the context of one-dimensional real or complex systems) was Sullivan's theorem stating that any two period-doubling quadratic-like maps are always quasiconformally conjugate.

For *critical circle maps* (with a single critical point of power-law type), the only invariant is the rotation number—this is a fundamental result due to J-C. Yoccoz. But there are additional difficulties. Since the first return map to a neighborhood of the critical point on the circle is discontinuous, one has to work with *commuting pairs* of maps—as introduced by O. Lanford. The notion of *holomorphic commuting pair*, introduced in EdF's thesis in 1992, opened the door for the use of Sullivan's holomorphic and quasi-conformal methods in the circle setting. Welington was very much interested in this work, and soon afterwards he and EdF embarked on a project for understanding more deeply the rigidity, universality and renormalization convergence of critical circle maps. Building on the complex bounds first obtained in EdF's thesis and later generalized by Yampolsky, they obtained the following result [32, 34].

Theorem 6 (2000, W. de Melo and EdF) *Any two real-analytic critical circle maps with the same rotation number of bounded type and the same (odd) power-law at their critical points are $C^{1+\alpha}$ conjugate.*

This may be regarded as the analogue of Herman's KAM theorem for circle diffeomorphisms. This result was later extended by Khanin and Teplinskii, Khmelev and Yampolsky to cover all rotation numbers, not just bounded type. The price to pay is that one can only deduce that the conjugacy is C^1. There are good reasons as to why this is the best one can expect: in [32], the authors constructed C^∞ counterexamples to $C^{1+\alpha}$-rigidity, and later Ávila constructed real-analytic counterexamples.

The main ingredient in the proof of Theorem 6 is to show that one has exponential convergence to zero of the C^0 distance between the successive renormalizations of both maps. This uses an adaptation of the *tower construction* of McMullen to holomorphic pairs. In fact, in his 1998 Fields Medal address, McMullen cited Theorem 6 as one of the main applications of his theory.

Theorem 6 was generalized for C^3 critical circle maps by P. Guarino in his thesis, written under Welington's supervision, and this resulted in a very nice paper by Guarino and Welington [46]. Subsequently, in [47], Guarino, Martens and Welington removed the hypothesis of bounded type on the rotation number, at the cost of requiring one more degree of differentiability for the maps. Thus, the final result can be stated as follows.

Theorem 7 (2018, W. de Melo, P. Guarino, M. Martens) *Any two C^4 critical circle maps with the same irrational rotation number and with a unique critical point of the same odd power-law type are conjugate by a C^1 diffeomorphism. Morevoer, there exists a full-measure set of rotation numbers (containing those with bounded type) for which the conjugacy is in fact $C^{1+\alpha}$.*

This can be regarded as the state of the art concerning renormalization and rigidity of critical circle maps.

1.8 Hyperbolicity of Renormalization: Unimodal Maps (1999–2006)

In the late 90s, Welington teamed up with EdF and Alberto Pinto in order to establish the global hyperbolicity of the renormalization operator in the context of unimodal maps having a finite degree of smoothness—this had been conjectured almost 20 years earlier by Lanford. The basic idea was to combine Lyubich's breakthrough concerning the hyperbolicity of renormalization in the space of quadratic-like germs with a strong generalization of certain non-linear functional-analytic techniques first developed by A. Davie in the context of Lanford's period-doubling renormalization. Several technical difficulties had to be overcome, mostly having to do with the fact that, in the world of C^r maps, the renormalization operator is not Fréchet differentiable—so that the notion of *hyperbolicity* a-priori does not even make sense. In the end, a full proof of Lanford's conjecture in the bounded combinatorics setting was obtained. In very informal terms, the result can be stated as follows (see [39]).

Theorem 8 (2006, W. de Melo, E. de Faria, A. Pinto) *If $r \geq 2 + \alpha$, where $0 < \alpha < 1$ is close to 1, the limit set of the renormalization operator in the space of C^r unimodal maps of bounded combinatorial type is a hyperbolic Cantor set where the operator acts as the full shift in a finite number of symbols. In addition, for points of such limit set,*

(i) *The local unstable manifolds are real analytic curves;*
(ii) *The local stable manifolds are of class C^1, and together they form a continuous lamination whose holonomy is $C^{1+\beta}$ for some $\beta > 0$;*

In [39], the authors also proved that the *global* stable manifolds of renormalization are of class C^1. This was done combining Theorem 8 with the implicit function theorem in Banach spaces, at the expenses of losing one more degree of differentiability (*i.e.*, assuming that $r \geq 3 + \alpha$ with α close to one).

1.9 Interval Maps: Stochasticity (2002–2003)

Another question is the following: How typical is it for a map from the quadratic family to be either

- stochastic, or
- hyperbolic.

A result in this direction was proved by Lyubich.

Theorem 9 (2002, M. Lyubich) *For Lebesgue almost all a, either $f_a(x) = ax(1 - x)$ is hyperbolic or a summability condition is satisfied.*

By a result of Martens and Nowicki this summability condition implies a summability criterion for the existence of absolutely continuous invariant measures due to Nowicki and SvS.

As an aside, no growth condition is needed. A recent paper by Bruin, Shen and SvS shows that for any map (possibly multimodal) the summability criterion can be replaced by something much weaker: for any f without periodic attractors there exists $C > 0$ so that if for any critical point c one has $|Df^n(f(c))| \geq C$ then f has an acip.

As one of the outcomes of a very successful collaboration, Artur Ávila, Welington and Misha Lyubich proved the following:

Theorem 10 (2003, A. Ávila, W. de Melo, M. Lyubich) *Within any non-trivial real-analytic family of quasiquadratic maps, for almost all parameters the above dichotomy holds.*

The proof is based on the following ingredients:

- The space of analytic maps is foliated by codimension-one analytic submanifolds, 'hybrid classes': maps in these submanifolds are topologically conjugate and, in the hyperbolic case, they have the same multiplier at attracting cycles.
- hybrid classes laminate a neighbourhood of any non-parabolic map.
- This allows the authors to transfer the regular or stochastic property of the quadratic family to any nontrivial real analytic family: The holonomy along the foliation occurs in a quasisymmetric manner, since it comes from holomorphic motions parametrized by a complex Banach ball. But in general it is not absolutely continuous. However, the holonomy respects the property of exponential decay of the parapuzzle geometry (and therefore the Martens-Nowicki criterion).
- The main point is a construction of a transverse vector field v at f (crossing hybrid classes transversally). A solution α of the equation

$$v(z) = \alpha(f(z)) - f'(z)\alpha(z)$$

yields an infinitesimal quasiconformal change of coordinates used to perform an infinitesimal change of f in the direction of the vector field v. One finds v such that the equation does not have a solution, i.e. v cannot be "horizontal" (tangent to hybrid classes).
- To construct v, they start with a smooth vector field v, holomorphic on the critical value puzzle piece U_1 and vanishing on other preimages of a central puzzle piece U_0 leading to U_1. The fact that one has large scales (since the critical point is quadratic) and a Key Lemma are used to show that this vector field cannot be horizontal.
- They then approximate v by a holomorphic vector field using Mergelyan's theorem.

1.10 Two More Books (2001–2010)

In the beginning of 2001, Welington took up the task of writing notes for a short course to be delivered at the 23rd Brazilian Math Colloquium at IMPA later that year. Soon after starting on his own, he decided to invite one of us (EdF) to join in the project, and the set of notes was published as [5]. A few years later, these notes were expanded and polished to a full-size book. The goal of the book was to present the most important mathematical tools of a holomorphic and/or quasiconformal nature that are used in the study of one-dimensional dynamical systems. This book was published by Cambridge in 2008 (see [7]).

In the meantime, Welington became interested in Mathematical Physics, more precisely in Quantum Field Theory (QFT), thanks primarily to several conversations with Dennis Sullivan. In order to learn more about the subject, he decided to teach a summer course about it. He eventually convinced EdF to embark on the project of writing a longish set of notes [6] for a course to be delivered at the 26rd Brazilian Math Colloquium at IMPA, in 2007. The idea was to write a book which would help clarify many of the difficult ideas appearing in QFT in a way that would be accessible and make sense to mathematicians. This was quite an adventure, since both authors were coming from outside Mathematical Physics. This book, too, was published by Cambridge, in 2010 (see [8]).

In fact, the story behind these two Cambridge books is a bit more non-linear than the previous two paragraphs suggest. In 2006, during the ICM held in Madrid, Welington met David Tranah, the editor at Cambridge University Press. In their conversation, Welington mentioned that he and EdF were writing a book on mathematical aspects of QFT. Tranah said that he might be interested in publishing such a book. Upon his return to Brazil, Welington urged EdF to write to Tranah. In the e-mail exchange that followed, EdF mentioned the first book (on holomorphic dynamics), and Tranah told him that he might be interested in publishing that one too.

1.11   Prizes

Welington was elected to the Brazilian Academy of Sciences in 1991, gave a talk at the International Congress of Mathematics in 1998 and became a member of the Third World Academy of Science in 2003.

2   Welington, the Person

Welington met his wife Gilza while he was still quite young. They were inseparable, truly adored each other and even felt comfortable squabbling in front of others. No doubt, Gilza was instrumental in Welington's success.

Usually Welington and Gilza lived relatively modestly. However, occasionally Welington would love to eat out in famous restaurants and talked about these memorable meals for years after.

Welington loved to be near the sea and in particular to sail and the sense of adventure and freedom. As soon as he could afford it, he and Gilza bought a weekend condominium in Angra dos Reis together with a sailing boat. Many, many weekends Gilza and Welington would invite visitors from IMPA to stay with them there. There was a strict ritual: to go sailing after breakfast, starting to drink beers at noon (and definitely not a minute before) and then to swim, snorkel or go for a hike on one of the beautiful islands in the bay. Lunch would be in one of the lovely small beach restaurants or a packed lunch with self-picked oysters. During these trips mathematics was always the main topic of discussion (Gilza would normally prefer to stay in the condominium), but the atmosphere would also be light hearted. At SvS's 60th birthday party he reminisced about the time that Henk Broer threw Dennis Sullivan (and SvS) overboard, but was unsure whether to tackle Welington, who after all was the captain.

In 1987, Welington made a trip of 700 nautical miles from Berkeley to the Marquesas islands. The idea of the trip came from Steve Smale in whose boat the trip was made. Steve was the captain, and Welington and Charles Pugh were the crew. Oftentimes Welington spoke with amazement about this extraordinary trip.

Welington's website contains a quote from Richard Bode's book *'First you have to row a little boat'* which no doubt describes his philosophy of life, one of embracing life as an exciting adventure:

> For the truth is that I already know as much about my fate as I need to know. The day will come when I will die. So the only matter of consequence before me is what I will do with my allotted time. I can remain on shore, paralyzed with fear, or I can raise my sails and dip and soar in the breeze.

3 Welington's Ph.D. Students

Welington was a mentor to many young mathematicians but took on relatively few Ph.D. students. He always had high standards and was very proud of his students. Perhaps unbeknownst to them, he followed their careers with great interest.

Maria José Pacífico (1980)
Antonio Augusto Gaspar Ruas (1982)
Edson Vargas (1989)
Artur Ávila (2001)
Daniel Smania (2001)
Alejandro Kocsard (2007)
Pablo Guarino (2012)

Publications of Welington de Melo

1. J. Palis Jr., W. de Melo, Introdução aos sistemas dinâmicos, in *Projeto Euclides [Euclid Project]*, vol. 6 (Instituto de Matemática Pura e Aplicada, Rio de Janeiro, 1978)
2. J. Palis Jr., W. de Melo, *Geometric Theory of Dynamical Systems* (Springer, Heidelberg, 1982) [An introduction] (Translated from the Portuguese by A.K. Manning)
3. Zh. Palis, V. di Melu, Геометрическаятеориядинамическихсистем. Современ-наяМатематика: ВводныеКурсы. [Contemporary Mathematics: Introductory Courses]. "Mir", Moscow, 1986. Bvedenie. [An introduction] (Translated from the English by V.N. Kolokoltsov, Translation edited and with notes and an afterword by D.V. Anosov)
4. W. de Melo, S. van Strien, One-dimensional dynamics, in *Ergebnisse der Mathematik und ihrer Grenzgebiete (3) [Results in Mathematics and Related Areas (3)]*, vol. 25 (Springer, Heidelberg, 1993)
5. E. de Faria, W. de Melo, *One Dimensional Dynamics: The Mathematical Tools*. Publicações Matemáticas do IMPA. [IMPA Mathematical Publications] (Instituto de Matemática Pura e Aplicada (IMPA), Rio de Janeiro, 2001). 23o Colóquio Brasileiro de Matemática [D23rd Brazilian Mathematics Colloquium]
6. E. de Faria, W. de Melo, *Mathematical Aspects of Quantum Field Theory*. Publicações Matemáticas do IMPA. [IMPA Mathematical Publications] (Instituto Nacional de Matemática

Pura e Aplicada (IMPA), Rio de Janeiro, 2007) (With a foreword by Dennis Sullivan, 26o Colóquio Brasileiro de Matemática) [26th Brazilian Mathematics Colloquium]

7. E. de Faria, W. de Melo, Mathematical tools for one-dimensional dynamics, in *Cambridge Studies in Advanced Mathematics*, vol. 115 (Cambridge University Press, Cambridge, 2008)

8. E. de Faria, W. de Melo, Mathematical aspects of quantum field theory, in *Cambridge Studies in Advanced Mathematics*, vol. 127 (Cambridge University Press, Cambridge, 2010) (With a foreword by Dennis Sullivan)

9. W. de Melo, Structural stability of diffeomorphisms on two-manifolds. Invent. Math. **21**, 233–246 (1973)

10. W. de Melo, Structural stability on two-manifolds, in *Dynamical systems (Proc. Sympos., Univ. Bahia, Salvador, 1971)* (Academic Press, New York, 1973) pp. 255–257

11. W. de Melo, On the structure of the pareto set of generic mappings. Bol. Soc. Brasil. Mat. **7**(2), 121–126 (1976)

12. W. de Melo, Stability and optimization of several functions. Topology **15**(1), 1–12 (1976)

13. W. de Melo, Accessibility of an optimum. Lect. Notes Math. **597**, 429–440 (1977)

14. C. Gutiérrez, W. de Melo, The connected components of Morse-Smale vector fields on two manifolds. Lect. Notes Math. **597**, 230–251 (1977)

15. A. Lins, W. de Melo, C.C. Pugh, On Liénard's equation. Lect. Notes Math. **597**, 335–357 (1977)

16. W. de Melo, Moduli of stability of two-dimensional diffeomorphisms. Topology **19**(1), 9–21 (1980)

17. W. de Melo, J. Palis, Moduli of stability for diffeomorphisms. In *Global Theory of Dynamical Systems (Proc. Internat. Conf., Northwestern Univ., Evanston, Ill., 1979)*. Lect. Notes Math. vol. 819 (Springer, Heidelberg, 1980), pp. 318–339

18. W. de Melo, J. Palis, S.J. van Strien, Characterising diffeomorphisms with modulus of stability one. In *Dynamical Systems and Turbulence, Warwick 1980 (Coventry, 1979/1980)*, Lect. Notes Math. vol. 898 (Springer, Heidelberg, 1981), pp. 266–285

19. W. de Melo, G.L. dos Reis, P. Mendes, Equivariant diffeomorphisms with simple recurrences on two-manifolds. Trans. Amer. Math. Soc. **289**(2), 793–807 (1985)

20. W. de Melo, A finiteness problem for one-dimensional maps. Proc. Amer. Math. Soc. **101**(4), 721–727 (1987)

21. W. de Melo, S.J. van Strien, Diffeomorphisms on surfaces with a finite number of moduli. Ergod. Theory Dyn. Syst. **7**(3), 415–462 (1987)

22. W. de Melo, F. Dumortier, A type of moduli for saddle connections of planar diffeomorphisms. J. Differ. Equ. **75**(1), 88–102 (1988)

23. W. de Melo, S. van Strien, One-dimensional dynamics: the Schwarzian derivative and beyond. Bull. Amer. Math. Soc. (N.S.) **18**(2):159–162 (1988)

24. W. de Melo, S. van Strien, A structure theorem in one-dimensional dynamics. Ann. Math. **129**(3), 519–546 (1989)

25. M. Martens, S. van Strien, W. de Melo, P. Mendes, On cherry flows. Ergod. Theory Dyn. Syst. **10**(3), 531–554 (1990)

26. M. Martens, W. de Melo, S. van Strien, Julia-Fatou-Sullivan theory for real one-dimensional dynamics. Acta Math. **168**(3–4), 273–318 (1992)

27. W. de Melo, C. Pugh, The C^1 Brunovský hypothesis. J. Differ. Equ. **113**(2), 300–337 (1994)

28. W. de Melo, Full families of circle endomorphisms. In *Dynamical Systems and Chaos (Hachioji, 1994)*, vol. 1 (World Sci. Publ., River Edge, NJ, 1995), pp. 25–27

29. W. de Melo, B.F. Svaiter, The cost of computing integers. Proc. Amer. Math. Soc. **124**(5), 1377–1378 (1996)

30. W. de Melo, On the cyclicity of recurrent flows on surfaces. Nonlinearity **10**(2), 311–319 (1997)

31. W. de Melo, Rigidity and renormalization in one-dimensional dynamical systems. In *Proceedings of the International Congress of Mathematicians*, vol. II (Berlin, 1998), number extra vol. II, pp. 765–778

32. E. de Faria, W. de Melo, Rigidity of critical circle mappings I. J. Eur. Math. Soc. (JEMS) **1**(4), 339–392 (1999)

33. W. de Melo, A.A. Pinto, Rigidity of C^2 infinitely renormalizable unimodal maps. Comm. Math. Phys. **208**(1), 91–105 (1999)
34. E. de Faria, W. de Melo, Rigidity of critical circle mappings II. J. Amer. Math. Soc. **13**(2), 343–370 (2000)
35. A. Avila, W. de Melo, M. Martens, On the dynamics of the renormalization operator. In *Global Analysis of Dynamical Systems* (Inst. Phys., Bristol, 2001), pp. 449–460
36. M. Martens, W. de Melo, Universal models for Lorenz maps. Ergod. Theory Dyn. Syst. **21**(3), 833–860 (2001)
37. A. Avila, M. Lyubich, W. de Melo, Regular or stochastic dynamics in real analytic families of unimodal maps. Invent. Math. **154**(3), 451–550 (2003)
38. W. de Melo, Bifurcation of unimodal maps. Qual. Theory Dyn. Syst. **4**(2):413–424 (2004)
39. E. de Faria, W. de Melo, A. Pinto, Global hyperbolicity of renormalization for C^r unimodal mappings. Ann. Math. **164**(3), 731–824 (2006)
40. W. de Melo, Rigidity in dynamics. Bull. Belg. Math. Soc. Simon Stevin **15**(5, Dynamics in perturbations), 789–796 (2008)
41. V.V.M.S. Chandramouli, M. Martens, W. de Melo, C.P. Tresser, Chaotic period doubling. Ergod. Theory Dyn. Syst. **29**(2), 381–418 (2009)
42. W. de Melo, Dynamics on the circle. In *Dynamics, Games and Science I*, Springer Proc. Math. vol. 1 (Springer, Heidelberg, 2011), pp. 651–659
43. W. de Melo, Fine structure of hyperbolic diffeomorphisms [book review of mr2464147]. Bull. Amer. Math. Soc. (N.S.) **48**(1), 131–136 (2011)
44. W. de Melo, Renormalization in one-dimensional dynamics. J. Differ. Equ. Appl. **17**(8), 1185–1197 (2011)
45. W. de Melo, P.A.S. Salomão, E. Vargas, A full family of multimodal maps on the circle. Ergod. Theory Dyn. Syst. **31**(5), 1325–1344 (2011)
46. P. Guarino, W. de Melo, Rigidity of smooth critical circle maps. J. Eur. Math. Soc. (JEMS) **19**(6), 1729–1783 (2017)
47. P. Guarino, M. Martens, W. de Melo, Rigidity of critical circle maps. Duke Math. J. **167**(11), 2125–2188 (2018)

Some Monoids of Pisot Matrices

Artur Avila and Vincent Delecroix

Abstract A matrix norm gives an upper bound on the spectral radius of a matrix. Knowledge on the location of the dominant eigenvector also leads to upper bound of the second eigenvalue. We show how this technique can be used to prove that certain semi-group of matrices arising from continued fractions have a Pisot spectrum: namely for all primitive matrices in this semi-group all eigenvalues except the dominant one is smaller than one in absolute value.

Keywords Pisot matrices · Spectral radius · Pisot spectrum

1 Introduction

A *dominant eigenvalue* of a real square matrix is an eigenvalue of maximum modulus. We call a square matrix *Pisot* if it has non-negative integer entries, its dominant eigenvalue is simple and all eigenvalues different from the dominant one have absolute values less than one. Recall that a non-negative square matrix A is primitive if there exists a positive integer n so that A^n has all its entries positive. In this article, we prove that several monoids of non-negative matrices enjoy the property that all of its primitive elements are Pisot.

Our first family of matrices is related to the so called *fully subtractive* (multidimensional) continued fraction algorithm. For an integer $d \geq 2$ we define for each $k = 1, \ldots, d$ the matrix $A_{FS,d}^{(k)}$ by

To the memory of Welington de Melo, dear teacher and friend.

A. Avila (✉)
Institut fur Mathematik, Universitat Zurich and Instituto de Matemática Pura e Aplicada-IMPA, Rio de Janeiro, Brazil
e-mail: artur.avila@gmail.com

V. Delecroix
Laboratoire Bordelais de Recherche en Informatique-LaBRI, Talence, France
e-mail: vincent.delecroix@labri.fr

$$(A^{(k)}_{FS,d})_{ij} = \begin{cases} 1 \text{ if } j = k \text{ or } i = j, \\ 0 \text{ otherwise} \end{cases}$$

For $d = 3$ this boils down to the three matrices

$$A^{(1)}_{FS,3} = \begin{pmatrix} 1 & 0 & 0 \\ 1 & 1 & 0 \\ 1 & 0 & 1 \end{pmatrix}, \qquad A^{(2)}_{FS,3} = \begin{pmatrix} 1 & 1 & 0 \\ 0 & 1 & 0 \\ 0 & 1 & 1 \end{pmatrix}, \qquad A^{(3)}_{FS,3} = \begin{pmatrix} 1 & 0 & 1 \\ 0 & 1 & 1 \\ 0 & 0 & 1 \end{pmatrix}.$$

All non-degenerate products of the matrices $A^{(k)}_{FS,d}$ satisfy the Pisot property.

Theorem 1 *Let $A = A^{(i_1)}_{FS,d} A^{(i_2)}_{FS,d} \cdots A^{(i_n)}_{FS,d}$ be a product of the fully subtractive matrices in dimension d. Then the matrix A is primitive if and only if all letters $\{1, \ldots, d\}$ appear in the sequence $(i_1, i_2, \ldots, i_n)$. Moreover, if the matrix A is primitive then it is Pisot.*

The case $d = 3$ of Theorem 1 was proved in [3]. The authors used an induction on characteristic polynomials and our approach is radically different.

The same result holds for another set of 3×3 matrices related to the Brun multidimensional continued fractions. Let

$$A^{(1)}_{Br} = \begin{pmatrix} 1 & 1 & 0 \\ 0 & 1 & 0 \\ 0 & 0 & 1 \end{pmatrix}, \qquad A^{(2)}_{Br} = \begin{pmatrix} 1 & 1 & 0 \\ 1 & 0 & 0 \\ 0 & 0 & 1 \end{pmatrix}, \qquad A^{(3)}_{Br} = \begin{pmatrix} 1 & 0 & 1 \\ 1 & 0 & 0 \\ 0 & 1 & 0 \end{pmatrix}.$$

Theorem 2 *Let $A = A^{(i_1)}_{Br} A^{(i_2)}_{Br} \cdots A^{(i_n)}_{Br}$ be a product of the 3×3 Brun matrices. Then, A is primitive if and only if the matrix $A^{(3)}_{Br}$ appears in the product. Moreover, if A is primitive then it is Pisot.*

This result was already known since the work of Brun [6].

The proofs of Theorems 1 and 2 only involves linear algebra and more precisely relations between eigenvalues and matrix norms. To some extent, it is very close to the following inequality that holds for a non-negative primitive $d \times d$ matrix A

$$\lambda_2 \leq \sup_{x \in v^\perp \setminus \{0\}} \frac{\|Ax\|}{\|x\|}.$$

where v is the Perron-Frobenius eigenvector of A, λ_2 the second largest absolute value and $\|.\|$ is any norm on $\mathbb{R}^d$. Our proof uses a partial information on v provided by the continued fraction algorithm associated to the matrices.

From a diophantine approximation point of view, the Pisot property is particularly interesting because it provides the so called exponential convergence of the continued fraction expansion for almost every vectors (see [9]). We show that the above results naturally extends to this situation in Sect. 6.

Beyond continued fractions, Pisot matrices are of special interest in substitutive dynamical systems. More precisely, replacing matrices with so called substitutions, one can build a uniquely ergodic subshift X of low complexity. The Pisot properties implies the existence of a rotation on a $d - 1$ dimensional torus that is a factor of X (interestingly this rotation only depends on the matrix). In many cases X can be proved to be measurably conjugate to this rotation (see [7] Chap. 7). This was the main motivation for the study of the fully subtractive matrices in [3].

We gratefully thank Eric Domenjoud and Milton Minervo for careful reading of earlier versions.

2   Fully Subtractive and Brun Continued Fractions

Our main focus are the fully subtractive and Brun algorithm. However, most of the definitions can be set in a more general context. To that purpose, let us consider a finite or countable set $\mathcal{A}$ that we call alphabet and for each $i \in \mathcal{A}$ a matrix $A^{(i)} \in \mathrm{SL}(d, \mathbb{R})$. In the case of the fully subtractive algorithm we have $\mathcal{A}_{FS,d} = \{1, 2, \ldots, d\}$ while for Brun algorithm $\mathcal{A}_{Br} = \{1, 2, 3\}$.

To the data $(\mathcal{A}, (A^{(i)})_{i \in \mathcal{A}})$ we associate the set of infinite words $\Delta = \mathcal{A}^{\mathbb{N}}$, the shift map $T : \Delta \to \Delta$ and a cocycle

$$\forall x \in \Delta, \forall n \geq 0, \quad A_n(x) = A^{(x_0)} A^{(x_1)} \ldots A^{(x_{n-1})}.$$

The maps $A_n : \Delta \to \mathrm{SL}(d, \mathbb{R})$ satisfy the so called *(transposed) cocycle property*: $A_{m+n}(x) = A_m(x) A_n(T^m x)$.

Recall that the projective space $\mathbb{P}(\mathbb{R}^d)$ is the quotient of $\mathbb{R}^d \backslash \{0\}$ by the relation $v \sim \alpha v$ for any $\alpha \neq 0$. In other words, it is the set of lines in $\mathbb{R}^d$. It can also constructed as a quotient of the $d - 1$-dimensional sphere. $\mathbb{P}(\mathbb{R}^d)$ is compact.

Our main tool is the following definition.

Definition 3 Let $(A^{(i)})_{i \in \mathcal{A}}$ be a set of matrices in $\mathrm{SL}(d, \mathbb{R})$ where $\mathcal{A}$ is a finite or countable alphabet. We say that a set $D \subset \mathbb{P}(\mathbb{R}^d)$ is *adapted* to these matrices if it is non-empty, it is the closure of its interior and for all $i \in \mathcal{A}$ we have $A^{(i)} D \subset D$.

For example $\mathbb{P}(\mathbb{R}^d)$ is always adapted. But we will be interested in the somewhat smallest adapted set in order to localize the dominant eigenvector.

Let

$$D_{FS,d} = \{(x_1, \ldots, x_d) \in \mathbb{P}(\mathbb{R}_+^d) : \forall i, j, k \quad x_i < x_j + x_k\}$$

and

$$D_{Br} = \{(x, y, z) \in \mathbb{P}(\mathbb{R}_+^3) : x > y > z\}.$$

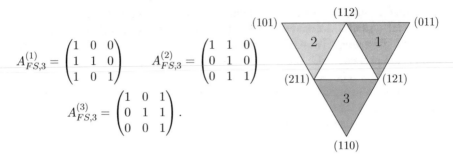

$$A_{FS,3}^{(1)} = \begin{pmatrix} 1 & 0 & 0 \\ 1 & 1 & 0 \\ 1 & 0 & 1 \end{pmatrix} \qquad A_{FS,3}^{(2)} = \begin{pmatrix} 1 & 1 & 0 \\ 0 & 1 & 0 \\ 0 & 1 & 1 \end{pmatrix}$$

$$A_{FS,3}^{(3)} = \begin{pmatrix} 1 & 0 & 1 \\ 0 & 1 & 1 \\ 0 & 0 & 1 \end{pmatrix}.$$

Fig. 1 Fully subtractive partition of the domains with $d = 3$. The points (xyz) in the picture corresponds to the point in $\mathbb{P}(\mathbb{R}_+^3)$ with coordinates (x, y, z). In other words, to the line $\mathbb{R}_+(x, y, z)$

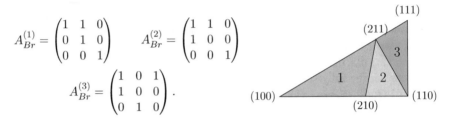

$$A_{Br}^{(1)} = \begin{pmatrix} 1 & 1 & 0 \\ 0 & 1 & 0 \\ 0 & 0 & 1 \end{pmatrix} \qquad A_{Br}^{(2)} = \begin{pmatrix} 1 & 1 & 0 \\ 1 & 0 & 0 \\ 0 & 0 & 1 \end{pmatrix}$$

$$A_{Br}^{(3)} = \begin{pmatrix} 1 & 0 & 1 \\ 1 & 0 & 0 \\ 0 & 1 & 0 \end{pmatrix}.$$

Fig. 2 The matrices and domains for the Brun algorithm

Then it is easily seen that $D_{FS,d}$ is adapted for the fully subtractive matrices in dimension d and D_{Br} is adapted for the Brun matrices.

Given a countable collection of matrices $(A^{(i)})_{i \in A}$ and $D \subset \mathbb{P}(\mathbb{R}^d)$ adapted, we define $D^{(i)} = A^{(i)}D$. In Figs. 1 and 2 one can see the projective picture of the domains $D^{(1)}$, $D^{(2)}$ and $D^{(3)}$. Note that in these cases, the domains $D^{(i)}$ are disjoint but that it is not a requirement in our definition. Moreover, one can see that in the Brun case the $D^{(i)}$ form a partition while it is not the case for the fully subtractive.

If the $D^{(i)}$ are disjoint one can define a continued fraction algorithm as follows. One defines a partial map $f : D \dashrightarrow D$ by setting $fx = (A^{(i)})^{-1}x$ on $D^{(i)}$.

One can compute that for Brun one has

$$f_{Br}(x, y, z) = \text{sort}(x - y, y, z)$$

where sort : $\mathbb{P}(\mathbb{R}_+^3) \to D$ is the map which permutes the coordinates in order to sort them. While for the fully subtractive one has

$$f_{FS,d}(x) = (x_1 - x_i, x_2 - x_i, \ldots, x_{i-1} - x_i, x_i, x_{i+1} - x_i, \ldots, x_d - x_i)$$
$$\text{if } x_i = \min(x_1, \ldots, x_d).$$

3 Strategy

The proofs of Theorems 1 and 2 follow a general strategy that we describe now.

Let $(A^{(i)})_{i\in\mathcal{A}}$ be a finite or countable set of matrices as in Sect. 2. Let also $\Delta = \mathcal{A}^{\mathbb{N}}$ and $D \subset \mathbb{P}(\mathbb{R}^d)$ be adapted. We define for a finite word $w = i_0 i_1 \ldots i_{n-1} \in \mathcal{A}^n$ the set $D^{(w)} = A^{(i_0)} A^{(i_1)} \ldots A^{(i_{n-1})} D$. It generalizes the definition $D^{(i)}$ that we already used for words of length 1 (identified to letters). In terms of continued fractions, it corresponds to the set of points in D corresponding to the cylinder $[w]$ induced by the coding by $\mathcal{A}$. Note that for any w the set $D^{(w)}$ is *not* empty. Given an infinite word $x = x_0 x_1 \ldots \in \Delta$ we also set $D_n(x) = A_n(x) D = D^{(x_0 x_1 \ldots x_{n-1})}$. In particular $D_0(x) = D$ and $D_1(x) = D^{(x_0)}$. Given x, the set $D_n(x)$ are nested and we define $D_{\infty}(x) = \bigcap_{n\geq 0} D_n(x)$.

We let $\|.\|$ be the L^{∞} norm on $\mathbb{R}^d$ and the associated operator norm on matrices. That is for a vector v and a matrix A

$$\|v\| = \max(|v_1|, \ldots, |v_d|) \quad \text{and} \quad \|A\| = \max_{i=1,\ldots,d} \sum_{j=1,\ldots,d} |a_{ij}|.$$

In this section the norm used on $\mathbb{R}^d$ has no importance. But it turns out that, to apply the results to continued fraction algorithms in Sects. 4 and 5 the most convenient one was twice the L^{∞} norm.

To a non-zero vector v in $\mathbb{R}^d$, we associate its dual hyperplane $H_v = \{z \in \mathbb{R}^d; \langle v, z \rangle = 0\}$. Given a non zero vector v in $\mathbb{R}^d$ we define the following semi-norm on $d \times d$ matrices

$$\|B\|_v = \sup_{z\in H_v\setminus\{0\}} \frac{\|Bz\|}{\|z\|} = \max_{\substack{\|z\|\leq 1 \\ z\in H_v}} \|Bz\|.$$

More generally, if $\Lambda \subset \mathbb{R}^d$ is a cone, we define

$$\|B\|_{\Lambda} = \sup_{v\in\mathbb{P}(\Lambda)} \|B\|_v.$$

These semi-norm will be applied to the transposed cocycle B of A that is $B^{(w)} = (A^{(w)})^*$ for finite words w and $B_n(x) = (A_n(x))^*$ for points $x \in D$ and $n \geq 0$. These matrices satisfy the *cocycle property* $B_{m+n}(x) = B_n(T^m x) B_m(x)$.

Lemma 4 *Let $(A^{(i)})_{i\in\mathcal{A}}$ be a finite or countable set of matrices in $\mathrm{SL}(d, \mathbb{R})$ and let $D \subset \mathbb{P}(\mathbb{R}^d)$ be adapted. Let $B^{(i)}$ (respectively $B_n(x)$) denote the transposed of $A^{(i)}$ (resp. $A_n(x)$). If for all $i \in \mathcal{A}$ we have*

$$\left\| B^{(i)} \right\|_{D^{(i)}} \leq 1. \tag{1}$$

Then for any point $x = x_0 x_1 \ldots \in \Delta$ *we have*

$$\|B_n(x)\|_{D_n(x)} \leq 1$$

where $D_n(x) = A_n(x)D$.

The corollary below shows that under the assumption (1), the generated monoid satisfies a "weak Pisot" condition.

Corollary 5 *Under the same assumption of Lemma 4, each matrix* $A^{(w)} = A^{(i_0)} A^{(i_1)}$ $\ldots A^{(i_{n-1})}$ *has at most one eigenvalue (with multiplicity) greater than one in absolute value.*

Proof of Lemma 4. We prove the Lemma by induction. The hypothesis corresponds to the case $n = 1$. Assume that this inequality holds for some n. By definition $A_{n+1}(x) = A_1(x)A_n(Tx)$ and $D_{n+1}(x) = A_1(x)D_n(Tx)$. Hence $v \in D_n(Tx)$ if and only if $A_1(x)v \in D_{n+1}(x)$. Let us choose $v \in D_n(Tx)$, then

$$\|B_{n+1}(x)\|_{A_1(x)v} = \|B_n(Tx)B_1(x)\|_{A_1(x)v} \leq \|B_n(Tx)\|_v \cdot \|B_1(x)\|_{A_1(x)v}.$$

This inequality shows that if $\|B_n(Tx)\|_v \leq 1$ then $\|B_{n+1}(x)\|_{A_1(x)v} \leq 1$ which concludes the proof of the Lemma. $\qquad\square$

Proof of Corollary 5. Let us consider the point $x = (i_0 i_1 \ldots i_{p-1})^\infty$ in Δ so that $B_p(x) = B^{(i_0)} \ldots B^{(i_{n-1})}$. Because $B_p(x)$ is simply the transposed of $A_p(x)$ we can simply prove the statement for B_p.

By Lemma 4 and because $D_\infty(x) \subset D_p(x)$ we have

$$\|B_p(x)\|_{D_\infty(x)} \leq 1.$$

Now the union $M = \bigcup_{v \in D_\infty(x)} H_v$ is preserved by $B_p(x)$. On the other hand, we have a direct sum $\mathbb{R}^d = E_u \oplus E_s$ where E_u (respectively E_s) is the direct sum of the eigenspaces $\ker(B_p(x) - \lambda I)^d$ with eigenvalue λ whose absolute value is greater than 1 (resp. smaller or equal than 1). By the above inequality, any eigenvector of $B_p(x)$ contained in M must have an eigenvalue of absolute less than one. More generally, E_u does not intersect M. Now, M contains a $d - 1$ vector space and hence E_u has dimension at most 1. $\qquad\square$

4   Pisot Property for Arnoux-Rauzy Matrices

In this section we prove Theorem 1. Let $d \geq 2$ be an integer and let e_1, e_2, $\ldots e_d$ be the canonical basis of $\mathbb{R}^d$. Let $e = e_1 + e_2 + \ldots + e_d$ and for $i = 1, \ldots, k$ let $f_i = e - e_i$. The domain $D_{FS,d}$ is the convex hull of the rays vectors $\mathbb{R}_+ f_i$.

Let as usual $B^{(w)} = (A^{(w)})^*$ and $B_n(x) = (A_n(x))^*$. We claim that we have the following stronger property than (1) from Lemma 4

$$\forall i = 1, \ldots, d, \qquad \|B^{(i)}\|_D \leq 1.$$

Let us prove this claim. Let $v \in D_\infty$, then we may write $v = \mu_1 f_1 + \mu_2 f_2 + \ldots + \mu_d f_d$ for some non-negative numbers μ_i that satisfy $\mu_1 + \mu_2 + \ldots + \mu_d = 1$. We hence have $v = e - \sum \mu_i e_i$ and

$$H_v = \{z \in \mathbb{R}^d; \langle z, v \rangle = 0\} = \{z \in \mathbb{R}^d; \sum z_j = \sum \mu_j z_j\}.$$

Given $z \in H_v$ we have $B^{(i)}z = (z_1, \ldots, z_{i-1}, \sum_j \mu_j z_j, z_{i+1}, \ldots, z_d)$, In other words, $B^{(i)}$ acts on H_v as a stochastic matrix $P(i, v)$ which is the identity except its i-th row which is $(\mu_1, \mu_2, \ldots, \mu_d)$. In particular $\|B^{(i)}\|_v = \|P(i, v)\|_v \leq 1$.

By the claim, the monoid generated by the matrices $A_{FS,d}^{(i)}$ satisfies the weak Pisot condition of Corollary 5. What remains to prove is that when a product is irreducible, there is no remaining eigenvalue of absolute value 1. To that purpose, let us consider a finite product $A = A_{FS,d}^{(i_0)} A_{FS,d}^{(i_1)} \cdots A_{FS,d}^{(i_{p-1})}$. If one of the letter $\{1, \ldots, d\}$ is missing in the sequence $(i_0, i_1, \ldots, i_{p-1})$ then $Ae_i = e_i$ and so the matrix is not primitive. On the other hand, if all letters appear it is easy to see that all entries in A are positive.

Now let $x = (x_0 x_1 \ldots x_{p-1})^\infty$ be a periodic point that contains all letters from $\mathcal{A}$. Because of positivity the dominant eigenvalue of $A_p(x)$ is simple and its associated eigenvector v_0 is so that $D_\infty(x) = \mathbb{R}_+ v_0$. Let $v_n = A_n(x)^{-1} v_0$ be the Perron-Frobenius eigenvector of $A_p(T^n x)$. Each v_n is positive and hence the coefficients $\mu_1, \mu_2, \ldots, \mu_d$ that appear in the stochastic matrices $P(x_i, v_i)$ are all positive. Now, the product $P(x_{p-1}, v_{p-1}) \ldots P(x_1, v_1) P(x_0, v_0)$ is a stochastic matrix with all its entries positive. Hence, its second eigenvalue, which is also the second eigenvalue of $A_p(x)$, is less than 1 in absolute value.

5   Pisot Property for Brun Algorithm (in Dimension 3)

We now turn to the proof of Theorem 2. Let $\mathcal{A} = \{1, 2, 3\}$ and $A_{Br}^{(1)}, A_{Br}^{(2)}, A_{Br}^{(3)}$ be the matrix of the Brun algorithm. We let $\Delta = \mathcal{A}^\mathbb{N}$ and denote by A and B respectively the cocycle and the transposed cocycle. We claim that, as in the case of the fully subtractive, we have the stronger property that

$$\forall i \in \{1, 2, 3\}, \qquad \|B^{(i)}\|_D \leq 1.$$

We only need to consider the matrix $B^{(1)} = \begin{pmatrix} 1 & 0 & 0 \\ 1 & 1 & 0 \\ 0 & 0 & 1 \end{pmatrix}$ since the other two are obtained by multiplying by a permutation matrix which will not change the L^∞-norm.

Let $v = \mu_1(1:0:0) + \mu_2(1:1:0) + \mu_3(1:1:1) \in D$ for some μ_1, μ_2, μ_3 such that $\mu_1 + \mu_2 + \mu_3 = 1$ and $H_v = \{z \in \mathbb{R}^d; z_1 + z_2 = \mu_1 z_2 - \mu_3 z_3\}$. Now, for any $z \in H_v$ we have $B^{(1)}(z_1, z_2, z_3) = (z_1, z_1 + z_2, z_3) = (z_1, \mu_1 z_2 - \mu_3 z_3, z_3)$. In particular $\|B^{(1)}(z_1, z_2, z_3)\| \le \|(z_1, z_2, z_3)\|$. As in the case of the fully subtractive algorithm we define $P(v, 1) = \begin{pmatrix} 1 & 0 & 0 \\ 0 & \mu_1 & -\mu_3 \\ 0 & 0 & 1 \end{pmatrix}$ that is so that the action of $B^{(1)}$ and $P(1, v)$ coincide on H_v. Similarly, we define $P(2, v)$ and $P(3, v)$ corresponding to $B^{(2)}$ and $B^{(3)}$. Note that, contrarily to the fully subtractive case, $P(i, v)$ is not a stochastic matrix.

Now, given a product $A = A_{Br}^{(i_0)} \ldots A_{Br}^{(i_{n-1})}$, if 3 does not appear in the sequence $(i_0, \ldots, i_{n-1})$ then $Ae_3 = e_3$ and hence the matrix A can not be irreducible. Conversely, if 3 appears then A^3 is easily seen to be positive.

As in the fully subtractive case we consider the sequence of eigenvectors $(v_n)_{n \ge 0}$ and the product $P = P(x_{p-1}, v_{p-1}) \ldots P(x_1, v_1) P(x_0, v_0)$. For a primitive product we got that the numbers μ_j appearing in the definition of each $P(x_i, v_i)$ are all positive. As a consequence, the absolute value of the sum of each row of the product P is strictly smaller than 1. In other words $\|P\|_{v_0} = \|B_p(x)\|_{v_0} < 1$ which prove that there is no eigenvalue 1.

6 Lyapunov Exponents

Let $(A^{(i)})_{i \in \mathcal{A}}$ be a finite or countable set of matrices. Let Δ, T, A, B denote as before the infinite words, the shift map the cocycle and the transposed cocycle. Let also D be adapted to these matrices.

The asymptotics of the cocycle (or the transposed cocycle) are studied through *Lyapunov exponents*. Given a T-invariant ergodic probability measure μ on Δ, we associate the real numbers $\gamma_1^\mu \ge \gamma_2^\mu \ge \ldots \ge \gamma_d^\mu$ defined by

$$\forall k \in \{1, 2, \ldots, d\}, \quad \gamma_1 + \gamma_2 + \ldots + \gamma_k = \lim_{n \to \infty} \int_\Delta \frac{\log \|\wedge^k A_n(x)\|}{n} d\mu(x).$$

In order to be well defined we assume that

$$\int_\Delta \max \left(\log \|A_1(x)\|, \log \|A_1(x)^{-1}\| \right) d\mu(x) < \infty \tag{2}$$

and we refer to this condition as the log-*integrability* of the cocycle. If the alphabet $\mathcal{A}$ is finite the cocycle is automatically log-integrable. If x is a periodic point of T and $\mu = (\delta_x + \delta_{Tx} + \ldots + \delta_{T^{n-1}x})/n$ is the sum of Dirac masses distributed along its orbit, then the associated Lyapunov exponents are the logarithms of the absolute values of eigenvalues of $A_n(x)$ where n is the period of x. In that sense, Lyapunov exponents generalize eigenvalues.

Given a measure μ for which the cocycle is log-integrable, we say that (Δ, T, A, μ) has *Pisot spectrum* if the associated Lyapunov exponents satisfy $\gamma_1^\mu > 0 > \gamma_2^\mu$. This property is related to the strong convergence of higher dimensional continued fraction algorithm [9].

Now we restate Lemma 4 in a more dynamical context.

Lemma 6 *Let* $(A^{(i)})_{i \in \mathcal{A}}$ *be a finite or countable set of non-negative matrices in* $SL(d, \mathbb{R})$. *Let* (Δ, T, A, B) *be the associated full shift with its cocycle and its transposed cocycle. Let also* D *be adapted. Assume that*

$$\forall i \in \mathcal{A}, \quad \left\| (A^{(i)}) \right\|_{D^{(i)}} \leq 1.$$

Let μ *be a* T-*invariant and ergodic measure on* D *so that*

- *the cocycle* A_n *is log-integrable,*
- *there exists a cylinder* $[w]$ *such that* $\mu([w]) > 0$, $A^{(w)}$ *is positive and* $\left\| B^{(w)} \right\|_{D^{(w)}} < 1$.

Then two first Lyapunov exponents of the cocycle A_n *for the measure* μ *satisfies* $\gamma_1^\mu > 0 > \gamma_2^\mu$.

Proof Let us first prove that $\gamma_1 > 0$.

Now, by definition, for μ-almost every x

$$\gamma_1 = \lim_{n \to \infty} \frac{\log \|A_n(x)\|}{n}$$

Let $m = |w|$ be the length of w and consider the positions which are multiple of m. For a μ-generic x we have by Birkhoff theorem that

$$\lim_{n \to \infty} \frac{\#\{i \leq n : T^i(x) \in [w]\}}{n} = \mu([w])$$

In other words, given a sequence of length n large enough we can find a linear number of disjoint occurrences of w (up to a sublinear error). Let k_n be the number of these occurrences, then necessarily each entry of $A_n(x)$ is larger than the corresponding one in C^{k_n} where C is the matrix which contains a 1 in every position. In particular $\gamma_1 > 0$.

From the existence of w it also follows that for a μ-generic x the cone $D_\infty(x)$ is reduced to a line contained in the interior of $\mathbb{R}_+^d$. We can hence define μ-almost everywhere a function $v : \Delta \to \mathbb{R}_+^d$ by $D_\infty(x) = \mathbb{R}_+ v(x)$ and $\|v(x)\| = 1$. We then have the following formulas which holds for μ-almost every x

$$\gamma_1 = \lim_{n \to \infty} \frac{-\log \|A_n(x)^{-1} v(x)\|}{n} \quad \text{and} \quad \gamma_2 = \lim_{n \to \infty} \frac{\log \|B_n(x)\|_{v(x)}}{n}.$$

It is then easy to derive the estimate for γ_2. The map $x \mapsto v(x)$ and the dual hyperplanes $H_{v(x)}$ satisfy the following covariance properties

$$D_\infty(Tx) = A(x)^{-1} D_\infty(x) \quad \text{and} \quad H_{A^{-1}v} = A^* H_v.$$

Hence as in the proof of Lemma 4, we deduce that for all $v \in D_\infty(x)$

$$\|B_{m+n}(x)\|_{v(x)} \leq \|B_m(x)\|_{v(x)} \|B_n(T^m x)\|_{v(T^m x)}$$

In particular, for μ-almost every $x \in [w]$, any large $n \geq |w|$ we get that $\|B_n(x)\|_{v(x)} < 1$. Let $\delta = \|B^{(w)}\|_{D^{(w)}} < 1$. Using the same argument as in the estimation of γ_1 we get that

$$\gamma_2 \leq \liminf_{n \to \infty} \frac{k_n \log \delta}{n}.$$

And the above limit is strictly negative. $\square$

References

1. P. Arnoux, Š. Starosta, The rauzy gasket, in *Further Developments in Fractals and Related Fields* (Trends Math. Birkhäuser/Springer, New York, 2013)
2. P. Arnoux, Un exemple de semi conjugaisaon entre un échanges d'intervalles et une translation sur le tore. Bull. SMF **116**(4), 489–500 (1988)
3. P. Arnoux, S. Ito, Pisot substitutions and rauzy fractals. Bull. Belg. Math. Soc. Simon Stevin **8**(2), 181–207 (2001)
4. P. Arnoux, G. Rauzy, Représentation géométriques des suites de complexité $2n + 1$. Bull. SMF **119**(2), 199–215 (1991)
5. A. Brentjes, Multidimensional continued fraction algorithms, in *Mathematical Centre Tracts*, vol. 145 (Mathematisch Centrum, Amsterdam, 1981)
6. V. Brun, Algorithmes euclidiens pour trois et quatre nombres. Congr. Math. Scand. XII **I**, 45–64 (1957)
7. N.P. Fogg, Substitutions in dynamics, arithmetic and combinatorics, in *Lecture Notes in Mathematics* (2002)
8. C. Kraaikamp, R. Meester, Ergodic properties of a dynamical system arising from percolation theory. Ergod. Th. Dynam. Sys. **15**, 653–661 (1995)
9. J.C. Lagarias, The quality of the diophantine approximations found by the Jacobi-Perron algorithm and related algorithms. Mh. Math. **115**, 299–328 (1993)

On the Statistical Attractors
and Attracting Cantor Sets for Piecewise
Smooth Maps

P. Brandão, J. Palis and V. Pinheiro

Abstract We discuss some aspects of the asymptotic behavior of the forward orbits of most points for piecewise smooth maps of the interval, specially for contracting Lorenz maps. We are particularly interested in the statistical aspects of most orbits, the existence of statistical attractors and how most orbits in the basin of attraction of an attracting Cantor sets approach the attractor.

Keywords Metrical attractors · Statistical attractors · Contracting Lorenz maps · Piecewise smooth maps of the interval · Inaccessible basin of attraction · Strong ergodicity

In Memoriam of Welington de Melo

Just after concluding his Ph.D. at Berkeley, Jacob Palis returned to Rio de Janeiro and started to work at IMPA, conducting a seminar mostly concerned with new results on Dynamics. The seminar was intense, promoting talks even on Saturday mornings.

In one of these occasions a young student asked Palis to allow him to participate in the seminar, following a suggestion by Elon Lima who met this student at the Brazilian Mathematical Colloquium of 1969 in Poços de Caldas. This fellow was Welington de Melo. He assured Palis that he would study a lot and so, would much profit from the seminar. Palis was surprised, specially because this young student had just been accepted to the Master Program at IMPA. Palis was so impressed by his determination that finally said yes to him. Welington's performance was indeed outstanding: he concluded his doctorate in two years after joining the seminar, publishing his thesis in the celebrated journal "Inventiones mathematicae". This episode shows

P. Brandão · J. Palis
Estrada Dona Castorina,110,22460-320 Impa, Rio de Janeiro, BrazilP. Brandão
e-mail: paulo@impa.br

J. Palis
e-mail: jpalis@impa.br

V. Pinheiro (✉)
Instituto de Matematica-UFBA, Av. Ademar de Barros, s/n, 40170-110
Salvador, Bahia, Brazil
e-mail: viltonj@ufba.br

© Springer Nature Switzerland AG 2019
M. J. Pacifico and P. Guarino (eds.), *New Trends in One-Dimensional Dynamics*, Springer Proceedings in Mathematics & Statistics 285,
https://doi.org/10.1007/978-3-030-16833-9_4

31

two of the main features of his personality: strong determination and mathematical competence.

Welington became one of the world leaders in one-dimensional dynamics. He collaborated with main names in the field, taking part in several main moments of the development of the subject and finally, he was the Ph.D. advisor of Artur Avila. We observe that Avila was the first mathematician to win a Fields medal whose initial career was fully developed in the Southern hemisphere.

Above all, Welington cultivated in his whole life a "joie de vivre", specially crossing seas with friends in his boat. We do miss him.

1   Introduction

Mathematicians, and scientists in general, have been using strategies to reduce the dynamics that they are studying (or at least, features of them) to some more treatable low dimensional dynamical models, particularly to dynamics generated by maps of the interval. When one tries to reduce a multidimensional smooth dynamical system to the dynamics generated by a map f of the interval, it is not so surprising that f may not carry the same smoothness of the original dynamics.

One example of that concerns *Lorenz flows* and its attractors. Lorenz flows are specific three dimensional flows that were first studied by the meteorologist Edward Lorenz. He was trying to study some unpredictable behavior normally associated to weather using Navier-Stokes equations, which is a model of fluid convection and an infinite dimensional dynamical system. After he published his remarkable paper "Deterministic non-periodic flows" (1963), Lorenz flows were studied by many mathematicians, physicists, engineers and others. Eventually, V.S. Afraimovich, V.V. Bykov, L.P. Shil'nikov, in [1], and Guckenheimer and Williams, in [17], introduced the *geometric Lorenz flows*, similar to Lorenz's ones and exhibiting the same peculiar characteristics of the original Lorenz flows. This model consists of considering a hyperbolic singularity $p = (0, 0, 0) \in \mathbb{R}^3$ with one dimensional unstable manifold such that, in a linearizable neighborhood $V = (-r, r)^3$ of p, the local stable and unstable manifolds are $W_r^s(p) = V \cap (\{0\} \times \mathbb{R}^2)$ and $W_r^u(p) = V \cap (\mathbb{R} \times \{(0, 0)\})$, the return of each branch of $W^u(p) \setminus \{p\}$ to this neighborhood cuts transversally the plane $z = constant$, with eigenvalues $\lambda_2 < \lambda_3 < 0 < \lambda_1$ (see Fig. 1), and the expanding condition $\lambda_3 + \lambda_1 > 0$. The dynamics of a geometric Lorenz flow can be reduced to the dynamics of a map of preserving orientation interval map with a single discontinuity, called a Lorenz map. It is worth to be told that the geometric Lorenz flows (and so, the Lorenz maps) were the appropriate approach to study Lorenz flows, as it was proved by W. Tucker [37], that the original Lorenz flows are indeed geometric Lorenz ones.

In [3], Arneodo, Coullet and Tresser began to study a model obtained in the same way as the one obtained by Guckenheimer and Williams, just modifying the relation between the eigenvalues of the singularity, taking $\lambda_3 + \lambda_1 < 0$. This model is known as the contracting Lorenz flow and the attractors associated to it are more unstable

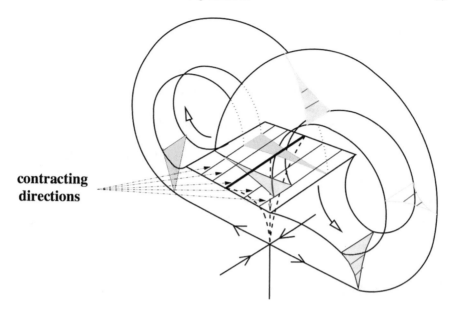

Fig. 1 Dynamics of the geometric Lorenz flows

than the Lorenz original ones, presenting far more diversified dynamics and many deferent types of attractors. In particular, the one dimensional maps associated to them "contain", as we will see later, all the possible dynamics of the Logistic maps (another very well known and studied class of dynamics).

The dynamics of smooth maps of the interval was studied exhaustively in the last four decades, in particular, properties as finiteness of attractors, non-existence of wandering intervals, classification of attractors, physical/SRB measures, generic families of unimodal maps with most parameters being stochastic or deterministic, the density of deterministic dynamics and so on. Nevertheless, several of the key techniques developed to smooths maps do not work for maps with discontinuities. Because of that, even for the simplest maps of the interval with one discontinuity, the Lorenz maps (particularly, the contractive ones), most of the equivalent results already obtained to smooth maps are not known.

Here we will try to point out some of the problems in which we are more interested in and that are open for maps with discontinuities. We mention a few open conjectures about such maps, some of them due to de Melo. To be concise and also because it is the simplest case, most of the definitions will be done in Sect. 2 only for contracting Lorenz maps, nevertheless everything can be set in a similar way to the general context of piecewise smooth maps of the interval. Still in Sect. 2, we use the concept of ergodicity for measures that are not necessarily invariant to define the (Ilyashenko) statistical attractors and compare them to Milnor attractors. In particular, we show in Proposition 2.5 that, if a contracting Lorenz map f has a cycle of intervals and the closure of the forward orbits of the critical values has empty interior, then f is ergodic with respect to Lebesgue and it has a single Milnor attractor containing a unique Ilyashenko statistical attractor.

Motivated by the questions raised in Sect. 2, we present in Sect. 3 the main result of the paper, Theorem 2, about how a typical orbit is attracted by an absorbing Cantor set. A point $p \in A$ is accessible from an A-outside point if $\exists q \notin A$ and a continuous curve γ such that $\gamma(0) = q, \gamma(1) = p$ and $\gamma((0, 1)) \cap A = \emptyset$. We say that p is asymptotically accessible from an A-outside point (for short, asymptotically accessible) if there exists $q \notin A$ such that for every $n \geq 0$ there is a $j \geq n$ and a continuous curve γ such that $\gamma(0) = q$, $\gamma(1) = f^j(p)$ and $\gamma((0, 1)) \cap A = \emptyset$. Every point of an attracting Cantor set is asymptotically inaccessible, i.e., it is not asymptotically accessible and Theorem 2 says that the same is true for almost every point in the basin of attraction of an attracting Cantor set. This is equivalent to say that the orbit of almost every point in the basin of attraction intersects a given gap of the Cantor set only a finite number of times.

2 Contracting Lorenz Maps

A *contracting Lorenz map* is a local difeomorphisrm $f : [0, 1] \setminus \{c\} \to [0, 1]$, $0 < c < 1$, such that 0 and 1 are fixed points, f has no repelling fixed points in $(0, 1)$ and $\lim_{x \to c} f'(x) = 0$. A C^3 interval map f has negative Schwarzian derivative if $Sf(x) < 0$ for every point x such that $f'(x) \neq 0$, where

$$Sf(x) = \frac{f'''(x)}{f'(x)} - \frac{3}{2}\left(\frac{f''(x)}{f'(x)}\right)^2 \tag{1}$$

A C^3 contracting Lorenz map $f : [0, 1] \setminus \{c\} \to \mathbb{R}$ is called *non-flat* if there exist $\varepsilon > 0$, constants $\alpha, \beta \geq 1$ and C^3 diffeomorphisms $\phi_0 : [c - \varepsilon, c] \to \mathrm{Im}(\phi_0)$ and $\phi_1 : [c, c + \varepsilon] \to \mathrm{Im}(\phi_1)$ such that

$$f(x) = \begin{cases} f(c_-) + \big(\phi_0(x - c)\big)^{\alpha} & \text{if } x \in (c - \varepsilon, c) \cap (0, 1) \\ f(c_+) + \big(\phi_1(x - c)\big)^{\beta} & \text{if } x \in (c, c + \varepsilon) \cap (0, 1) \end{cases}$$

In the remaining of this section (Sect. 2), f is a C^3 contracting Lorenz map $f : [0, 1] \setminus \{c\} \to [0, 1]$, $c \in (0, 1)$, with negative Schwarzian derivative.

The pre-orbit of a set $U \subset [0, 1]$ is $\mathcal{O}_f^-(U) := \bigcup_{j \geq 0} f^{-j}(U)$. If $x \notin \mathcal{O}_f^-(c)$, the forward orbit of x is $\mathcal{O}_f^+(x) = \{f^j(x)\,;\ j \geq 0\}$. On the other hand, if $x \in f^{-n}([0, 1])$ and $f^n(x) = c$, then $\mathcal{O}_f^+(x) = \{x, \cdots, f^n(x)\}$. The ω-*limit set* of a point $x \notin \mathcal{O}_f^-(c)$, denoted by $\omega_f(x)$, is the set of accumulating points of the forward orbit of x. That is,

$$\omega_f(x) := \{y \in [0, 1]\,;\ y = \lim_{j \to \infty} f^{n_j}(x) \text{ for some sequence } n_j \to +\infty\}.$$

Consider also the *lateral* ω-*limit sets* $\omega_f(x_-)$ and $\omega_f(x_+)$ as

$$\omega_f(x_\pm) := \{y \in [0, 1]; \ y = \lim_{j \to \infty} f^{n_j}(x_\pm) \text{ for some sequence } n_j \to +\infty\},$$

where $f^{n_j}(x_-) = \lim_{0 < \varepsilon \to 0} f^{n_j}(x - \varepsilon)$ and $f^{n_j}(x_+) = \lim_{0 < \varepsilon \to 0} f^{n_j}(x + \varepsilon)$. The *nonwandering set* of f, denoted by $\Omega(f)$, is the set of points p such that $\#\{j \geq 1; \ (p - \varepsilon, p + \varepsilon) \cap f^{-j}((p - \varepsilon, p + \varepsilon)) \neq \emptyset\} = \infty$ for every $\varepsilon > 0$. It is easy to see that $\omega_f(x) \subset \Omega(f)$ for every $x \in [0, 1] \setminus \mathcal{O}_f^-(c)$.

Since Poincaré's theory for circle homeomorphisms, it was observed that the existence of wandering intervals makes the classification and analysis of one dimensional dynamical systems far more complicated. We note that wandering intervals are not just intervals outside $\Omega(f)$, as one can see in the definition below, and this is the reason that some authors have called them *strongly wandering intervals*. A *wandering interval* is an open interval $I = (a, b) \subset [0, 1]$ such that (1) $f^n|_I$ is a homeomorphism between I and $f^n(I)$ for every $n \geq 1$, (2) I does not intersect the basin of attraction of a periodic-like attractor and (3) $f^n(I) \cap f^m(I) = \emptyset$ for every $0 \leq n < m$.

Due to the complexity of most dynamical systems, Poincaré has suggested that instead of attempting to describe the asymptotic behavior of all orbits of a dynamics, one can focus on the behavior of most orbits, thus avoiding the most pathological or peculiar. Because of that, attractors play a fundamental role in the study of dynamical systems for the understanding of future evolution of typical (not uncommon) initial states.

Definition 2.1 (Milnor [30]) A compact set A is called a (metrical) *attractor* for f if its *basin of attraction* $\beta_f(A) := \{x; \ \omega_f(x) \subset A\}$ has positive Lebesgue measure and there is no compact set $A' \subsetneq A$ so that $\beta_f(A')$ is the same as $\beta_f(A)$ up to a zero measure set.

Theorem 1 below characterizes the attractors (in Milnor sense) for contracting Lorenz maps. In particular, it says that either f has periodic-like attractors (at most two of them) and the the union of their basins of attraction has full Lebesgue measure or f has a single (non-periodic) attractor A containing almost every point in its basin of attraction and $\omega_f(x) = A$ for almost every x.

Theorem 1 ([9], see also [19, 32]) *Let f be a C^3 contracting Lorenz map f : $[0, 1] \setminus \{c\} \to [0, 1]$, $c \in (0, 1)$, with negative Schwarzian derivative. If f does not have periodic-like attractors, then f has an attractor A such that $\omega_f(x) = A$ for almost every $x \in [0, 1]$. In particular, $\mathrm{Leb}(\beta_f(A)) = 1$.*

Furthermore, f can have at most two periodic-like attractors. If f has a single periodic attractor, its basin of attraction has full Lebesgue measure. In the case that f has two periodic-like attractors, the union of their basins of attraction has full Lebesgue measure.

If f does not have periodic-like attractors, then A is either a cycle of intervals or a transitive Cantor set.

If A is a Cantor set, then $A = \omega_f(c_-)$ or $\omega_f(c_+)$. Moreover, if f is non-flat and A is a Cantor set, then c_- and $c_+ \in A = \omega_f(c_-) = \omega_f(c_+)$.

An attractor A is called a **periodic-like attractor** when A is a finite set and, given $p \in A$ we have $A = \bigcup_{j=1}^n f^j(p_-)$ or $\bigcup_{j=1}^n f^j(p_+)$, where $n = \#A$. Periodic-like attractors generalizes the concept of attracting periodic orbits in the context of piecewise smooth maps.

A **cycle of intervals** is a finite union of disjoint closed intervals $I_1 \cup \cdots \cup I_n$ such that $f|_{I_1 \cup \cdots \cup I_n}$ is transitive. For any continuous map $h : [0, 1] \to [0, 1]$ a cycle of intervals is always a chaotic attractor, in the sense that the topological entropy of f restricted to the cycle of interval is always positive. In contrast, contracting Lorenz maps can have both: chaotic cycle of intervals and cycle of intervals with zero topological entropy (as for Cherry attractors below).

If the attractor A is a Cantor set, it is called an **attracting Cantor set**. An attracting Cantor set can be a solenoid, a Cherry attractor or a wild attractor. A **solenoid attractor** appears when f is infinitely many times renormalizable. That is, when there exits a nested sequence of intervals $[a_n, b_n]$ with $a_1 < \cdots < a_n \nearrow c \nwarrow b_n < \cdots < b_1$ such that the first return map to each $[a_n, b_n]$ is conjugated to a contracting Lorenz map. In [8], Brandão shows that the topological attractor A of an infinitely renormalizable contracting Lorenz map is a **minimal set**, i.e., every orbit of A is dense in A. Notice that, as $c \in A$ and f cannot be extended continuously through c, the argument in [8] is more delicate than the usual argument for continuous maps.

A **topological attractor** A is a compact set such that $\beta_f(A) := \{x ; \omega_f(x) \subset A\}$ is not a **meager** set and there is no compact set $A' \subsetneq A$ so that $\beta_f(A')$ is the same as $\beta_f(A)$ up to a meager set. Recall that a set Γ is called **meager** if it is contained in a countable union of compact set with empty interior. In this context of subsets of the interval, $\Gamma \subset [0, 1]$ is not meager if and only if Γ contains a residual subset of some open set $U \subset [0, 1]$.

A **Cherry attractor** [13, 23] appears when there exists an interval $[a, b]$, with $a < b < c$ such that F, the first return map by f to $[a, b]$, is conjugate to an injective map of the circle $h : S^1 \setminus \{1\} \to S^1$ with an irrational rotation number, $S^1 = \{z \in \mathbb{C} ; |z| = 1\}$. As it has irrational rotation number, it follows from the theory of circle maps that $\omega_h(x) = \Omega(h)$ for every $x \in S^1 \setminus \{1\}$, where $\Omega(h)$ is the nonwandering set of h. Furthermore, $A_h := \Omega(h)$ is a minimal set and A_h is either the whole circle or an attracting Cantor set (with all its gaps being wandering intervals). By the conjugacy with the first return map F, the attractor A_h induces an attractor A_F for F and also an attractor A for f, called a Cherry attractor for f. This Cherry attractor is always a minimal set for f and it is either a cycle of intervals or an attracting Cantor set. If the Cherry attractor is a Cantor set, then some of its gaps are wandering intervals.

A far as we know, the wandering intervals associated to Cherry attractors are the only known examples of wandering intervals for C^2 non-flat contracting Lorenz maps. The existence of this kind of intervals for contracting Lorenz map that are not associated to Cherry attractors is an open problem with very little progress in the past 20 years. For C^2 diffeomorphism of the circle, the non-existence of wandering intervals was established in 1930 by Denjoy [15] and, in 1989, for non-flat C^2 maps of the interval, by de Melo and van Strien [27]. In the context of Lorenz maps, what is expected to happen in terms of wandering intervals is the following:

Conjecture 1 (Martens and de Melo [24]) *Let $f : [0, 1] \setminus \{c\} \to [0, 1]$ be a C^2 non-flat contracting Lorenz maps. If f has a wandering interval then f has a Cherry attractor.*

Even for contracting Lorenz maps that are C^3 and have negative Schwarzian derivative, Martens and de Melo's conjecture remains almost untouched. We may note that the fact that f cannot be extended continuously through c together with $f'(c_{\pm})$ being 0 plays a main role to make this conjecture hard to be proved. Indeed, if $f'(c_{\pm}) \neq 0$, the conjecture is true, proved by Mestel and Berry [29] in the beginning of 1990's decade.

An attractor A is called a **wild attractor** (also called an **absorbing Cantor set**) for the contracting Lorenz map f if it is properly contained in some compact transitive set Λ, i.e., $A \subsetneqq \Lambda$. If f does not admit wandering intervals, Λ must be a circle of intervals.

For continuous maps of the interval, a wild attractor occurs when one has a (metrical) attractor properly contained in a topological one. The question about the existence of wild attractors for S-unimodal maps was asked by Milnor [30] and proved by Bruin, Keller, Nowicki and van Strien in [11].

A C^1 map $g : [0, 1] \to [0, 1]$ is called a **symmetric unimodal map** if $g(0) = g(1) = 0$, $g'(x) \neq 0$ for $x \neq 1/2$ and g satisfies the symmetry $g(x) = g(1 - x) \, \forall x$. A particularly famous class of examples of symmetric unimodal maps is the one of the **Logistic maps** $g_t : [0, 1] to [0, 1]$ defined by $g_t(x) = 4tx(1 - x)$, where t is a parameter varying on $[0, 1]$. Given a symmetric unimodal map g, let $L : [0, 1] \setminus \{1/2\} \to [0, 1]$ be the Lorenz map defined by

$$L(x) = \begin{cases} g(x) & \text{if } x < 1/2 \\ 1 - g(x) & \text{if } x > 1/2 \end{cases}.$$

Note that $g \circ L = g \circ g$. Indeed, If $x < 1/2$ then $L(x) = g(x)$ and so, $g(L(x)) = g^2(x)$. If $x > 1/2$ then $L(x) = 1 - g(x)$ and by the symmetric, $g \circ L(x) = g(1 - g(x)) = g(g(x)) = g^2(x)$. Thus, $g \circ L(x) = g^2(x)$ for every $x \in [0, 1] \setminus \{1/2\}$. This means that g and L are semi-conjugated maps, with a at most two to one semi-conjugation and so, g and L have essentially the same dynamics. Moreover, as $|L'(x)| = |g'(x)|$, we get that

$$|(L^n)'(x)| = |L'(L^{n-1}(x))||(L^{n-1})'(x)| = |g'(L^{n-1}(x))||(L^{n-1})'(x)| =$$

$$= |(g \circ L^{n-1})'(x)| = |(g^{n-1} \circ g)'(x)| = |(g^n)'(x)|$$

for every $n \geq 1$ and $x \notin \mathcal{O}_g^-(1/2)$. That is, up to a sign, the derivative of points along the orbits of g and L are the same (Fig. 2).

$$g \qquad\qquad\qquad L$$

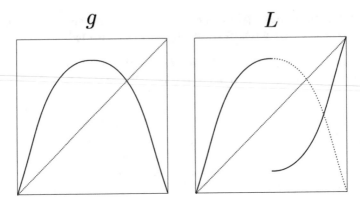

Fig. 2 A symmetric unimodal map g and its associated (symmetric) Lorenz map L

As the S-unimodal maps presented in [11] are symmetric and with wild attractors contained in cycle of intervals, one can use the contracting Lorenz maps associated to them to show, for Lorenz maps, the existence of wild attractors $A \subsetneq \Lambda$ with Λ being a cycle of intervals. In contrast with the wild attractors that come from symmetric unimodal maps, if there exists a contracting Lorenz map having wandering intervals as well as a wild attractor, $A \subsetneq \Lambda$, then the transitive set Λ is a Cantor set.

The wild attractor generated by symmetric unimodal maps is always a minimal set. But it is not expected that these are the only possible wild attractors that a contracting Lorenz map can display, even if the map does not have wandering intervals. Hence, one can ask about the minimality (as a set) of the wild attractors for contracting Lorenz maps, see Question 2.8.

An attractor A is called a ***physical attractor*** if it is the support of an ergodic f-invariant probability μ, called a ***physical measure***, such that Lebesgue almost every point $x \in \beta_f(A)$ is a μ-generic point, that is, $\frac{1}{n} \sum_{j=0}^{n-1} \delta_{f^j(x)}$ converges in the weak* topology to the measure μ. The simplest example of a physical attractor is a periodic-like attractor A, in this case $\mu = \frac{1}{\#A} \sum_{p \in A} \delta_p$ is the physical measure. When the physical measure is absolutely continuous with respect to Lebesgue, it is called a Sinai-Ruelle-Bowen (SRB) measure.

Milnor's definition of an attractor A deals only with the topological aspects of the asymptotical behavior of the orbits of the most points in $\beta_f(A)$. In contrast, if A is a physical attractor, the physical measure μ gives an accurate information about the average of time that a point will spend in a given open set V. Precisely, given $x \in [0, 1] \setminus \mathcal{O}_f^-(c)$ and $V \subset [0, 1]$, define the ***visiting frequency*** of x to V as

$$\tau_x(V) = \tau_{x,f}(V) = \limsup_{n \to \infty} \frac{1}{n} \#\{0 \le j < n \; ; \; f^j(x) \in V\},$$

then, for almost every point $x \in \beta_f(A)$ and any open set V, we have that $\tau_x(V) \ge \mu(V)$. Moreover, if $\mu(\partial V) = 0$ then $\tau_x(V) = \mu(V)$ for almost every $x \in \beta_f(A)$.

Nevertheless, even symmetric S-unimodal maps may not have physical attractors (see [18]). Hence, the same is true to contracting Lorenz maps. A concept more flexible than physical attractor, and that also gives informations about the visiting frequency, is the concept of statistical attractors proposed by Ilyashenko. For that, define the ***statistical ω-limit set*** of a point $x \in [0, 1] \setminus \mathcal{O}_f^-(c)$ as

$$\omega_f^*(x) = \{y \, ; \, \tau_x(B_\varepsilon(y)) > 0 \text{ for all } \varepsilon > 0\},$$

where $B_\varepsilon(y) = (y - \varepsilon, y + \varepsilon)$ is the ball of radius ε and center y.

Definition 2.2 (*Ilyashenko, see page 148 of* [4]) A compact set A is called a ***statistical attractor*** for f if its ***statistical basin of attraction*** $\beta_f^*(A) = \{x \, ; \, \omega_f^*(x) \subset A\}$ has positive Lebesgue measure and there is no compact set $A' \subsetneq A$ so that $\beta_f^*(A')$ is the same as $\beta_f^*(A)$ up to a zero measure set.

A (not necessarily invariant) probability μ is called ***ergodic*** if $\mu(U) = 0$ or 1 for every invariant set U, i.e., $f^{-1}(U) = U$. It was proved by Blokh and Lyubich [7] that every S-unimodal map without periodic attractors is ergodic with respect to Lebesgue measure. Let us show that ergodicity implies the existence of a single attractor, a single statistical attractor and a good behavior of the visiting frequency.

Proposition 2.3 *Let $\mathbb{X}$ be a compact metric space, $V \subset \mathbb{X}$ a Borel set and $g : V \to \mathbb{X}$ a measurable map. If g is ergodic with respect to a Borel probability μ, then there are compact sets $A^* \subset A \subset \mathbb{X}$ such that $\omega_g(x) = A$ and $\omega_g^*(x) = A^*$ for almost every $x \in \bigcap_{n \geq 0} g^{-n}(V)$. Furthermore, $x \mapsto \tau_{x,g}(U)$ is constant for almost every $x \in \bigcap_{n \geq 0} g^{-n}(V)$ and every measurable set $U \subset \mathbb{X}$.*

Proof As $\mathbb{X}$ is compact, for every $x \in X := \bigcap_{j \geq 0} g^{-1}(\mathbb{X})$, we have that $\omega_g(x)$ and $\omega_g^*(x)$ are nonempty compact sets. Furthermore, note that $\omega_g(x) = \omega_g(g(x))$ and $\omega_g^*(x) = \omega_g^*(g(x))$ for every $x \in X$. We may assume that $\mu(X) > 0$. As $g^{-1}(X) = X$, it follows from the ergodicity of μ that $\mu(X) = 1$. Let $\mathbb{K}$ be the set of all non-empty compact subsets of $\mathbb{X}$ with the Hausdorff metric $d_H(A, B) = \inf\{\varepsilon > 0 \, ; \, A \subset B_\varepsilon(B) \text{ and } B \subset B_\varepsilon(A)\}$, where $B_r(C) := \bigcup_{x \in C} B_r(x)$ is the ball of radius $r > 0$ with "center" $C \subset \mathbb{X}$. As $\mathbb{X}$ is a compact metric space, it is well known that $(\mathbb{K}, d_H)$ is also a compact metric space.

The proof of the existence of compact sets A and A^* such that $\omega_g(x) = A$ and $\omega_g^*(x)$ for μ almost every $x \in \mathbb{X} (= X \mod \mu)$ follows from the Lemma 2.4 below applied to the maps $\varphi, \psi : X \to \mathbb{K}$, where $\varphi(x) = \omega_g(x)$ and $\psi(x) = \omega_g^*(x)$. As $\omega_g(x) \supset \omega_g^*(x)$ always, we get that $A^* \subset A$. Finally, given a Borel set $U \subset \mathbb{X}$, consider $\xi : X \to [0, 1]$ defined by $\xi(x) = \tau_{x,g}(U)$. As $\xi(g(x)) = \xi(x) \, \forall x \in X$, it follows from Lemma 2.4 that ξ is constant for μ almost every $x \in X$, proving the proposition. $\qquad\qquad\square$

Although Lemma 2.4 is a well known fact for invariant measures, as we are not assuming the invariance of μ, we are providing a brief proof of it.

Lemma 2.4 *Let $\mathbb{X}$ and $\mathbb{Y}$ be separable metric spaces, $g : \mathbb{X} \to \mathbb{X}$ a measurable map, μ a Borel probability on $\mathbb{X}$ and $W \subset \mathbb{X}$ a measurable set. If μ is a g-ergodic probability (not necessarily invariant) with $\mu(W) = 1$ and $\varphi : W \to \mathbb{Y}$ is g-invariant (i.e., $\varphi \circ g = \varphi$), then φ is μ-almost constant, i.e., $\exists\, y \in \mathbb{Y}$ such that $\varphi(x) = y$ for μ-almost every x.*

Proof Consider $\nu = \varphi_* \mu = \mu \circ \varphi^{-1}$, the push-forward of μ. Suppose that the support of ν, supp ν, has more than one point. Say, $p, q \in$ supp ν with $p \neq q$. Take $\varepsilon > 0$ such that $B_\varepsilon(p) \cap B_\varepsilon(q) = \emptyset$. As φ is g-invariant, we get that $V_0 := \varphi^{-1}(B_\varepsilon(p))$ and $V_1 := \varphi^{-1}(B_\varepsilon(q))$ are f-invariant sets with $V_0 \cap V_1 = \emptyset$. As $p, q \in$ supp ν, we have that $\mu(V_0) = \nu(B_\varepsilon(p)) > 0$ and $\mu(V_1) = \nu(B_\varepsilon(q)) > 0$, contradicting the ergodicity of μ. $\qquad\square$

So, it follows from Blokh and Lyubich [7] and Proposition 2.3 that every S-unimodal map g without periodic attractors has only one attractor A_g and only one statistical attractor A^*. Furthermore, $\omega_g(x) = A_g$ and $\omega_g^*(x) = A_g^*$ for almost every $x \in [0, 1]$ and $\tau_{x,g}(U) = \tau_{y,g}(U)$ for almost every x and $y \in [0, 1]$ and every measurable $U \subset [0, 1]$. Unfortunately, it is not clear that Blokh and Lyubich's result can have some extension to contracting Lorenz maps in general. For instance, the existence of wandering intervals is an obstruction to the Lebesgue ergodicity of a map. Indeed, if $J = (a, b)$ is a wandering interval, then one can consider any $p \in (a, b)$ and $V = \bigcup_{n \geq 0} f^{-n}\big(\bigcup_{j \geq 0} f^j((a, p))\big)$. As V is f invariant and $0 < \mathrm{Leb}(V) < 1$, because $(p, b) \cap V = \emptyset$, we get that f cannot be ergodic. At least in one case, we can show that f is ergodic with respect to Lebesgue measure, see Proposition 2.5 below (as the proof of this proposition is a bit more technical, we left it to the Appendix).

Proposition 2.5 *If $\overline{\mathcal{O}_f^+(c_-) \cup \mathcal{O}_f^+(c_+)}$ has empty interior and the attractor of f is a cycle of intervals, then f is ergodic with respect to Lebesgue.*

Hence, under the hypotheses of Proposition 2.5, f has a unique statistical attractor $A^* \subset A$ such that $\omega_f^*(x) = A^*$ for almost every x. Furthermore, $x \mapsto \tau_x(U)$ is constant for almost every x and every $U \subset [0, 1]$.

As $f(\omega_f^*(x) \setminus \{c\}) = \omega_f^*(x)$, that is, $\omega_f^*(x)$ is a forward invariant set, another way to show that the contracting Lorenz map f has a unique statistical attractor is to prove that its attractor A is a minimal set. Indeed, as $\omega_f^*(x) \subset \omega_f(x) = A$ for almost every x, if A is a minimal set, it follows from the forward invariance of $\omega_f^*(x)$ that $\omega_f^*(x) = A$ for almost every x. In this direction we have the following result.

Proposition 2.6 *If f is infinitely renormalizable and non-flat, then the solenoid attractor $A = \omega_f(c_-) = \omega_f(c_+)$ is a Cantor set that is a minimal set. Furthermore, $\omega_f(x) = \omega_f^*(x) = A$ for every $x \in \beta_f(A)$, $\beta_f(A)$ is a residual subset of $[0, 1]$ and $\mathrm{Leb}(\beta_f(A)) = 1$.*

Proof As f is infinitely many times renormalizable, the (measurable) attractor A cannot be a cycle of intervals neither a periodic-like attractor. Thus, it follows Theorem 1 that the (measurable) attractor is a Cantor set $A = \omega_f(c_-)$ or $A = \omega_f(c_+)$.

On the other hand, by Theorem A of [8], the topological attractor A_{top} of f is a minimal Cantor set and it contains c. Thus, $\omega_f(c_-) = \omega_f(c_+) = A_{top}$ and so, $A = \omega_f(c_-) = \omega_f(c_+) = A_{top}$ is a minimal set. As commented before, it follows from $\omega_f^*(x) \subset \omega_f(x) = A$, for almost every x, together with the minimality of A and the forward invariance of $\omega_f^*(x)$ that $\omega_f^*(x) = A$ for almost every x. That is, $A^* = A$ is the statistical attractor of f and $\text{Leb}(\beta_f^*(A^*)) = 1$. Finally, the fact that $\beta_f(A)$ is a residual set comes from Theorem D of [8]. □

Our first question is about the existence and finiteness of statistical attractors for general contracting Lorenz maps.

Question 2.7 *Is almost every point $x \in [0, 1]$ contained in the basin of attraction of a statistical attractor? Is the number of statistical attractors of Contracting Lorenz maps finite?*

As noted before, one can use the minimality of an attractor A to conclude that A is also a statistical attractor. So, our second question is about the minimality of A as a set.

Question 2.8 *For contracting Lorenz maps, are all attracting Cantor sets minimal sets?*

Although this is not the focus of this paper, we want end this section mentioning that there are many other aspects of Contracting Lorenz maps that have been studied by many authors in the past 30 years. Some of them are on combinatorial properties [6, 20, 25, 32, 33, 35, 36] or bifurcation and features of the parameter space of two parameter families [14, 33, 35, 36] and characterization of complete families [25]. There are papers about existence of attractors of positive Lyapunov exponents [34], SRB measures [26], statistical stability of SRB measures [2], about entropy [6], thermodynamic formalism for contracting Lorenz maps and flows [31], generalizing chaotic contracting Lorenz maps to higher dimensions [5], etc. Finally, we want to mention [12] where one can have a discussion about some variations of the definition of statistical attractors.

3 Piecewise Smooth Maps of the Interval

If a map f of the interval is piecewise C^3 with negative Schwarzian negative (Theorem 1) or if f is a non-flat piecewise C^2 maps (see [10]), then there is a finite number of non-periodic attractors attracting almost every point that does not belong to the basin of attraction of some periodic-like attractor. Furthermore, if A is one of those non-periodic attractors then $\omega_g(x) = A$ for almost every x in the basin of attraction of A. Hence, A is somehow transitively approximated from outside for almost every point of its basin of attraction. Nevertheless, it is not clear that if, or when, A is a transitive set. As we saw in Sect. 2, all attracting Cantor sets of contracting Lorenz maps are transitive. What we don't know, in the Lorenz case, is if they are minimal sets. For general non-flat piecewise C^2 maps, even the transitivity of an attracting Cantor set is not known.

An initial approach by A. Avila and V. Pinheiro aiming to build an example of a non-transitive attracting Cantor set was the following. Let $f : [0, 1] \to [0, 1]$ be a non-flat C^2 map of the interval. Suppose that f has an attracting Cantor set Λ such that

$$\text{Leb}(\{x \in \beta_f(\Lambda) \, ; \, \#(\mathcal{O}_f^+(x) \cap J) = \infty\}) > 0$$

for some gap $J = (a, b)$ of Λ. For instance, you can assume that f is an unimodal map. A **gap of a Cantor set** $\Lambda \subset [0, 1]$ is a connected component of $[0, 1] \setminus \Lambda$. If such Λ and J exists, one can consider a new map $g : [0, 2] \setminus \mathcal{C}_g \to [0, 2]$ given by

$$g(x) = \begin{cases} f(x) & \text{if } x \in [0, 1) \setminus [a, b] \\ f(x) + 1 & \text{if } x \in J \\ f(x - 1) + 1 & \text{if } x - 1 \in (0, 1] \setminus [a, b] \\ f(x - 1) & \text{if } x - 1 \in J \end{cases}.$$

where $\mathcal{C}_g = \mathcal{C}_f \cup \{a, b, 1, a + 1, b + 1\}$. Note that g is a non-flat piecewise C^2. Given a set $U \supset [0, 1]$, let $1 + U := \{1 + x \, ; \, x \in U\}$. Note that, $\Lambda_g := \Lambda \cup (1 + \Lambda)$ is an attracting Cantor set for g with $\omega_g(x) = \Lambda_g$ for almost every $x \in \beta_g(\Lambda_g) = \beta_f(\Lambda) \cup (1 + \beta_f(\Lambda))$. Nevertheless, $g(\Lambda) = \Lambda$ and $g(1 + \Lambda) = \Lambda$. That is, Λ_g is not a transitive set. So, the existence of such Λ and J implies the existence of non-transitive attracting Cantor sets. However, it turns out that this strategy can not be implemented, as one can see in Corollary 3.2.

We say that the orbit $\mathcal{O}_f^+(x)$ of a point $x \in \beta_f(A)$ is **asymptotically inaccessible** if given any $p \notin A$ there is a $n_0 \geq 0$ such that $\{(1 - t)p + tf^n(x) \, ; \, t \in (0, 1)\}$ contains a point of A for every $n \geq n_0$.

Definition 3.1 An attracting Cantor set A has an **asymptotically inaccessible basin of attraction** if the forward orbit of almost every point in $\beta_f(A)$ is asymptotically inaccessible. This is equivalent to say that $\#(\mathcal{O}_f^+(x) \cap J) < \infty$ for almost every $x \in \beta_f(A)$ and every gap J of A.

Theorem 2 Let f be a non-flat piecewise C^2 map and A an attracting Cantor set of f. For almost every point $x \in \beta_f(A)$, either x belongs to a wandering interval or the forward orbit of x is asymptotically inaccessible.

As commented before, non-flat C^2 maps of the interval do not admit wandering intervals [27]. Hence, we get the following corollary.

Corollary 3.2 The basin of attraction of any attracting Cantor set for a non-flat C^2 map is always asymptotically inaccessible.

To prove Theorem 2, we need Lemma 3.3 and Proposition 3.4 below. A (not necessarily invariant) probability μ is called **strongly ergodic** with respect to a measurable map f if $\mu(U) = 0$ or 1 for every forward invariant set U, i.e., $f(U) \subset U$.

As every invariant set is a forward invariant one, it follows that a strongly ergodic measure is an ergodic one.

Lemma 3.3 *Let $\mathbb{X}$ be a compact metric space, μ a Borel probability on $\mathbb{X}$ and $f : \mathbb{X} \to \mathbb{X}$ a measurable map. If μ is strongly ergodic with respect to f and μ is f-non-singular, i.e., $\mu \circ f^{-1} \ll \mu$, then $\mu(\omega_f(x)) = 1$ for μ almost every $x \in \mathbb{X}$.*

Proof As μ is also ergodic, it follows from Proposition 2.3 that there is a compact set $A \subset \mathbb{X}$ such that $\omega_f(x) = A$ for μ almost every $x \in \mathbb{X}$. Suppose that $\mu(A) = 0$. In this case one can choose an open neighborhood V of A such that $\mu(V) < 1/2$. Given $n \geq 1$, let $U_n = \{x \in \mathbb{X};\ f^j(x) \in V \,\forall\, j \geq n\}$. As $\omega_f(x) = A$ for almost every x, we have that $\mu(\bigcup_{n\geq 1} U_n) = 1$. Hence, there is $n_0 \geq 1$ such that $\mu(U_{n_0}) > 0$. Note that $\mu(f^{n_0}(U_{n_0})) > 0$ because $\mu \circ f^{-1} \ll \mu$. As U_{n_0} is a forward invariant set, $f^{n_0}(U_{n_0})$ is also forward invariant. Thus, it follows from the strong ergodicity of μ that $\mu(f^{n_0}(U_{n_0})) = 1$, which is a contradiction with $f^{n_0}(U_{n_0}) \subset V$ and $\mu(V) < 1/2$. As a consequence, $\mu(A) > 0$. Thus, as $f(A) = A$, it follows again from the strong ergodicity of μ that $\mu(A) = 1$.

Proposition 3.4 *Let $f : [0, 1] \setminus C_f \to [0, 1]$ be a non-flat C^2 local diffeomorphism, with $C_f \subset (0, 1)$ being a finite set. Let $\mathbb{B}(f)$ denote the union of all wandering intervals of f with the basin of attraction of all periodic-like attractors of f. Let $F : U \to (a, b)$ be the first return map by f to an interval (a, b), where $U = \{x \in (a, b) ;\ \mathcal{O}_f^+(f(x)) \cap (a, b) \neq \varnothing\}$. If $F(P) = (a, b)$ for every connected component of U, then*

$$\mathrm{Leb}\left(\bigcap_{n\geq 0} F^{-n}((a, b)) \setminus (\mathbb{B}(f) \cup \mathcal{O}_f^-(\mathrm{Per}(f)))\right) \textit{ is either } 0 \textit{ or } b - a.$$

Furthermore, if $\mathrm{Leb}\left(\bigcap_{n\geq 0} F^{-n}((a, b)) \setminus (\mathbb{B}(f) \cup \mathcal{O}_f^-(\mathrm{Per}(f)))\right) = b - a$ *then* $\mathrm{Leb}\,|_{(a,b)}$ *is F strongly ergodic and $\omega_F(x) = [a, b]$ for almost every $x \in (a, b)$.*

Proof Write $V_0 = \bigcap_{n\geq 0} F^{-n}((a, b)) \setminus (\mathbb{B}(f) \cup \mathcal{O}_f^-(\mathrm{Per}(f)))$. As there is nothing to prove when $\mathrm{Leb}(V_0) = 0$, we may assume that $\mathrm{Leb}(V_0) > 0$.

Given $n \geq 1$, let $\mathcal{P}_n$ be the set of connected components of $F^{-n}((a, b))$. Given $x \in F^{-n}((a, b))$, let $\mathcal{P}_n(x)$ be the element of $\mathcal{P}_n$ containing x. Note that $F^n(\mathcal{P}_n(x)) = (a, b)$ for every $n \geq 1$ and $x \in F^{-n}((a, b))$.

Consider any F forward invariant set V, i.e., $F(V) \subset V$, with positive measure and contained in V_0. One example of such a set is V_0 itself. Let $p_0 \in V$ be a Lebesgue density point of V, i.e.,

$$\lim_{\delta \to 0} \frac{\mathrm{Leb}(V \cap B_\delta(p_0))}{\mathrm{Leb}(B_\delta(p_0))} = 1.$$

As $V \subset V_0$, $\mathcal{P}_n(p_0)$ is well defined for every $n \geq 1$ and $F^n(\mathcal{P}_n(p_0)) = (a, b)$.

Claim 3.5 diameter$(\mathcal{P}_n(p_0)) \to 0$ when $n \to \infty$.

Proof of the claim As $\mathcal{P}_1(p_0) \supset \mathcal{P}_2(p_0) \supset \mathcal{P}_3(p_0) \supset \cdots$, if $\lim_n$ diameter$(\mathcal{P}_n(p_0))$
> 0, then $(\alpha, \beta) := $ interior$(\bigcap_{n \geq 1} \mathcal{P}_n(p_0))$ is an interval with $\alpha < \beta$ and such that
$F^n|_{(\alpha, \beta)}$ is a diffeomorphism for every $n \geq 1$. This means that a is a homterval
(see Appendix). This leads to a contradiction. Indeed, by the Homterval Lemma
(Lemma 3.9 in Appendix), $(\alpha, \beta) \subset \mathbb{B}(f) \cup \mathcal{O}_f^-(\mathrm{Per}(f))$, and on the other hand,
Leb$((\alpha, \beta) \setminus (\mathbb{B}(f) \cup \mathcal{O}_f^-(\mathrm{Per}(f)))) \geq$ Leb$((\alpha, \beta) \cap V)) > 0$ (because p_0 is a density point of V).

Given $\varepsilon > 0$, choose $a < a' < p_0 < b' < b$ so that Leb$((a, b) \setminus (a', b')) < \varepsilon/2$.
Let us write $J := (a', b') \subset (a, b) =: I$. As f is a first return map, we have that
$f^j(\mathcal{P}_n(p_0)) \cap f^i(\mathcal{P}_n(p_0)) = \emptyset$ for every $0 \leq j < i < n$. As a consequence, both
$\max\{|f^j(\mathcal{P}_n(p_0))|; 0 \leq j \leq n\}$ and $\sum_{j=0}^{n-1} |f^j(\mathcal{P}_n(p_0))|$ are smaller or equal to 1.
Thus, it follows from item 1 of Proposition 3.11 in Appendix that there is $K > 0$,
depending only on f, $\frac{a'-a}{b-a}$ and $\frac{b-a'}{b-a}$ (in particular, not depending on n), such that

$$\left| \frac{(F^n)'(p)}{(F^n)'(q)} \right| \leq K, \forall p, q \in J_n(p_0) \text{ and } n \geq 1, \tag{2}$$

where $J_n(p_0) := (F^n|_{\mathcal{P}_n(p_0)})^{-1}(J)$ is the connected component of $F^{-n}(J)$ containing
p_0.

So, it follows from the bounded distortion given by (2) and the forward invariance
of V that Leb$((a', b') \setminus V) = 0$. Indeed, as $(a', b') = F^n(J_n(p_0))$, we get from (2)
that

$$\mathrm{Leb}((a', b') \setminus V) \overset{\star}{\leq} \mathrm{Leb}(F^n(J_n(p_0) \setminus V)) = \mathrm{Leb}((a', b')) \frac{\mathrm{Leb}(F^n(J_n(p_0) \setminus V))}{\mathrm{Leb}(F^n(J_n(p_0)))} \leq$$

$$\leq \mathrm{Leb}((a', b')) K \frac{\mathrm{Leb}(J_n(p_0) \setminus V)}{\mathrm{Leb}(J_n(p_0))} \to 0$$

Here the inequality $(\star)$ follows from the fact that $V \supset F^n(J_n(p_0) \cap V)$ and, then,
$(a', b') \setminus V \subset (a', b') \setminus F^n(J_n(p_0) \cap V) = F^n(J_n(p_0) \setminus V)$.

Finally, as we can take a' as close to a and b' as close to b as wished, we conclude
that Leb$(I \setminus V) = 0$. That is, Leb$(V) = 1$ for every forward invariant set with positive measure, proving that Leb is strongly ergodic. In particular, Leb$(V_0) = b - a$.
Applying Lemma 3.3, we get that Leb$(\omega_F(x)) = b - a$ for almost every x and, by
the compactness of the ω-limit, $\omega_F(x) = [a, b]$ for Lebesgue almost every $x \in [a, b]$

Now, we can prove Theorem 2.

Proof of Theorem 2 As f is piecewise C^2, there is a finite set $\mathcal{C}_f$ such that (1)
$g := f|_{[0,1] \setminus \mathcal{C}_f}$ is a C^2 local diffeomorphism and that (2) g cannot be extended through
$c \in \mathcal{C}_f$ as a C^2 local diffeomorphism.

As $\mathcal{O}_f^-(\mathcal{C}_f)$ is a countable set (so, with zero Lebesgue measure) and $\mathcal{O}_f^+(x) =$
$\mathcal{O}_g^+(x)$ for every $x \in [0, 1] \setminus \mathcal{O}_f^+(\mathcal{C}_f)$, we may consider g instead of f. That is, let us
assume that $f : [0, 1] \setminus \mathcal{C}_f \to [0, 1]$ is a non-flat C^2 local diffeomorphism. Without

loss of generality, we may also assume that $\{f(0), f(1)\} \subset \{0, 1\}$ (see, for instance, Remark 11 in [10]).

Let $\mathcal{W}(f)$ be the union of all wandering intervals of f. We may assume that $\mathrm{Leb}(\beta_f(A) \setminus \mathcal{W}(f)) > 0$, otherwise there is nothing to be proved. Consider a gap J of A and suppose by contradiction that there is $L \subset \beta_f(A) \setminus \mathcal{W}(f)$ with $\mathrm{Leb}(L) > 0$ and such that $\#(\mathcal{O}_f^+(x) \cap J) = \infty$ for every $x \in L$.

We say that $c_- \in \omega_f(x)$ if there is a sequence $n_j \to \infty$ such that $f^{n_j}(x) < c\ \forall j$ and $c = \lim_{j \to \infty} f^{n_j}(x)$. Similarly, we define $c_+ \in \omega_f(x)$. It follows from Theorem 1 in [10] that

$$\omega_f(x) = \bigcup_{\substack{c_\pm \in \omega_f(x) \\ c \in \mathcal{C}_f}} \overline{\mathcal{O}_f^+(c_\pm)} := \bigcup_{\substack{c_- \in \omega_f(x) \\ c \in \mathcal{C}_f}} \overline{\mathcal{O}_f^+(c_-)} \cup \bigcup_{\substack{c_+ \in \omega_f(x) \\ c \in \mathcal{C}_f}} \overline{\mathcal{O}_f^+(c_+)},$$

for almost every $x \in [0, 1] \setminus \big(\mathbb{B}_0(f) \cup \mathbb{B}_1(f) \cup \mathcal{O}_f^-(\mathrm{Per}(f))\big)$, where $\mathbb{B}_0(f)$ is the union of the basins of attraction of all attracting periodic-like orbits of the map f and $\mathbb{B}_1(f)$ is the set of all points x such that $\omega_f(x)$ is a cycle of intervals.

As a consequence,

$$A = \omega_f(x) = \bigcup_{\substack{c_\pm \in \omega_f(x) \\ c \in \mathcal{C}_f}} \overline{\mathcal{O}_f^+(c_\pm)},$$

for almost every $x \in \beta_f(A)$.

Let $\mathcal{C}^- = \{c \in \mathcal{C}_f ; c_- \notin \omega_f(x) \text{ for almost every } x \in \beta_f(A)\}$ and $\mathcal{C}^+ = \{c \in \mathcal{C}_f ; c_+ \notin \omega_f(x) \text{ for almost every } x \in \beta_f(A)\}$.

Remark 3.6 Note that $c_- \in \omega_f(x)$ for almost every $x \in \beta_f(x)$ and $c \in \mathcal{C}_f \setminus \mathcal{C}^-$. Similarly, $c_+ \in \omega_f(x)$ for almost every $x \in \beta_f(x)$ and $c \in \mathcal{C}_f \setminus \mathcal{C}^+$.

Given $r > 0$ let

$$B(r) := \bigcup_{c \in \mathcal{C}^-} (c - r, c) \cup \bigcup_{c \in \mathcal{C}^+} (c, c + r)$$

and set, for $n \geq 1$,

$$U_n = \{x \in \beta_f(A) ; \overline{\mathcal{O}_f^+(x)} \cap B(1/n) = \emptyset\}.$$

As $\mathrm{Leb}(\beta_f(A) \setminus \bigcup_{n \geq 1} U_n) = 0$ and $\mathrm{Leb}(L) > 0$, let $\ell \geq 1$ be so that $\mathrm{Leb}(L \cap U_\ell) > 0$.

Consider any non-flat C^2 map $g : [0, 1] \setminus \mathcal{C}_f \to [0, 1]$ such that $g(x) = f(x)$ for every $x \notin U_\ell$ and that $g(c_-) \in \{0, 1\}$ for every $c \in \mathcal{C}^-$ and $g(c_+) \in \{0, 1\}$ for every $c \in \mathcal{C}^+$ (in Fig. 3 we have an illustration of such a map g).

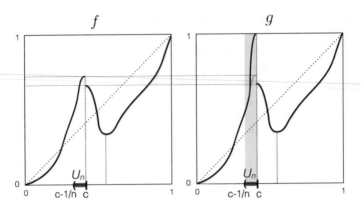

Fig. 3 The map g on the right is the deformation in the "lateral neighborhood" $(c - 1/n, c)$ of a critical point $c \in C^-$ of the map f, on the left side of the picture, as asked in the proof of Theorem 2

Note that $\mathcal{O}_g^+(x) = \mathcal{O}_f^+(x)$ for every $x \in L$ and so $\omega_g(x) = A$ for every $x \in L$. It follows from Remark 3.6 that $g^j(c_-) = f^j(c_-) \in A$ for every $j \geq 1$ and $c \in \mathcal{C}_f \setminus \mathcal{C}^-$. Also, $g^j(c_+) = f^j(c_+) \in A$ for every $j \geq 1$ and $c \in \mathcal{C}_f \setminus \mathcal{C}^+$. As $g^j(c_-) \in \{0, 1\}$ for every $j \geq 1$ and $c \in \mathcal{C}^-$ and, as $g^j(c_+) \in \{0, 1\}$ for every $j \geq 1$ and $c \in \mathcal{C}^+$, we get that $g^j(c_\pm) \notin J$ for every $j \geq 1$ and $c \in \mathcal{C}_f$. Furthermore, as $\partial J \subset A \cup \{0, 1\}$, we get that $g^j(\partial J) \cap J = \emptyset$ for every $j \geq 1$, that is, J is a **nice interval** for g. Therefore, if $U = \{x \in J ; \mathcal{O}_g^+(g(x)) \cap J \neq \emptyset\}$ and $G : U \to J$ is the first return map to J by g, we get that $G(P) = J$ for every connected component P of U (here, we are using that J is nice together with fact that forward orbit of critical values, $g(c_\pm)$ with $c \in \mathcal{C}_f$, does not intersect J).

As

$$L \cap U_\ell \subset \bigcap_{n \geq 0} G^{-n}(J) \setminus (\mathbb{B}(g) \cup \mathcal{O}_g^-(\mathrm{Per}(g))),$$

where $\mathbb{B}(g) = \mathbb{B}_0(g) \cup \mathcal{W}(g)$, it follows from Proposition 3.4 that $\omega_G(x) = \overline{J}$ for every $x \in L \cap U_\ell$, but this is a contradiction, as A is a Cantor set and $\omega_G(x) \subset \omega_g(x) = \omega_f(x) = A$ for every $x \in L \cap U_\ell$.

Now, we present two questions about piecewise smooth maps of the interval. Question 3.7 below is a natural one, if one has in mind the conjecture of Martens and de Melo (Conjecture 1). We say that A is a **Cherry-like attractor** if $\#A = \infty$ and there is an open set U such that $A \cap U \neq \emptyset$ and $\omega_f(x) = A$ for every $x \in U$.

Question 3.7 *Are all the wandering intervals of a non-flat piecewise C^2 map contained in the basin of attraction of Cherry-like attractors?*

The question below was proposed by de Melo to the authors during the *International Conference on Dynamical Systems*, Búzios, 2016.

Question 3.8 *Is there an upper bounded for the period of non expanding periodic points of a given non-flat piecewise C^2 map of the interval?*

Recall that a periodic point p is called **expanding** if $|(f^n)'(p)| > 1$, where n is the period of p. For C^2 non-flat maps of the interval (that is, maps without discontinuities), the answer of the question above was proved by Welington de Melo himself in a joint work with Sebastian van Strien [27].

Appendix

A **homterval** is an open interval $I = (a, b)$ such that $f^n|_I$ is a homeomorphism for $n \geq 1$. This is equivalent to assume that $I \cap \mathcal{O}_f^-(\mathcal{C}_f) = \emptyset$. A homterval I is called a *wandering interval* if $I \cap \mathbb{B}_0(f) = \emptyset$ and $f^j(I) \cap f^k(I) = \emptyset$ for all $1 \leq j < k$, where $\mathbb{B}_0(f)$ the union of the basins of attraction of all periodic-like attractors.

Lemma 3.9 (Homterval Lemma, see [28]) *Let $I = (a, b)$ be a homterval of f. If I is not a wandering interval, then $I \subset \mathbb{B}_0(f) \cup \mathcal{O}_f^-(\mathrm{Per}(f))$. Furthermore, if f is C^3 with $Sf < 0$, and I is not a wandering interval, then the set $I \setminus \mathbb{B}_0$ has at most one point.*

Lemma 3.10 (Koebe's Lemma [28]) *For every $\varepsilon > 0$, $\exists K > 0$ such that the following holds: Let M, T be intervals in $\mathbb{R}$ with $M \subset T$ and denote respectively by L and R the left and right components of $T \setminus M$. If $f : T \to f(T) \subset \mathbb{R}$ is a C^3 diffeomorphism with negative Schwarzian derivative and*

$$|f(L)| \geq \varepsilon |f(M)| \text{ and } |f(R)| \geq \varepsilon |f(M)|,$$

then $\frac{|Df(x)|}{|Df(y)|} \leq K$ for $x, y \in M$.

Proposition 3.11 [See Proposition 3 in [10] and Proposition 2 of [38]] *Let $\mathcal{C}_f \subset (0, 1)$ be a finite set. If $f : [0, 1] \setminus \mathcal{C}_f \to [0, 1]$ is a C^2 non-flat local diffeomorphism, then there exists a function $O(\varepsilon)$ with $O(\varepsilon) \to 0$ as $\varepsilon \searrow 0$ with the following property: Let $J \subset T$ be an interval, R, L the connected components of $T \setminus J$ and $\delta := \min\{|R|/|J|, |L|/|J|\}$. Let n be an integer and $T_0 \supset J_0$ intervals such that $f^n|_{T_0}$ is a diffeomorphism, $f^n(T_0) = T$ and $f^n(J_0) = J$. If $\varepsilon := \max\{|f^j(T_0)|; 0 \leq j \leq n\}$, then*

(1) $\left|\frac{Df^n(x)}{Df^n(y)}\right| \leq \left(\frac{1+\delta}{\delta}\right)^2 e^{O(\varepsilon) \sum_{i=0}^{n-1} |f^i(J_0)|}$, *for all $x, y \in J_0$;*

(2) $\exists \delta' > 0$ *depending only on ε and $\sum_{i=0}^{n-1} |f^i(J_0)|$ such that, for all $x, y \in J_0$ and $1 \leq j \leq n$, we have $\left|\frac{Df^j(x)}{Df^j(y)}\right| \leq \left(\frac{1+\delta'}{\delta'}\right)^2 e^{O(\varepsilon) \sum_{i=0}^{n-1} |f^i(J_0)|}$;*

(3) $\frac{|Df^n(x)|}{|Df^n(y)|} \leq \exp\left(\frac{2}{\delta|J|}|f^n(x) - f^n(y)| + O(\varepsilon) \sum_{i=0}^{n-1} |f^i(x) - f^j(y)|\right)$ *for every $x, y \in J_0$.*

Proposition 3.12 (Proposition 3.4 for maps with negative Schwarzian derivative) *Let U be an open subset of an interval (a, b) and $F : U \to (a, b)$ be a C^3 local* *diffeomorphism with negative Schwarzian derivative and such that $F(P) = (a, b)$ for every connected component of U. If $\mathbb{B}(F)$ denotes the union of all wandering intervals of F with the basin of attraction of all periodic-like attractors of F, then*

$$\mathrm{Leb}\left(\bigcap_{n \geq 0} F^{-n}((a, b)) \setminus \mathbb{B}(F)\right) \text{ is either } 0 \text{ or } b - a.$$

Furthermore, if $\mathrm{Leb}\left(\bigcap_{n \geq 0} F^{-n}((a, b)) \setminus \mathbb{B}(F)\right) = b - a$ *then* $\mathrm{Leb}|_{(a,b)}$ *is* F *strongly ergodic and* $\omega_F(x) = [a, b]$ *for almost every* $x \in (a, b)$. $\qquad\square$

Proof Similar to the proof of Proposition 3.4. Essentially, one only need to replace Proposition 3.11 by Koebe's Lemma (Lemma 3.10).

Proof of Proposition 2.5 As the attractor A is a cycle of intervals, we get from Theorem 1 that $\omega_f(x) = A$ for almost every $x \in [0, 1]$. In particular, this forbids f to have wandering intervals and periodic attractors. As $\overline{\mathcal{O}_f^+(c_-)} \cup \overline{\mathcal{O}_f^+(c_+)}$ has empty interior, we can consider a connected component $J \neq \emptyset$ of $A \setminus (\overline{\mathcal{O}_f^+(c_-)} \cup \overline{\mathcal{O}_f^+(c_+)})$. As $\overline{\mathcal{O}_f^+(c_-)} \cup \overline{\mathcal{O}_f^+(c_+)} \neq A$, A is not a minimal set and so it cannot be a Cherry attractor. Hence, it follows from Theorem D of [8] that A is a chaotic cycle of intervals. In particular, this implies that $\overline{\mathrm{Per}(f) \cap A} = A$. Thus, choose $q, p \in \mathrm{Per}(f) \cap \mathrm{interior}(J)$ with $p < q$. Let I be any connected component of $(p, q) \setminus \overline{\mathcal{O}_f^+(c_-)} \cup \overline{\mathcal{O}_f^+(c_+)}$. Note that I is a nice interval, that is, $\mathcal{O}_f^+(\partial I) \cap I = \emptyset$. Let $I^* = \{x \in I ; \mathcal{O}_f^+(f(x)) \cap I \neq \emptyset\}$ and $F : I^* \to I$ the first return map, by f, to I. As $\omega_f(x) \supset I$ for almost every $x \in [0, 1]$ and $\mathrm{Leb} \circ f^{-1} \ll \mathrm{Leb}$, we get that $\mathrm{Leb}(I) = \mathrm{Leb}(I^*) = \mathrm{Leb}(\bigcap_{n \geq 0} F^{-1}(I^*))$. Thus, it follows from Proposition 3.12 that $\mathrm{Leb}|_I$ is strongly ergodic with respect to F, in particular, F is ergodic.

The ergodicity of $\mathrm{Leb}|_I$ with respect to f implies that Leb is also ergodic with respect to f. Indeed, if $V = f^{-1}(V)$ and $\mathrm{Leb}(V) > 0$ then, as $\mathrm{Leb} \circ f^{-1} \ll \mathrm{Leb}$ and $\mathcal{O}_f^+(x) \cap I \neq \emptyset$ for Leb almost all $x \in [0, 1]$, we get that $\mathrm{Leb}(V \cap I) > 0$. As F is the first return map to I, it follows that $F^{-1}(V \cap I) = V \cap I$. Thus, by the ergodicity with respect to F, $\mathrm{Leb}(V \cap I) = \mathrm{Leb}(I)$. That is, $\mathrm{Leb}(I \setminus V) = 0$. Using that $\mathrm{Leb} \circ f^{-1} \ll \mathrm{Leb}$, we get that $\mathrm{Leb}([0, 1] \setminus V) = 0$, proving that Leb is ergodic with respect to f.

References

1. V.S. Afraimovich, V.V. Bykov, L.P. Shil'nikov, On the appearance and structure of the Lorenz attractor. Dokl. Acad. Sci. USSR **234**, 336–339 (1977)
2. J.F. Alves, M. Soufi, Statistical stability and limit laws for Rovella maps. Nonlinearity **25**(12), 3527–3552 (2012)
3. A. Arneodo, P. Coullet, C. Tresser, A possible new mechanism for the onset of turbulence. Phys. Lett. A **81**(4), 197–201 (1981)

4. V.I. Arnold, V.S. Afrajmovich, Y. Ilyashenko, L.P. Shilnikov, *Bifurcation Theory and Catastrophe Theory* (Springer-Verlag, Berlin, Heidelberg, 1999)
5. V. Araujo, A. Castro, M.J. Pacifico, V. Pinheiro, Multidimensional Rovella-like attractors. J. Differ. Equations 3163–3201 (2011)
6. L. Alsedà, J. Llibre, M. Misiurewicz, C. Tresser, Periods and entropy for Lorenz-like maps. Ann. Inst. Fourier **39**, 929–952 (1989)
7. A. Blokh, M. Lyubich, Measurable dynamics of S-unimodal maps of the interval. Ann. Sci. École Norm. Sup. **24**, 545–573 (1991)
8. P. Brandão, On the structure of contracting Lorenz maps, to appear on Annales de l'Institut Henri Poincaré. Analyse Non Linéaire (2018). https://doi.org/10.1016/j.anihpc.2017.12.001
9. P. Brandão, J. Palis, V. Pinheiro, On the finiteness of attractors for one-dimensional maps with discontinuities (2013). arXiv:1401.0232
10. P. Brandão, J. Palis, V. Pinheiro, On the finiteness of attractors for piecewise C^2 maps of the interval. Erg. Theo. Dyn. Sys. **261**, 1–21 (2017)
11. H. Bruin, G. Keller, T. Nowicki, S.J. van Strien, Wild Cantor attractors exist. Ann. Math. **143**, 97–130 (1996)
12. E. Catsigeras, On Ilyashenko's statistical attractors. CDSS. **29**, 78–97 (2013)
13. T. Cherry, Analytic quasi-periodic curves of discontinuous type on the torus. Proc. London Math. Soc. **44**, 175–215 (1938)
14. P. Collet, P. Coullet, C. Tresser, Scenarios under constraint. J. Phyique Lett. **46**, 143–147 (1985)
15. A. Denjoy, Sur les courbes definies par les équations différentielles à la surface du tore. J. de Mathématiques Pures et Appliquées **11**, 333–375 (1930)
16. J.M. Gambaudo, C. Tresser, Dynamique réguliére ou chaotique. Applications du cercle ou de l'intervalle ayant une discontinuité. Comptes Rendus Acad. Sci. Paris, Ser I. **300**, 311–313 (1985)
17. J. Guckenheimer, R.F. Williams, Structural stability of Lorenz attractors. Publ. Math. IHES **50**, 59–72 (1979)
18. F. Hofbauer, G. Keller, Quadratic maps without asymptotic measure. Commun. Math. Phys. **127**, 319–337 (1990)
19. G. Keller, M.S. Pierre, Topological and measurable dynamics of Lorenz maps, in *Ergodic Theory, Analysis, and Efficient Simulation of Dynamical Systems*, ed. by B. Fiedler (Springer, Berlin, Heidelberg, 2001)
20. R. Labarca, C.G. Moreira, Essential dynamics for Lorenz maps on the real line and the lexicographical world. Annales de l'Institut Henri Poincaré. Analyse Non Linéaire 23, 683694 (2006)
21. M. Lyubich, Non-existence of wandering intervals and structure of topological attractors of one dimensional dynamical systems: Schwarzian derivative. Ergod. Theory Dyn. Syst. **9**, 737–749 (1989)
22. R. Mañé, R. Mane, Hyperbolicity, sinks and measure in one dimensional dynamics. Commun. Math. Phys. **100**(4), 495–524, Commun. Math. Phys. (1987) **112**, 721–724 (Erratum) (1985)
23. M. Martens, P. Mendes, W. de Melo, S. van Strien, On Cherry flows. Ergod. Theory Dyn. Syst. **10**, 531–554 (1990)
24. M. Martens, W. de Melo, Universal models for Lorenz maps (pre-print version, 1996), de Melo's homepage at IMPA. http://w3.impa.br/%7Edemelo/lorenz.ps
25. M. Martens, W. de Melo, Universal models for Lorenz maps. Ergod. Theory Dyn. Syst. **21**, 833–860 (2001)
26. R. Metzger, Sinai-Ruelle-Bowen measures for contracting Lorenz maps and flows (Annales de l'Institut Henri Poincaré, Analyse Non Linéaire, 2000)
27. W. de Melo, S. van Strien, A structure theorem in one dimensional dynamics. Ann. Math. **129**, 519–546 (1989)
28. W.C. de Melo, S.V. Strien, *One Dimensional Dynamics* (Springer 1993)
29. B. Mestel, D. Berry, Wandering intervals for Lorenz maps with bounded nonlinearity. Bull. Lond. Math. **23**, 183–189 (1991)
30. J. Milnor, On the concept of attractor. Commun. Math. Phys. **99**, 177–195 (1985)

31. M.J. Pacifico, M.J. Todd, Thermodynamic formalism for contracting Lorenz flows. Stat. Phys. **139**(1), 159–176 (2010)
32. M. St. Pierre, Topological and measurable dynamics of Lorenz maps. Dissertationes Mathematicae,**17** (1999)
33. I. Procaccia, S. Thomas, C. Tresser, Frist-return maps as a unifield renormalization scheme for dynamical systems. Phys. Rev. A. **35**, 1–17 (1986)
34. A. Rovella, The dynamics of perturbations of contracting Lorenz maps. Bul. Soc. Brazil Mat. (N.S.) **24**(2), 233–259 (1993)
35. C. Tresser, Nouveaux types de transitions vers une entropie topologique positive. Comptes Rendus Acad. Sci. Paris, Ser I. **296**, 729–732 (1983)
36. C. Tresser, R.F. Williams, Splitting words and Lorenz braids. Phys. D **62**, 15–21 (1993)
37. W. Tucker, The Lorenz attractor exists. C. R. Acad. Sci. Paris, Série I **328**, 1197–1202 (1999)
38. S. van Strien, E. Vargas, Real bounds, ergodicity and negative Schwarzian for multimodal maps. J. Am. Math. Soc. **17**, 749–782 (2004)

Correspondences in Complex Dynamics

Shaun Bullett, Luna Lomonaco and Carlos Siqueira

*In memory of Welington de Melo, whose intellectual honesty
inspired us all.*

Abstract This paper surveys some recent results concerning the dynamics of two
families of holomorphic correspondences, namely $\mathcal{F}_a : z \to w$ defined by the relation

$$\left(\frac{aw-1}{w-1} \right)^2 + \left(\frac{aw-1}{w-1} \right) \left(\frac{az+1}{z+1} \right) + \left(\frac{az+1}{z+1} \right)^2 = 3,$$

and

$$\mathbf{f}_c(z) = z^\beta + c, \text{ where } 1 < \beta = p/q \in \mathbb{Q},$$

which is the correspondence $\mathbf{f}_c : z \to w$ defined by the relation

$$(w-c)^q = z^p.$$

Both can be regarded as generalizations of the family of quadratic maps $f_c(z) = z^2 + c$. We describe dynamical properties for the family $\mathcal{F}_a$ which parallel properties
enjoyed by quadratic polynomials, in particular a Böttcher map, periodic geodesics

S. Bullett
School of Mathematical Sciences, Queen Mary University of London, London E1 4NS, UK
e-mail: s.r.bullett@qmul.ac.uk

L. Lomonaco
Department of Applied Mathematics, University of São Paulo, São Paulo CEP 05508-900, Brazil
e-mail: lluna@ime.usp.br

C. Siqueira (✉)
Department of Mathematics, University of São Paulo, São Paulo CEP 05508-900, Brazil

Department of Mathematics, Federal University of Bahia, Salvador CEP 40170-115, Brazil
e-mail: carlos.siqueira@ufba.br

© Springer Nature Switzerland AG 2019
M. J. Pacifico and P. Guarino (eds.), *New Trends in One-Dimensional
Dynamics*, Springer Proceedings in Mathematics & Statistics 285,
https://doi.org/10.1007/978-3-030-16833-9_5

and Yoccoz inequality, and we give a detailed account of the very recent theory of holomorphic motions for hyperbolic multifunctions in the family $\mathbf{f}_c$.

Keywords Holomorphic correspondences · Holomorphic motions · Yoccoz inequality · Hausdorff dimension · Matings · External rays

1 Introduction

A holomorphic correspondence on the Riemann sphere is a relation $z \mapsto w$ given implicitly by a polynomial equation $P(z, w) = 0$. Any rational map is an example of a holomorphic correspondence. Indeed, if $f(z) = p(z)/q(z)$, then $w = f(z)$ iff $P(z, w) = 0$, where $P(z, w) = wq(z) - p(z)$. In particular, the family of quadratic polynomials $f_c(z) = z^2 + c$ (parametrized by $c \in \mathbb{C}$) can be regarded as an analytic family of holomorphic correspondences. The grand orbits of any finitely generated Kleinian group can also be regarded as those of a holomorphic correspondence.

This paper is concerned with two families of holomorphic correspondences which generalize quadratic polynomials in different ways. The first is the family $\mathcal{F}_a : z \to w$ defined by

$$\left(\frac{aw - 1}{w - 1}\right)^2 + \left(\frac{aw - 1}{w - 1}\right)\left(\frac{az + 1}{z + 1}\right) + \left(\frac{az + 1}{z + 1}\right)^2 = 3, \tag{1}$$

where $a \in \mathbb{C}$ and $a \neq 1$, introduced in the early nineties by Bullett and Penrose [4]. They proved:

Theorem 1.1 *For every a in the real interval [4,7], the correspondence $\mathcal{F}_a$ is a mating between some quadratic map $f_c(z) = z^2 + c$ and the modular group $\Gamma = \mathrm{PSL}(2, \mathbb{Z})$,*

and conjectured that the connectedness locus for this family is homeomorphic to the Mandelbrot set.

The second family is

$$\mathbf{f}_c(z) = z^\beta + c, \quad c \in \mathbb{C}, \tag{2}$$

where $\beta > 1$ is a rational number and $z^\beta = \exp \frac{1}{q}(\log z^p)$. If $\beta = p/q$ in lowest terms, then each member of the family (2) of multifunctions is a holomorphic correspondence, defined by the relation $(w - c)^q = z^p$. Hence $\mathbf{f}_c$ maps every $z \neq 0$ to a set consisting of q points. If p and q are not relatively prime, we shall use the notation $z^{p/q} + c$ to express the holomorphic correspondence $(w - c)^q = z^p$. Thus $z^2 + c$ and $z^{4/2} + c$ denote different correspondences.

In this paper we describe the dynamics of holomorphic correspondences from various perspectives, exploring the concepts of hyperbolicity and holomorphic motions for (2) and describing results concerning a Böttcher map, periodic geodesics, and a Yoccoz inequality for the family of matings (1). As we shall see, the techniques involved in the two studies are independent, but as we have already noted, both families can be viewed as generalizations of the quadratic family, and our techniques for studying them are motivated by the notions of hyperbolicity, external rays, Yoccoz inequalities and local connectivity, which are inextricably related to one another in the study of quadratic polynomials $f_c(z) = z^2 + c$. For this reason, it will be convenient to start by recalling some well known facts, techniques and open questions concerning this celebrated family of maps. Excellent sources for details are the books of Milnor [19] and de Faria and de Melo [8]. An overview of a century of complex dynamics is presented in the article by Rees [25].

1.1 Dynamics of Quadratic Maps

Consider the action of $f_c(z) = z^2 + c$ on the Riemann sphere $\widehat{\mathbb{C}}$. For any polynomial of degree $d \geq 2$ acting on $\widehat{\mathbb{C}}$, the point $z = \infty$ is a superattracting fixed point. Let $\mathcal{A}_c$ denote its basin of attraction. The filled Julia set $K_c = K_{f_c}$ is the set of points with bounded orbit, that is $K_c = \widehat{\mathbb{C}} \setminus \mathcal{A}_c$. The Julia set $\mathcal{J}_c = \mathcal{J}_{f_c}$ is the common boundary of these regions: $\mathcal{J}_c = \partial K_c = \partial \mathcal{A}_c$. The *Mandelbrot set* M is the connectedness locus of the family $f_c(z) = z^2 + c$; that is the set of all parameters $c \in \mathbb{C}$ such that $\mathcal{J}_c$ is connected.

On a neighbourhood of ∞, the quadratic polynomial f_c is conformally conjugate to $f_0(z) = z^2$ by the *Böttcher map* φ_c, which is the conjugacy tangent to the identity at infinity. In the case $\mathcal{J}_c$ (or equivalently K_c) is connected, the Böttcher map extends to a conformal conjugacy:

$$\varphi_c : \mathbb{C} \setminus K_c \to \mathbb{C} \setminus \overline{\mathbb{D}}_1$$

(An analogue of this map for the family $\mathcal{F}_a$ will appear in Sect. 2.4.) The *external ray* $R_\theta^c \in \mathbb{C} \setminus K_c$ with argument $\theta \in \mathbb{R}/\mathbb{Z}$ is the preimage under the Böttcher map φ_c of the half-line $te^{2\pi i\theta} \in \mathbb{C} \setminus \overline{\mathbb{D}}_1$, with $t \in (1, \infty)$. When

$$\lim_{t \to 1^+} \varphi_c^{-1}(te^{2\pi i\theta}) = z,$$

we say that R_θ^c lands at z. We know that *rational rays land* [9, 19], and that *repelling and parabolic periodic points are landing points* of at least one and at most finitely many rays [19]. By Carathéodory's theorem, if $\mathcal{J}_c$ is locally connected, then every external ray lands. We remark that the Böttcher map and external rays can also be defined for degree d polynomials, and in this case as well rational rays land and repelling and parabolic periodic points are landing points [19]. (*Hyperbolic geodesics* play an analogous role for the family $\mathcal{F}_a$ and enjoy similar properties to external rays, see Sect. 2.4).

Using the Böttcher map, Douady and Hubbard constructed a conformal homeomorphism between the complement of the Mandelbrot set and the complement of the closed unit disk:

$$\Phi : \mathbb{C} \setminus M \to \mathbb{C} \setminus \overline{\mathbb{D}}_1$$

$$c \quad \to \quad \varphi_c(c),$$

proving that the Mandelbrot set is compact and connected [9]. This isomorphism also allows the definition of *parameter space external rays*: the parameter ray of argument θ is $\mathcal{R}_\theta = \Phi^{-1}(R_\theta^0)$. The celebrated conjecture that M is locally connected (known as MLC) implies that every parameter space external ray lands. MLC is crucial in one dimensional complex dynamics, since it has been proved ([10]) to imply density of hyperbolicity (Conjecture 1.1) for the family of quadratic polynomials. A rational map is called *hyperbolic* when all its critical points are attracted to attracting cycles. Hyperbolic maps are among the best understood rational maps. Indeed, if the quadratic polynomial f_c is hyperbolic then (i) every orbit in the interior of the filled Julia set K_c (if non-empty) converges to the finite attracting cycle (which is unique since f_c is quadratic); (ii) every orbit outside K_c converges to ∞; and (iii) f_c is expanding and topologically mixing on the Julia set $\mathcal{J}_c = \partial K_c$. A major conjecture in holomorphic dynamics is:

Conjecture 1.1 (Density of hyperbolicity) *The set of hyperbolic rational maps is open and dense in the space of rational maps* Rat_d *of the same degree.*

A version of this conjecture dates back to Fatou, and for this reason Conjecture 1.1 is often known as the *Fatou conjecture*. Note that it concerns density of hyperbolicity, since openness of the set of hyperbolic maps is known.

Strongly related to hyperbolicity is the concept of structural stability. A map f_a is *structurally stable* if f_c is topologically conjugate to f_a, for every c in an open set containing a. For rational maps on the Riemann sphere J-*stability*, which roughly speaking means stability on a neighborhood of the Julia set, is usually considered [25]. Mañé, Sad and Sullivan [21] have shown that the set of J-structurally stable rational maps is open and dense in the space of rational maps Rat_d of the same degree. Since in any family of holomorphic maps the set of hyperbolic parameters forms an open and closed subset of the J-stable parameters, Conjecture 1.1 is equivalent to the following (see [18]):

Conjecture 1.2 *A J-stable rational map of degree d is hyperbolic.*

For quadratic polynomials, Conjecture 1.1 claims that the set of c such that $f_c(z) = z^2 + c$ is hyperbolic is an open and dense subset of the complex plane. On the other hand, density of J-stability implies that each of the infinitely many components U of $\mathbb{C} \setminus \partial M$ is the parameterization domain of a holomorphic motion $h_c : \mathcal{J}_a \to \mathcal{J}_c$, $c \in U$ (holomorphic motions are defined in Sect. 3.1), with base point $a \in U$ arbitrarily fixed, and every h_c being a *quasi-conformal conjugacy*. If U is a component of $\mathbb{C} \setminus \partial M$ having one point a for which $f_a(z) = z^2 + a$ is hyperbolic, then f_c is hyperbolic for every c in U, and thus in the quadratic setting density of hyperbolicity is equivalent to conjecturing that *every component of* $\mathbb{C} \setminus \partial M$ *is hyperbolic*. Note that, since $\mathcal{J}_0 = \mathbb{S}^1$, it follows that $\mathcal{J}_c$ is a *quasicircle* (image of $\mathbb{S}^1$ under a quasiconformal homeomorphism) for every c close to zero (more precisely, for every c in the same hyperbolic component as $c = 0$). (A generalization of this fact for $z^\beta + c$ is given by Theorem 3.4).

In the late eighties J.-C. Yoccoz made a major contribution towards the MLC conjecture, proving that MLC holds at every point $c \in \partial M$ such that f_c is not infinitely renormalizable. A key ingredient is what is now known as the *Yoccoz inequality*. It can be shown that if z is a repelling fixed point for a degree d polynomial P with connected filled Julia set, then just finitely many external rays γ_i, say q', land at z. Each γ_i is periodic with the same period, and there exists $p' < q'$ such that $P \circ \gamma_i = \gamma_{i+p'}$ for any i. The number of cycles of rays landing at z is $m = \gcd(p', q')$, and $\theta = p/q = (p'/m)/(q'/m)$ is called the *combinatorial rotation number* of P at z.

Theorem 1.2 (Yoccoz-Pommerenke-Levin inequality [13, 15, 22]) *If z is a repelling fixed point of a degree d polynomial P with connected filled Julia set, and $\theta = p/q$ is its combinatorial rotation number in lowest terms, then*

$$\frac{\operatorname{Re}\tau}{|\tau - 2\pi i\theta|^2} \geq \frac{mq}{2\log d}, \tag{3}$$

for some branch τ of $\log P'(z)$.

(A Yoccoz inequality for the family $\mathcal{F}_a$ is developed by the first two authors in [3]; see Theorem 1.3. While the original Yoccoz inequality is proven for degree d polynomials, and so applies to iterates of degree 2 polynomials and hence to periodic orbits, an inequality of the form presented in Theorem 1.3 has so far only been proved for repelling fixed points.)

In 1994, C. McMullen made a deep contribution toward MLC, by proving that every component of the interior of the Mandelbrot set meeting the real axis is hyperbolic [18]. In the late nineties, M. Lyubich [17], and independently Graczyk and Swiatek [12] proved density of hyperbolicity for the real quadratic family. About ten years later Kozlovski, Shen and van Strien proved it for real polynomials of higher degree, by proving that any real polynomial can be approximated by hyperbolic real polynomials of the same degree [14]. However, density of hyperbolicity for degree d rational maps on $\widehat{\mathbb{C}}$ is still open.

1.2 Dynamics of Holomorphic Correspondences

We now outline our main results described in this paper, concerning the families (1) and (2): these involve generalizations of the concepts presented in Sect. 1.1. *Readers who want to see the proofs—as Welington always did—can find those concerning family (1) in [2, 3], and those concerning family (2) in [26–29].*

Part I. We start with an abstract definition of matings between quadratic maps and PSL$(2, \mathbb{Z})$ (Sect. 2.1) with the help of Minkowski's question mark function. This description dates back to 1994, when the first author together with Bullett and Penrose [4] started investigating the family $\mathcal{F}_a$. The formal definitions of limit sets and the connectedness locus $\mathcal{C}_\Gamma$ for this family are given in Sect. 2.2. There we also define a *mating* between the modular group and a map in the parabolic quadratic family

$$\text{Per}_1(1) = \{P_A(z) = z + 1/z + A \mid A \in \mathbb{C}\}/(A \sim -A),$$

and present a result which is a significant advance on Theorem 1.1, namely that for any $a \in \mathcal{C}_\Gamma$, the correspondence $\mathcal{F}_a$ is a mating between PSL$(2, \mathbb{Z})$ a parabolic map in $\text{Per}_1(1)$ (see Theorem 2.1, and Figs. 4 and 5).

We open Sect. 2.4 by recalling the existence of a Böttcher map for the family $\mathcal{F}_a$ when $a \in \mathcal{C}_\Gamma$ (see Theorem 2.2), and we then use it to construct periodic geodesics on the regular domain of $\mathcal{F}_a$ (an analogue of periodic external rays). These land (see Theorem 2.3), analogously to the rational external rays for the quadratic family of polynomials.

By a quite technical and deep argument [3] it can be shown that when a is in $\mathcal{C}_\Gamma$ every repelling fixed point z of $\mathcal{F}_a$ is the landing point of exactly one periodic cycle of geodesics. It follows, as for polynomials, that z has a well-defined combinatorial rotation number $\theta = p/q$. A geodesic in the cycle is stabilized by a Sturmian word $W_{p/q}$, in α and β, of rotation number p/q (Sturmian words are defined in Sect. 2.5: $W_{p/q}$ is unique up to cyclic permutation for any given p/q).

Theorem 1.3 (Yoccoz inequality) *Let $a \in \mathcal{C}_\Gamma$ and z be a repelling fixed point of f_a whose combinatorial rotation number is $\theta = p/q$ in lowest terms. Then there is a branch τ of $\log f_a'(z)$ such that*

$$\frac{\text{Re}\,\tau}{|\tau - 2\pi i\theta|^2} \geq \frac{q^2}{4p \, \log(\lceil q/p \rceil + 1)}, \quad \text{if } \theta \leq 1/2; \text{ and}$$

$$\frac{\text{Re}\,\tau}{|\tau - 2\pi i\theta|^2} \geq \frac{q^2}{4(q-p) \, \log(\lceil q/(q-p) \rceil + 1)}, \quad \text{if } \theta > 1/2.$$

The inequalities of both Theorems 1.2 and 1.3 have geometric interpretations as restricting the logarithm of the derivative at a repelling fixed point to a round disk for each p/q. See Fig. 1 for illustrations.

Theorem 1.3 provides a key step in the strategy of the first two authors to prove that the part $M_\Gamma = \mathcal{C}_\Gamma \cap \{z : |z - 4| \leq 3\}$ of the connectedness locus $\mathcal{C}_\Gamma$ of the family (1) is homeomorphic to the connectedness locus M_1 of the parabolic family $\{z \mapsto z + \frac{1}{z} + A : A \in \mathbb{C}\}/(A \sim -A)$. With the result announced by Carsten Peterson and Pascale Roesch that M_1 is homeomorphic to the Mandelbrot set M [23], this will finally prove the long-standing conjecture that M_Γ (pictured in Fig. 6) is homeomorphic to M.

Fig. 1 Disks in the τ-plane
permitted by the Yoccoz
inequality: on the left for the
matings $\mathcal{F}_a$, and on the right
for the classical case of
quadratic polynomials. (In
each case the disks plotted
correspond to all
$p/q \in [0, 1/2]$ with $q \le 8$,
and to 1/16)

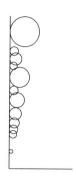

Part II. Section 3 describes the dynamics of hyperbolic correspondences in the family
(2). We start by defining Julia sets (see Fig. 7 for an example). The main subject is the
generalization of holomorphic motions, which involves the construction of a solenoid
associated to the Julia set of $\mathbf{f}_c(z) = z^\beta + c$ (Theorem 3.4). For parameters c close to
zero, the dynamics of $z^\beta + c$ on its Julia set $\mathcal{J}_c$ is the projection of a (single-valued)
dynamical system $f_c : U \to U$ given by as holomorphic map defined on a subset
$U \subset \mathbb{C}^2$. The maximal invariant set of f_c is a solenoid whose projection is $\mathcal{J}_c$. The
projection of the holomorphic motion in $\mathbb{C}^2$ yields a branched holomorphic motion on
the plane, as defined by Lyubich and Dujardin [11] for polynomial automorphisms
of $\mathbb{C}^2$. Branched holomorphic motions are described in greater generality for the
family (2) in [29].

The advantage of the solenoid construction is that it makes possible to apply
certain techniques of Thermodynamic Formalism to the family of maps $f_c : U \to U$
and use them to estimate the Hausdorff dimension of $\mathcal{J}_c$. For example,

Theorem 1.4 (Hausdorff dimension) *If $q^2 < p$ then for every c sufficiently close to
zero,*

$$\dim_H \mathcal{J}_c < 2,$$

where $\dim_H$ denotes the Hausdorff dimension of $\mathcal{J}_c$.

In the family of Fig. 2 we have $p = 5$ and $q = 2$. Since $2^2 < 5$, it follows that
$\mathcal{J}_c$ is the projection of a solenoid having zero Lebesgue measure. The assumption
$q^2 < p$ may not be sharp. The essential idea is that $\dim_H \mathcal{J}_c \to 2$ as $\beta \to 1$, which
is supported by many experiments.

Notation and Terminology.

1. Holomorphic correspondences are denoted by $\mathcal{F}, \mathcal{G}, \ldots$ in the context of matings,
   or by $\mathbf{f}, \mathbf{g}, \ldots$ when studying hyperbolic multifunctions.
2. By the term *multifunction* we mean any multivalued map. Every multifunction
   maps points to subsets.

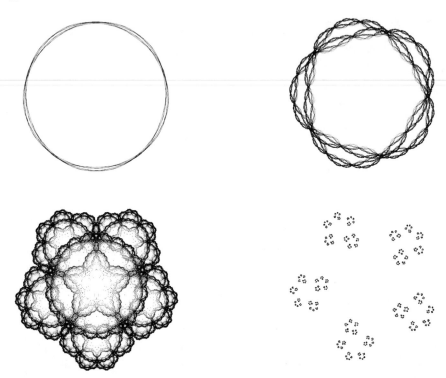

Fig. 2 Julia sets of $z^\beta + c$, where $\beta = 5/2$ is fixed. The values of c are, respectively, 0.05, $(1 + i)/5$, 0.7 and $2 + i$, read from upper-left to bottom-right. $\mathcal{J}_c$ is a circle at the singularity $c = 0$, but the first figure reveals that $\mathcal{J}_c$ is the *shadow* of a solenoid for every c close to zero with $c \neq 0$. As we perform the branched motion, more bifurcations are added to $\mathcal{J}_c$. The complexity increases up to a certain moment (third to fourth steps) when the process reverses and $\mathcal{J}_c$ becomes a Cantor dust. The first three are connected and the fourth is a Cantor set. In this family, $\mathcal{J}_c$ is a Cantor set for $|c|$ sufficiently large

3. $\mathbb{S}^1 = \{z \in \mathbb{C} : |z| = 1\}$, $\widehat{\mathbb{C}} = \mathbb{C} \cup \{\infty\}$, $\mathbb{H} = \{z = x + iy \in \mathbb{C} : y > 0\}$, and $f^n = \underbrace{f \circ \cdots \circ f}_{n}$.

4. $\Gamma = \mathrm{PSL}(2, \mathbb{Z})$ is the modular group consisting of all Möbius transformations

$$z \mapsto \frac{az + b}{cz + d},$$

where $ad - bc = 1$ and $a, b, c, d \in \mathbb{Z}$. The operation is the standard composition $\circ$. The *generators* of the modular group that we shall use are the maps

$$\alpha(z) = z + 1 \text{ and } \beta(z) = \frac{z}{z + 1}.$$

Consider

$$P(z, w) = (w - (z + 1))(w(z + 1) - z) = 0. \tag{4}$$

The grand orbits of $\mathrm{PSL}(2, \mathbb{Z})$ on $\mathbb{H}$ are identical to those of the holomorphic correspondence $\mathcal{H} : \widehat{\mathbb{C}} \to \widehat{\mathbb{C}}$ determined by $P(z, w) = 0$.

2 Mating Quadratic Maps with PSL(2, $\mathbb{Z}$)

Recall that in the case of hyperbolic quadratic polynomials $f_c(z) = z^2 + c$, the *topological mating* between f_c and $f_{c'}$ is the map

$$g : \frac{K_c \cup K_{c'}}{\sim} \to \frac{K_c \cup K_{c'}}{\sim}$$

induced by f_c and $f_{c'}$ on the quotient space, where $\sim$ is the smallest closed relation such that $\varphi_c(z) \sim \varphi_{c'}(\bar{z})$, for every $z \in \mathbb{S}^1$ (φ_c is the boundary extension of the Böttcher coordinate and K_c is a copy of the filled Julia set). The two maps are *matable* if the quotient space is a sphere, and g can be realized as a rational map. By applying Thurston's characterization of rational maps among critically finite branched coverings of the sphere, Tan [30] and Rees [24] proved that two quadratic polynomials f_c, $f_{c'}$ with periodic critical points are matable if and only if c and c' do not belong to complex conjugate limbs of the Mandelbrot set.

Matings can also be constructed between Fuchsian groups: by applying the Bers Simultaneous Uniformization Theorem certain Fuchsian groups can be mated with (abstractly isomorphic) Fuchsian groups to yield *quasifuchsian* Kleinian groups. (See [7] for a discussion of matings in various contexts in conformal dynamics.) What is a surprise when first encountered is that certain Fuchsian groups can be mated with polynomial maps (see Sect. 2.1). This is achieved in a larger category of conformal dynamical systems, containing both rational maps and finitely generated Kleinian groups, the category of holomorphic correspondences on the Riemann sphere. These are multifunctions $\mathcal{F} : \widehat{\mathbb{C}} \to \widehat{\mathbb{C}}$, for which there is a polynomial $P(z, w)$ in two complex variables such that $\mathcal{F}(z) = \{w \in \widehat{\mathbb{C}} : P(z, w) = 0\}$.

2.1 Mating Quadratic Polynomials with PSL(2, $\mathbb{Z}$)

Examples of matings between quadratic polynomials and the modular group were discovered by the first author and Christopher Penrose in the early '90s. To understand their existence we first consider how one can construct an abstract (topological) model (see also [1, 4] for more details).

Topogical Mating: Minkowski's Question Mark Function. Let

$$h : \hat{\mathbb{R}}_{\geq 0} \to [0, 1]$$

denote the homeomorphism which sends $x \in \mathbb{R}$ represented by the continued fraction

$$[x_0; x_1, x_2, \ldots] = x_0 + \cfrac{1}{x_1 + \cfrac{1}{x_2 + \cfrac{1}{x_3 + \ldots}}}$$

to the binary number

$$h(x) = 0.\underbrace{1 \ldots 1}_{x_0} \underbrace{0 \ldots 0}_{x_1} \underbrace{1 \ldots 1}_{x_2} \ldots$$

This is a version of Minkowski's question mark function [20]. It conjugates the pair of maps $\alpha : x \to x + 1$, $\beta : x \to x/(x + 1)$ to the pair of maps $t \to t/2$, $t \to (t + 1)/2$ (the inverse binary shift).

If the Julia set $J(f_c)$ of $f_c : z \to z^2 + c$ is connected and locally connected then the Böttcher map $\varphi_c : \widehat{\mathbb{C}} \setminus \mathbb{D} \to \widehat{\mathbb{C}} \setminus K(f_c)$ extends to a continuous surjection $S^1 \to J(f_c)$, which semi-conjugates the map $z \to z^2$ on S^1 (the binary shift) to the map f_c on $J(f_c)$. We deduce that we may use the homeomorphism h described above to glue the action of f_c^{-1} on $J(f_c)$ to that of α, β on $\hat{\mathbb{R}}_{\geq 0}/\{0 \sim \infty\}$. Equally well we can glue the action of f_c^{-1} on $J(f_c)$ to that of α^{-1}, β^{-1} on $\hat{\mathbb{R}}_{\leq 0}/\{0 \sim -\infty\}$.

We now take two copies K_- and K_+ of the filled Julia set K_c of f_c and glue them together at the boundary point of external angle 0 to form a space $K_- \vee K_+$. Each point $z \in K_c$ has a corresponding z' defined by $f_c(z') = f_c(z)$. Consider the $(2 : 2)$ correspondence defined on $K_- \vee K_+$ by sending

- $z \in K_-$ to $f_c(z) \in K_-$ and to $z' \in K_+$;
- $z \in K_+$ to $f_c^{-1}(z) \in K_+$.

It is an elementary exercise to check that this correspondence on $K_- \vee K_+$ can be glued to the correspondence defined by α and β on the complex upper half-plane using the homeomorphisms $\hat{\mathbb{R}}_{\geq 0}/\{0 \sim \infty\} \to \partial K_-$ and $\hat{\mathbb{R}}_{\leq 0}/\{0 \sim -\infty\} \to \partial K_+$ defined above. Thus we have a topological mating between the action of the modular group on the upper half-plane and our $(2 : 2)$ correspondence on $K_- \vee K_+$.

Holomorphic Mating. Reassured by the existence of this topological construction, we define a (holomorphic) mating between a quadratic polynomial f_c, $c \in M$ and $\Gamma = PSL(2, \mathbb{Z})$ to be a $(2 : 2)$ holomorphic correspondence $\mathcal{F}$ such that:

1. there exists a completely invariant open simply-connected region Ω and a conformal bijection $\varphi : \Omega \to \mathbb{H}$ conjugating $\mathcal{F}|_\Omega$ to $\alpha|_\mathbb{H}$ and $\beta|_\mathbb{H}$;
2. $\widehat{\mathbb{C}} \setminus \Omega = \Lambda = \Lambda_- \cup \Lambda_+$, where $\Lambda_- \cap \Lambda_+ = \{P\}$ (a single point) and there exist homeomorphisms $\phi_\pm : \Lambda_\pm \to K_c$ conjugating respectively $\mathcal{F}|_{\Lambda_-}$ to $f_c|_{K_c}$ and $\mathcal{F}|_{\Lambda_+}$ to $f_c^{-1}|_{K_c}$

In 1994 the first author and C. Penrose proved that for all parameters a in the real interval [4, 7], the correspondence $\mathcal{F}_a$ is a mating between a quadratic polynomial $f_c(z) = z^2 + c$, $c \in [-2, +1/4] \subset \mathbb{R}$ and the modular group $\Gamma = PSL(2, \mathbb{Z})$ (see [4]).

2.2   The Regular and Limit Sets of $\mathcal{F}_a$

Consider the family of holomorphic correspondences $\mathcal{F}_a : \widehat{\mathbb{C}} \to \widehat{\mathbb{C}}$, defined by the polynomial equation (1). The change of coordinate $\phi_a : \widehat{\mathbb{C}} \to \widehat{\mathbb{C}}$ given by

$$\phi_a(z) = \frac{az + 1}{z + 1}$$

conjugates $\mathcal{F}_a$ to the correspondence

$$J \circ \mathrm{Cov}_0^Q, \tag{5}$$

where J is the (unique) conformal involution fixing 1 and a, and Cov_0^Q is the deleted covering correspondence of the function $Q(z) = z^3$, that is to say, the correspondence defined by the relation

$$\frac{Q(w) - Q(z)}{w - z} = 0, \text{ i.e. } z^2 + zw + w^2 = 3.$$

So $\mathcal{F}_a$ and $J \circ \mathrm{Cov}_0^Q$ are the same correspondence in different coordinates, and in that sense we write $\mathcal{F}_a = J \circ \mathrm{Cov}_0^Q$.

By a *fundamental domain* for Cov_0^Q (respectively J) we mean any maximal open set U which is disjoint from $\mathrm{Cov}_0^Q(U)$ (respectively $J(U)$). We require our fundamental domains to be simply-connected and bounded by Jordan curves (see Fig. 3).

Klein Combination Locus. Let $P = 1$ denote the common fixed point of Cov_0^Q and J. The point P is a *parabolic fixed point*. The *Klein combination locus* $\mathcal{K}$ is the subset of $\mathbb{C}$ consisting of all a for which there are fundamental domains Δ_{Cov} and Δ_J of Cov_0^Q and J, respectively, such that

$$\Delta_{Cov} \cup \Delta_J = \widehat{\mathbb{C}} \setminus \{P\}.$$

We call such a pair of fundamental domains a *Klein Combination pair*.

In [2] we show that $\{a \in \mathbb{C} : |a - 4| \leq 3, a \neq 1\} \subset \mathcal{K}$, and that when a is in the interior of this disk the *standard* fundamental domains (see Fig. 3) are a Klein combination pair. More generally we prove that for every $a \in \mathcal{K}$, we can always choose a Klein combination pair whose boundaries $\partial\Delta_{Cov}$ and $\partial\Delta_J$ are transversal to the attracting-repelling axis at P.

Fig. 3 Standard
fundamental domains for
Cov_0^Q and J. The curve in
blue is $\text{Cov}_0^Q((-\infty, -2])$.
The region to the right of this
curve is a fundamental
domain Δ_{Cov} of Cov_0^Q. The
unbounded region
determined by the red circle
is a fundamental domain Δ_J
of the involution J. The
parabolic fixed point P is the
point 1

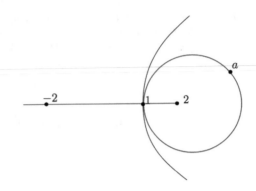

Now suppose $a \in \mathcal{K}$ and let Δ_{Cov} and Δ_J be a corresponding pair of fundamental domains of Cov_0^Q and J such that $\partial \Delta_{Cov}$ and $\partial \Delta_J$ are transversal to the attracting-repelling axis at P. It follows that $P \in \mathcal{F}_a^n(\overline{\Delta_J})$, for every n, and $\mathcal{F}_a(\Delta_J)$ is compactly contained in $\Delta_J \cup \{P\}$. By definition,

$$\Lambda_{a,+} = \bigcap_{n=1}^{\infty} \mathcal{F}_a^n(\overline{\Delta_J}) \tag{6}$$

where $\mathcal{F}_a = J \circ \text{Cov}_0^Q$, is the *forward limit set* of $\mathcal{F}_a$. Similarly, since Δ_{Cov} is forward invariant, the complement of Δ_{Cov} is invariant under $\mathcal{F}_a^{-1}$ and

$$\Lambda_{a,-} = \bigcap_{n=1}^{\infty} \mathcal{F}_a^{-n}(\widehat{\mathbb{C}} \setminus \Delta_{Cov}) \tag{7}$$

is the *backward limit set* of $\mathcal{F}_a$. The sets $\Lambda_{a,-}$ and $\Lambda_{a,+}$ have only one point in common, the point P. Their union, Λ_a, is the *limit set* of $\mathcal{F}_a$. An example of a plot of a limit set of $\mathcal{F}_a$ is displayed in Fig. 4. (In this plot we use the original coordinate system of (1), so $P = 0$ and J is the involution $z \leftrightarrow -z$.)

We have $\mathcal{F}_a^{-1}(\Lambda_{a,-}) = \Lambda_{a,-}$, and the restriction of $\mathcal{F}_a$ to this set is a $(2:1)$ single-valued holomorphic map denoted by f_a. The involution J maps $\Lambda_{a,-}$ onto $\Lambda_{a,+}$ and determines a conjugacy from f_a to

$$\mathcal{F}_a^{-1} : \Lambda_{a,+} \to \Lambda_{a,+}.$$

The *regular domain* of $\mathcal{F}_a$ is $\Omega_a = \widehat{\mathbb{C}} \setminus \Lambda_a$. This set is completely invariant under $\mathcal{F}_a$ (forward and backwards). By the Klein Combination Theorem it can be shown that if Ω_a contains no critical points it is tiled by copies of the intersection of any pair of Klein combination domains, [6].

Fig. 4 A connected limit set for $\mathcal{F}_a$, where $a = 4.56 + 0.42i$

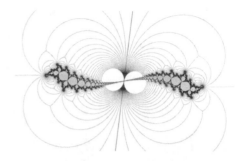

Fig. 5 Julia set of the hybrid equivalent member of $\mathrm{Per}_1(1)$

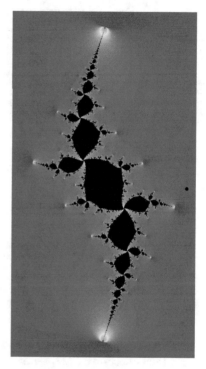

Connectedness Locus. The *connectedness locus* $\mathcal{C}_\Gamma$ of the family $\mathcal{F}_a$ is the subset of $\mathcal{K}$ consisting of all a such that the limit set Λ_a is connected. When $a \in \mathcal{C}_\Gamma$, the regular domain Ω_a contains no critical points, and moreover is simply connected.

Bullett and Penrose [4] conjectured that for every $a \in \mathcal{C}_\Gamma$, the correspondence $\mathcal{F}_a$ is a mating between some quadratic map $f_c(z) = z^2 + c$ and the modular group $\mathrm{PSL}(2, \mathbb{Z})$. More recently, this conjecture was settled affirmatively by Bullett and Lomonaco [2], provided the quadratic family is replaced by a quadratic family of *parabolic* maps (see Figs. 4 and 5).

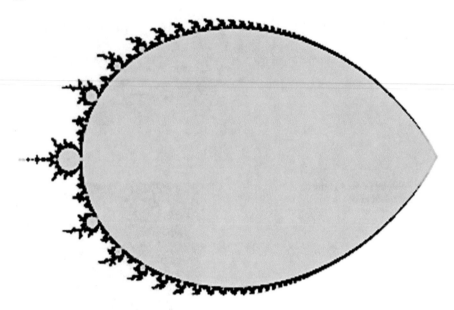

Fig. 6 A plot of $M_\Gamma = C_\Gamma \cap \{z : |z - 4| \leq 3\}$, which is conjecturally homeomorphic to the Mandelbrot set [4]

2.3 Mating Parabolic Maps with PSL(2, ℤ).

The family $\mathrm{Per}_1(1)$ consists of quadratic rational maps of the form $P_A(z) = 1 + 1/z + A$, where $A \in \mathbb{C}$. The maps in $\mathrm{Per}_1(1)$ all have a persistent parabolic fixed point at ∞ and critical points at ± 1. The connectedness locus for the family $\mathrm{Per}_1(1)$ is the *parabolic Mandelbrot set* M_1, which has been proved to be homeomorphic to the Mandelbrot set by Petersen and Roesch [23]. We say that $\mathcal{F}_a$ is a *mating* between P_A and $\mathrm{PSL}(2, \mathbb{Z})$ if:

1. on the completely invariant open simply-connected region Ω_a there exists a conformal bijection $\varphi_a : \Omega_a \to \mathbb{H}$ conjugating $\mathcal{F}_a : \Omega_a \to \Omega_a$ to $\alpha|_\mathbb{H}$ and $\beta|_\mathbb{H}$; and
2. the (2 : 1) branch of $\mathcal{F}_a$ which fixes $\Lambda_{a,-}$ (given by the holomorphic map f_a) is hybrid equivalent to P_A on the backward limit set $\Lambda_{a,-}$.

In [2], using the theory of parabolic-like maps developed by the second author (see [16]), the first two authors proved the following (see Figs. 4 and 5):

Theorem 2.1 *For every* $a \in C_\Gamma$, *the correspondence* $\mathcal{F}_a$ *is mating between a parabolic map in* $\mathrm{Per}_1(1)$ *and* $\mathrm{PSL}(2, \mathbb{Z})$.

The following conjecture has been open for at least 20 years [4] (Figs. 6, 7, 8):

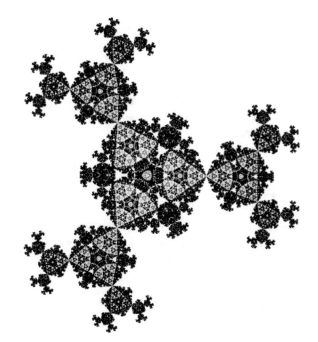

Fig. 7 The Julia set of $z \mapsto z^{\frac{3}{2}} + 2$

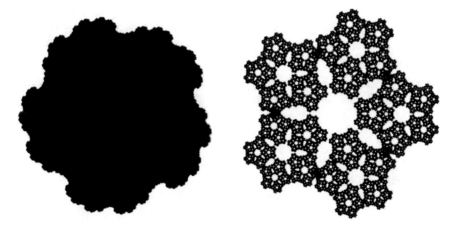

Fig. 8 Filled Julia sets in the family $\mathbf{f}_c(z) = z^{\beta} + c$, where $\beta = 5/4$. In the first figure (left), K_c is a full compact set corresponding to $c = 3 + 2i$. In the second we have a Carpet for $c = 26$. If $|c|$ is sufficiently large, K_c is a Cantor repeller. Since the Mandelbrot set M is contained in a disk of radius 2 around the origin, the fact that $c = 26$ and $\mathcal{J}_c$ is still connected seems odd. However, this is one of the main features of the family (2), and it is experimentally clear that $\mathcal{M}_{\beta}$ tends to cover the plane as $\beta \to 1+$

Conjecture 2.1 *The Mandelbrot set is homeomorphic to M_Γ.*

The first two authors have developed a detailed strategy for proving that M_Γ is homeomorphic to M_1. This, together with the proof by Petersen and Roesch that M_1 is homeomorphic to M, would finally prove Conjecture 2.1. A key step in the strategy to prove that M_Γ is homeomorphic to M_1 makes use of a Yoccoz inequality for matings, which we prove using a generalization of the technique of external rays (the subject of the next section).

2.4 Periodic Geodesics

Böttcher Coordinates. Consider the holomorphic correspondence $\mathcal{H}$ on the upper half-plane obtained from the generators $\alpha(z) = z + 1$ and $\beta(z) = z/(z+1)$ of $PSL(2, \mathbb{Z})$, i.e. defined by the polynomial equation (4). As part of the proof of Theorem 2.1 it is shown in [2] that:

Theorem 2.2 (Böttcher map) *If $a \in \mathcal{C}_\Gamma$, there is a unique conformal homeomorphism $\varphi_a : \Omega_a \to \mathbb{H}$ such that*

$$\mathcal{H} \circ \varphi_a = \varphi_a \circ \mathcal{F}_a.$$

By the Schwarz lemma, the Böttcher map is an isometry with respect to the hyperbolic metric, and maps geodesics to geodesics. Geodesics in Ω_a, or equivalently in $\mathbb{H}$, play a role for the correspondences $\mathcal{F}_a$ analogous to the role played by external rays for quadratic polynomials f_c.

Periodic Geodesics Land. We call a geodesic γ in the hyperbolic plane *periodic* if $W\gamma = \gamma$ for some W in the semi-group generated by α and β, that is

$$W = g_1 \circ g_2 \circ \cdots \circ g_n$$

where $n \geq 1$ and each $g_i \in \{\alpha, \beta\}$. (Note that the decomposition of W must include both α and β, since α^n and β^n, being parabolic, do not preserve any geodesic.)

Since $\mathbb{H}$ is geodesically complete, γ is a curve $\mathbb{R} \to \mathbb{H}$, and the limits

$$\gamma(-\infty) := \lim_{t \to -\infty} \gamma(t), \quad \gamma(\infty) := \lim_{t \to \infty} \gamma(t)$$

are by definition the *landing points* of γ. Every periodic geodesic in the hyperbolic plane lands: its landing points are in $\mathbb{R} \setminus \{0\}$.

If $a \in \mathcal{C}_\Gamma$, the regular domain is a hyperbolic Riemann surface, that is, it has a unique complete metric of constant curvature -1 determining its geometry. A geodesic $\hat{\gamma}$ in Ω_a is called *periodic* if $\varphi_a \circ \hat{\gamma}$ is a periodic geodesic of $\mathbb{H}$.

We say that a periodic geodesic $\hat{\gamma} : \mathbb{R} \to \Omega_a$ *lands* if the limits $\hat{\gamma}(\infty)$ and $\hat{\gamma}(-\infty)$ exist. They are the *right and left landing points,* respectively.

Theorem 2.3 *If $a \in C_\Gamma$, then every periodic geodesic lands. The left landing point belongs to $\Lambda_{a,-}$ and the right landing point is in $\Lambda_{a,+}$.*

As a corollary, the Böttcher map extends to all landing points of periodic geodesics. Indeed it extends to all landing points of preperiodic geodesics, and moreover these correspond under φ_a to the set of all quadratic irrationals in $\mathbb{R}$ (the set of real numbers with preperiodic continued fraction expansions).

2.5 Repelling Fixed Points, and Sturmian Sequences

The following result is again analogous to a result for quadratic polynomials, but the proof is quite technical and deep (even more so than in the case of polynomials, which is already difficult, see [3]), and at present we only have a proof for repelling fixed points, whereas for polynomials it is known for repelling and parabolic cycles:

Theorem 2.4 *A repelling fixed point in $\Lambda_-(\mathcal{F}_a)$ of a correspondence $\mathcal{F}_a$ with $a \in C_\Gamma$ is the landing point of exactly one periodic cycle of geodesics.*

This theorem has the consequence that to a repelling fixed point $z \in \Lambda_-$ of a correspondence $\mathcal{F}_a$ with $a \in C_\Gamma$ we can associate a periodic geodesic $\hat{\gamma}$ which lands there, and a finite word W in α and β which fixes $\varphi_a \circ \hat{\gamma}$. Letting f_a denote the (locally defined) branch of $\mathcal{F}_a$ which fixes z, we deduce that since f_a is locally a homeomorphism the cyclic order of the images of $\hat{\gamma}$ around z is preserved by f_a. Thus f_a has a well-defined *combinatorial rotation number* around z, and this number is rational.

Sturmian Sequences. Recall that a sequence $(s_i) \in \{0,1\}^{\mathbb{N}}$ is *Sturmian* if, for every n, the number of $1's$ in any two blocks of length n differs by at most one. There is an obvious equivalent definition for bi-infinite sequences.

If (s_i) is Sturmian, then the points of the orbit of $x = 0.s_1 s_2 \ldots$ (binary) under $f(z) = z^2$ on the unit circle are necessarily in the same order as the points of some rigid rotation R_θ, and vice versa. This θ is uniquely determined; it is by definition the *rotation number* of (s_i). Equivalently, θ is the limiting frequency of $1's$ in the sequence [5].

For each rational p/q (modulo 1) in lowest terms, there is a unique (up to cyclic permutation) finite word $W_{p/q} = (s_i) \in \{0,1\}^q$ such that the orbit of $x = 0.\overline{s_1 \ldots s_q}$ under $f(z) = z^2$ is in the same order around the circle as the points of an orbit of the rigid rotation $R_{p/q}$ (here $\overline{s_1 \ldots s_q}$ denotes a recurring block). For example $W_{1/3} = 001$, and $W_{2/5} = 00101$.

We call $W_{p/q}$ the finite Sturmian word of rotation number p/q, since the bi-infinite sequence made up of repeated copies of $W_{p/q}$ is the unique (up to shift) periodic Sturmian sequence of rotation number p/q. Finally we remark that there is nothing special about the symbols 1 and 0: identical terminology for Sturmian sequences and words may be applied if we replace 1 and 0 by α and β respectively.

We now return to the situation that $a \in C_\Gamma$, and z is a repelling fixed point of $f_a : \Lambda_{a,-} \to \Lambda_{a,-}$. If $\hat{\gamma}$ is a periodic geodesic landing at z, it has a combinatorial rotation number p/q (by Theorem 2.4), and any finite word W in α and β which fixes $\varphi_a \circ \hat{\gamma}$ is Sturmian, hence (a cyclic permutation of) a power of $W_{p/q}$. By establishing and applying bounds for the eigenvalues of the Sturmian words $W_{p/q}$ in α and β, we prove our Yoccoz inequality, Theorem 1.3 (see [3]).

3 Hyperbolic Correspondences

We now turn to the study of the one parameter family of holomorphic correspondences defined by (2). This family is perhaps the simplest generalization of the quadratic family as a multifunction.

It will be useful to recall some well-known facts directly related to the dynamics of $\mathbf{f}_c(z) = z^\beta + c$ when $\beta > 1$ is a rational number.

Hyperbolic Quadratic Maps. The notion of hyperbolicity can be given in several equivalent forms. According to the simplest one, $f_c(z) = z^2 + c$ is *hyperbolic* if $f_c^n(0)$ converges to an attracting cycle. Since the fixed point at infinity is an attracting cycle, every parameter in the complement of the Mandelbrot set corresponds to a hyperbolic map.

Since every finite attracting cycle attracts the orbit of a critical point, the map f_c can have at most one finite attracting cycle. Any quadratic map with a finite attracting cycle corresponds to a point in the interior of the Mandelbrot set M, and an equivalent form of the Fatou conjecture states that this is the only possibility for a quadratic map in the interior of M.

The closure of attracting cycles is denoted by $\mathcal{J}_c^*$. It turns out that f_c *is hyperbolic iff the basin of attraction of $\mathcal{J}_c^*$ is $\widehat{\mathbb{C}} \setminus \mathcal{J}_c$*. For this reason, we call $\mathcal{J}_c^*$ the dual Julia set of f_c.

This equivalent definition of hyperbolicity should be preserved in any generalization, mainly because of its intrinsic dynamical significance.

We shall use this equivalent property to define hyperbolic correspondences and centers in the family $\mathbf{f}_c(z) = z^\beta + c$, but first we need to extend the concepts of orbit, Julia set and multiplier of a cycle.

Cycles. Consider the family (2). Every sequence $(z_i)_0^\infty$ for which the points satisfy $z_{i+1} \in \mathbf{f}_c(z_i)$ is a *forward orbit*. A backward orbit is characterized by $z_{i+1} \in \mathbf{f}_c^{-1}(z_i)$. If $\varphi : U \to \mathbb{C}$ is an injective holomorphic map from a region U of the plane such that $\varphi(z) \in \mathbf{f}_c(z)$, for every z in U, then φ is a *univalent branch* of $\mathbf{f}_c$. By a cycle of length n we mean any periodic forward orbit with minimal period n. The quantity

$$\lambda = \prod_0^{n-1} \varphi_i'(z_i),$$

where φ_i is the unique univalent branch taking z_i to z_{i+1}, is the *multiplier* of the cycle. If $z = 0$ then there is no univalent branch defined at z; if some point of the cycle is 0, then by definition $\lambda = 0$.

The cycle is *repelling* if $|\lambda| > 1$, and attracting if $|\lambda| < 1$.

Julia Sets, Hyperbolic Correspondences. The Julia set of $\mathbf{f}_c$, denoted by $\mathcal{J}_c$, is the closure of the union of all repelling cycles of $\mathbf{f}_c$. Similarly, the dual Julia set $\mathcal{J}_c^*$ is the closure of the union of all attracting cycles in $\mathbb{C}$.

The ω-limit set of a point z, denoted $\omega(z)$, consists of every ζ such that $z_{i_k} \to \zeta$ as $k \to \infty$, for some bounded forward orbit (z_i) starting at $z_0 = z$, and some subsequence (z_{i_k}). We may use $\omega(z, \mathbf{f}_c)$ to make explicit the dependence on the dynamics of $\mathbf{f}_c$.

The dual Julia set is a *hyperbolic attractor* if $\mathcal{J}_c^*$ is forward invariant and supports an attracting conformal metric $\rho(z)|dz|$, in the sense that

$$\sup_{z,\varphi} \|\varphi'(z)\|_\rho < 1,$$

where the sup is taken over all $z \in \mathcal{J}_c^*$ and all univalent branches φ of $\mathbf{f}_c$ at z. It is implicit in this definition that $\mathcal{J}_c^*$ does not contain the critical point, for then no univalent branch is defined at 0.

If $\mathcal{J}_c^*$ is a hyperbolic attractor, then the *basin of attraction* of $\mathcal{J}_c^* \cup \{\infty\}$ is well defined and consists of all z such that $\omega(z) \subset \mathcal{J}_c^* \cup \{\infty\}$.

Definition 3.1 We say that $\mathbf{f}_c$ is *hyperbolic* if $\mathcal{J}_c^*$ is a hyperbolic attractor and the basin of attraction of $\mathcal{J}_c^* \cup \{\infty\}$ is $\hat{\mathbb{C}} \setminus \mathcal{J}_c$.

Connectedness Locus. For every c there is bounded disk B centered at 0 whose complement is invariant under $\mathbf{f}_c$, and every forward orbit of a point in $\mathbb{C} \setminus B$ converges exponentially fast to ∞.

We define

$$K_c = \bigcap_{n>0} \mathbf{f}_c^{-n}(B) \tag{8}$$

as the *filled Julia set* of $\mathbf{f}_c$. A point z belongs to K_c iff there is at least one bounded forward orbit under $\mathbf{f}_c$ starting at z.

A connected compact subset of the plane is *full* if its complement in the Riemann sphere is connected.

Theorem 3.1 *If $\beta = p/q$ and p is prime, then K_c is either connected or totally disconnected. If $0 \in K_c$, then K_c is full.*

The connectedness locus $\mathcal{M}_\beta$ of the family $\mathbf{f}_c$ is by definition the set of all parameters c for which K_c is connected.

Another important subset of the parameter space is

$$\mathcal{M}_{\beta,0} = \{c \in \mathbb{C} : 0 \in K_c\}. \tag{9}$$

Notice that both sets generalize the definition of Mandelbrot set for the quadratic family, but if β is not an integer, there is no reason to believe that $\mathcal{M}_\beta = \mathcal{M}_{\beta,0}$. In view of Theorem 3.1,

$$\mathcal{M}_{\beta,0} \subset \mathcal{M}_\beta,$$

if p is prime.

What is the structure of K_c when c belongs to $\mathcal{M}_\beta \setminus \mathcal{M}_{\beta,0}$? Below we answer this question.

Carpets. A set $\Lambda \subset \mathbb{C}$ is a hyperbolic repeller of $\mathbf{f}_c$ if (i) $\mathbf{f}_c^{-1}(\Lambda) = \Lambda$; and (ii) Λ supports an expanding conformal metric defined on a neighborhood of Λ. (See [29]). A filled Julia set K_c is a *Carpet* if (i) K_c is connected but not full; and (ii) K_c is a hyperbolic repeller.

Intuitively, every Carpet presents holes, and by the contraction of the branches of $\mathbf{f}_c^{-1}$, every hole comes with infinitely many small copies.

We say that K_c is a *Cantor repeller* if K_c is a hyperbolic repeller and also a Cantor set. In this case, $\mathcal{J}_c = K_c$.

Theorem 3.2 *If $\beta = p/q$ and p is prime, then K_c can be classified as*

1. *full if $c \in \mathcal{M}_{\beta,0}$;*
2. *a Carpet with $K_c = \mathcal{J}_c$, if c belongs to $\mathcal{M}_\beta \setminus \mathcal{M}_{\beta,0}$; and*
3. *a Cantor repeller if c is in $\mathbb{C} \setminus \mathcal{M}_\beta$.*

Centers. A center is a point c of the parameter space such that

$$\mathbf{g}_c^n(0) = \{0\},$$

for some $n > 0$, where $\mathbf{g}_c : K_c \to K_c$ is the restriction of the correspondence $\mathbf{f}_c$ to the set K_c ($\mathbf{g}_c$ is well-defined, since every point of K_c has at least one image in K_c).

This definition is motivated by a well-known fact from the quadratic family, where every bounded hyperbolic component U has a center [9, 10] defined as the unique point $c \in U$ for which the multiplier of the finite attracting cycle of f_c is zero.

Hence, in the case of the quadratic family, the number of bounded hyperbolic components is countably infinite, and every such component is encoded by a solution of $f_c^n(0) = 0$, for some $n > 0$.

Simple Centers. A center is called *simple* if there is only one orbit of 0 under $\mathbf{g}_c$, and this orbit is necessarily a cycle containing 0.

Let $\mathcal{S}_d = \{a \in \mathbb{C} : a^{d-1} = -1\}$, for $d > 1$. For every pair (d, a) in the infinite set

$$\bigcup_{d>1} \{d\} \times \mathcal{S}_d,$$

the point a is a simple center of family of the holomorphic correspondences $\mathbf{f}_c : z \mapsto w$ given by $(w - c)^2 = z^{2d}$. Indeed, it was shown in [29] that the first two iterates of 0 under $\mathbf{f}_a$ are $0 \mapsto a \mapsto a^n + a = 0$ and $0 \mapsto a \mapsto -a^n + a = -2a^n$, where $-2a^n$ is a point in the basin of infinity of $\mathbf{f}_a$.

Open Problems. A fundamental program for the family $\mathbf{f}_c(z) = z^\beta + c$ is given by the following problems:

I. Show that every perturbation of a center corresponds to a hyperbolic correspondence;

II. Show that the set $\mathcal{M}'_\beta$ of hyperbolic parameters is indeed open and every component of $\mathcal{M}'_\beta$ is encoded by a center;

III. Decide if the set of parameters for which $c \mapsto \mathcal{J}_c$ is continuous in the Hausdorff topology is open and dense (computer experiments seem to support this statement);

IV. Show that every component of $\mathbb{C} \setminus \partial \mathcal{M}_\beta$ is hyperbolic.

V. Classify Julia sets with zero Lebesgue measure.

The first Problem I can be solved with a generalization of the proof of Theorem 3.3 (see [27] for a detailed exposition); the second is very realistic but still unresolved; the third is in many aspects a generalization of the celebrated work of Mañé, Sad and Sullivan [21] (see also [29] and Sect. 3.1 for a discussion of holomorphic motions in the family (2)); and the fourth and fifth may be as difficult as the Fatou conjecture (which has been open for a century). Indeed, the Fatou conjecture is equivalent to the following assertion [21]: if c is in the interior of the Mandelbrot set, then the Julia set of $f_c(z) = z^2 + c$ has zero Lebesgue measure. Theorem 1.4 is perhaps the first result towards this classification.

Theorem 3.3 (Hyperbolicity) *If c is in the complement of $\mathcal{M}_{\beta,0}$, or c is sufficiently close to a simple center, then $\mathbf{f}_c$ is hyperbolic.*

3.1 Holomorphic Motions

Quasiconformal deformations of Julia sets in the family $\mathbf{f}_c$ can be explained by the theory of branched holomorphic motions introduced by Lyubich and Dujardin [11] for polynomial automorphisms of $\mathbb{C}^2$. For more details, see [29].

First, let us recall some classical facts about holomorphic motions.

Let $\Lambda \subset \mathbb{C}^n$ and $U \subset \mathbb{C}$ be an open set. A family of injections $h_c : \Lambda \to \mathbb{C}^n$ is a *holomorphic motion* with base point $a \in U$ if (i) h_a is the identity, and (ii) $c \mapsto h_c(z)$ is holomorphic on U, for every z fixed in Λ.

Branched Holomorphic Motions. Let Λ and U be subsets of $\mathbb{C}$ and suppose U open and nonempty. A branched holomorphic motion with base point $a \in U$ is a multifunction $\mathbf{h} : U \times \Lambda \to \mathbb{C}$ with the following properties: (i) $\mathbf{h}(a, z) = \{z\}$, for every $z \in \Lambda$. In other words, $\mathbf{h}_a = \mathbf{h}(a, \cdot)$ is the identity; and (ii) there is a family $\mathcal{F}$ of holomorphic maps $f : U \to \mathbb{C}$ such that

$$\bigcup_{z \in \Lambda} G_z(\mathbf{h}) = \bigcup_{f \in \mathcal{F}} G(f),$$

where $G(f) = \{(z, fz); z \in U\}$ is the graph of f and $G_z(\mathbf{h})$ is the graph of $c \mapsto \mathbf{h}_c(z)$.

The key difference in the definitions of branched and (non-branched) holomorphic motion is that bifurcations are allowed in the branched family, so that $\mathbf{h}_c(z)$ is a set instead of a single point.

3.2 Solenoidal Julia Sets

Recently, Siqueira and Smania have presented another way of interpreting branched holomorphic motions on the plane as projections of (non-branched) holomorphic motions on $\mathbb{C}^2$. The method is general and applies to every hyperbolic Julia set [29], but we shall restrict to bifurcations near $c = 0$.

There is a family of holomorphic maps $f_c : U_0 \to V_0$ such that U_0 and V_0 are open subsets of $\mathbb{C}^2$, the closure of U_0 is contained in V_0, and the maximal invariant set

$$S_c = \bigcap_{n=1}^{\infty} f_c^{-n}(V_0)$$

is the closure of periodic points of f_c. (All periodic points are repelling in a certain generalized sense, see [29]). This description holds for every c in a neighborhood of zero. The dynamics of $\mathbf{f}_c$ on $\mathcal{J}_c$ is a topological quotient of $f_c : S_c \to S_c$, in the sense that $\pi(S_c) = \mathcal{J}_c$ and π sends two points in S_c related by f_c to two points in $\mathcal{J}_c$ related by $\mathbf{f}_c$: $\pi f_c(x)$ is an image of $\pi(x)$ under $\mathbf{f}_c$, for every $x \in S_c$.

Let $\pi_c : S_c \to \mathcal{J}_c$ denote the projection $(z, w) \mapsto z$.

Theorem 3.4 (Holomorphic motions) *There is a holomorphic motion $h_c : S_0 \to \mathbb{C}^2$ with base point $c = 0$ such that*

1. $h_c(S_0) = S_c$ *and h_c is a conjugacy (homeomorphism) from $f_0 : S_0 \to S_0$ to $f_c : S_c \to S_c$.*
2. *the projected motion $\mathbf{h}_c(z) = \pi_c \circ f_c \circ \pi_0^{-1}(z)$ is a branched holomorphic motion mapping $\mathcal{J}_0 = \mathbb{S}^1$ to $\mathcal{J}_c = \mathbf{h}_c(\mathbb{S}^1)$.*
3. S_0 *is a solenoid, and $\mathbf{f}_c$ is hyperbolic, for every c in U.*

See [29] for the solenoidal description of S_0 (indeed, S_0 is the Williams-Smale solenoid for certain values of p and q).

In Fig. 2, the motion of $\mathcal{J}_c$ is illustrated in four steps.

3.3   Conformal Iterated Function Systems

Dual Julia sets $\mathcal{J}_c^*$ in the family (2) often appear as limit sets of conformal iterated function systems (CIFS). This phenomenon is easy to explain when c is close to zero, and very convenient to motivate further generalizations.

Indeed, using the contraction of $\mathbf{f}_c$ around $z = 0$ one can prove that for every $c \neq 0$ close to zero, there is an open disk D such that $D_1 = \mathbf{f}_c(D)$ is another disk avoiding zero and compactly contained in D.

Since D_1 is simply connected, there are q conformal branches $f_j : D_1 \to \mathbb{C}$ such that $\mathbf{f}_c(z) = \{f_j(z)\}_j$, for every $z \in D_1$. Moreover, the images $f_j(D_1)$ are disjoint disks. It follows that

$$\mathbf{f}_c(D_1) \subset \mathbf{f}_c(D) = D_1;$$

and the family of maps $f_j : D_1 \to D_1$ is a CIFS. The limit set of this CIFS is $\Lambda = \cap_n \omega^n(D_1)$, where

$$\omega(A) = \bigcup_{j=1}^{q} f_j(A),$$

for any $A \subset D_1$. The most important fact derived from this construction is that Λ is the closure of attracting periodic orbits: $\Lambda = \mathcal{J}_c^*$.

This analysis has many generalizations, including holomorphic motions and Hausdorff dimension. Theorem 1.4, for example, is stated in great generality in [28].

In [27] we give a general account establishing a rigidity result which states that $\mathcal{J}_c^*$ is finite at simple centers, but any perturbation of c yields a hyperbolic correspondence whose dual Julia set is a Cantor set. In the case of c close to zero, for example, $\mathcal{J}_c^*$ is either a Cantor set if $c \neq 0$ (indeed, Λ comes from a CIFS without overlaps) or a single point set $\mathcal{J}_0^* = \{0\}$.

Acknowledgements The authors would like to thank the Fundação de amparo a pesquisa do estado de São Paulo, which has supported the first two authors by the grant FAPESP 2016/50431-6, and the third by FAPESP 2016/16012-6. C.S. is very grateful to Edson de Faria for the hospitality at IME-USP, to Sylvain Bonnot for suggesting the investigation of some interesting problems relating holomorphic correspondences to automorphisms of $\mathbb{C}^2$, and to Daniel Smania for many discussions and key ideas on the dynamics of hyperbolic correspondences, specially those concerning Gibbs states and Hausdorff dimension. L.L. and C.S. would like to express their sincere gratitude to the scientific committee and organizers of the conference *New trends in one-dimensional dynamics*, on the occasion of Welington de Melo's seventieth birthday, specially to Pablo Guarino and Maria J. Pacifico (significant content of this paper has been previously announced in this conference).

References

1. B. Branner, N. Fagella, *Quasiconformal Surgery in Holomorphic Dynamics*, vol. 141 (Cambridge University Press, 2014)
2. S. Bullett, L. Lomonaco, *Mating Quadratic Maps with the Modular Group II.* arxiv.org/abs/1611.05257 (2016)
3. S. Bullett, L. Lomonaco, *Dynamics of Modular Matings.* arxiv.org/abs/org/abs/1707.04764 (2017)
4. S. Bullett, C. Penrose, Mating quadratic maps with the modular group. Invent. Math. **115**(1), 483–511 (1994)
5. S. Bullett, P. Sentenac, Ordered orbits of the shift, square roots, and the devil's staircase. Math. Proc. Cambridge Philos. Soc. **115**(3), 451–481 (1994)
6. S. Bullett, A combination theorem for covering correspondences and an application to mating polynomial maps with Kleinian groups. Conform. Geom. Dyn. **4**, 75–96 (2000)
7. S. Bullett, Matings in holomorphic dynamics, in *Geometry of Riemann Surfaces*, ed. by F.P. Gardiner, G. Gonzales-Diez, C. Kourouniotis. London Mathematical Society Lecture Notes, vol. 368 (Cambridge University Press, 2010), pp. 88–119
8. E. de Faria, W. de Melo, *Mathematical Tools for One-Dimensional Dynamics*, in Cambridge Studies in Advanced Mathematics, vol. 115 (Cambridge University Press, 2008)
9. A. Douady, J.H. Hubbard, Étude dynamique des polynômes complexes. Partie I. Publications Mathématiques d'Orsay **84**(2), 75 (1984)
10. A. Douady, J.H. Hubbard, Étude dynamique des polynômes complexes. Partie II. (with collaboration of P. Lavaurs, Tan Lei and P. Sentenac). Publications Mathématiques d'Orsay **85**(4), v-154 (1985)
11. R. Dujardin, M. Lyubich, Stability and bifurcations for dissipative polynomial automorphisms of $\mathbb{C}^2$. Invent. Math. **200**(2), 439–511 (2015)
12. J. Graczyk, G. Światek, *The Real Fatou Conjecture*, in Annals of Mathematics Studies, vol. 144 (NJ, 1998)
13. J.H. Hubbard, *Local Connectivity of Julia Sets and Bifurcation Loci; Three Theorems of J.-C. Yoccoz*, in Topological Methods in Modern Mathematics (Stony Brook, NY, 1991), pp. 467–511
14. O. Kozlovski, W. Shen, S. van Strien, Density of hyperbolicity in dimension one. Ann. Math. **166**(1), 145–182 (2007)
15. G.M. Levin, On Pommerenke's inequality for eigenvalues of fixed points. Colloq. Math. **62**(1), 167–177 (1991)
16. L. Lomonaco, On parabolic-like maps. Ergod. Theory Dyn. Syst. **35**, 2171–2197 (2015)
17. M. Lyubich, Dynamics of quadratic polynomials I, II. Acta Math. **178**(2), 185–247 (1997)
18. C.T. McMullen, *Complex Dynamics and Renormalization*, in Annals of Mathematics Studies, vol. 135 (NJ, 1994)
19. J. Milnor, *Dynamics in One Complex Variable*, in Annals of Mathematics Studies, vol. 160 (Princeton University Press, 2006)
20. H. Minkowski, Zur geometrie der zahlen, in *Verhandlungen des III. Internationalen Mathematiker-Kongresses* (Heidelberg, Berlin, 1904), pp. 164–173
21. R. Mañé, P. Sad, D. Sullivan, On the dynamics of rational maps. Annales scientifiques de l'École Normale Supérieure **16**(2), 193–217 (1983)
22. C. Pommerenke, On conformal mapping and iteration of rational functions. Complex Variables Theory Appl. 117–126 (1986)
23. C. Petersen, P. Roesch, *Personal Conversation* (2017)
24. M. Rees, *Realization of Matings of Polynomials as Rational Maps of Degree Two* (1986) (Manuscript)
25. M. Rees, One hundred years of complex dynamics. Proc. Roy. Soc. A 472 (2016)
26. C. Siqueira, Dynamics of holomorphic correspondences. Ph.D. thesis, University of São Paulo, ICMC, Digital library USP, 8 2015
27. C. Siqueira, *Dynamics of Hyperbolic Correspondences* (2017)

28. C. Siqueira, *Hausdorff Dimension of Julia Carpets* (2018) (In preparation)
29. C. Siqueira, D. Smania, Holomorphic motions for unicritical correspondences. Nonlinearity **30**(8), 3104 (2017)
30. L. Tan, Mating quadratic polynomials. Ergod. Theory Dyn. Syst. **12**, 589–620 (1991)

Braid Equivalence in the Hénon Family I

A. de Carvalho, T. Hall and P. Hazard

Abstract We give two general constructions of braid equivalences which exist between certain deformations of the 2-branched Horsehoe map. We then give numerical evidence suggesting that these constructions of braid equivalences are always realised in the Hénon family.

Keywords Combinatorial dynamics · Braids · Surface homeomorphisms

1 Introduction

During the last three decades of the twentieth century, much effort was devoted to the study of families of low-dimensional dynamical systems depending on parameters. There is today a very thorough theory explaining the dynamics of families of one-dimensional (real and complex) endomorphisms. The dynamics of the real quadratic family $f_a(x) = a - x^2$, for example, is nearly completely understood [12]. In the 1970s Hénon introduced the family[1] which now bears his name, a two-dimensional analog of the quadratic family: $F_{a,b}(x, y) = (f_a(x) - by, x)$. This is a family of plane diffeomorphisms depending on two parameters which, for $b = 0$, degenerates

---

[1]Hénon considered a different parametrisation of this family,

$$H_{a,b}(x, y) = (-y + 1 - ax^2, bx).$$

A. de Carvalho (✉) · P. Hazard
University of São Paulo-USP, São Paulo, Brazil
e-mail: andresallesdecarvalho@gmail.com

P. Hazard
e-mail: pete@ime.usp.br

T. Hall
University of Liverpool, Liverpool, England
e-mail: T.Hall@liverpool.ac.uk

© Springer Nature Switzerland AG 2019
M. J. Pacifico and P. Guarino (eds.), *New Trends in One-Dimensional Dynamics*, Springer Proceedings in Mathematics & Statistics 285,
https://doi.org/10.1007/978-3-030-16833-9_6

to the quadratic family. In contrast to the quadratic family, and despite the existence of several beautiful results about it, our understanding of the Hénon family is still rather rudimentary. While there are many similarities between the two families which make it possible to use knowledge of the former to help in understanding the latter, there are also many fundamental differences which demand that different techniques be developed.

One of the most basic aspects of the dynamics of a parametrized family is the way in which the periodic orbit structure changes as the parameters vary. This article is concerned with periodic orbits in Hénon family in the parameter regions close to degeneration, and exploits both similarities and differences between the quadratic and Hénon families.

Periodic orbits of endomorphisms of the real line are specified by their associated cyclic permutation: the way that their points, ordered on the line, are permuted by the endomorphism. For homeomorphisms of the plane such as Hénon maps, the analogous specification, introduced by Boyland [2, 3], is the braid type. If $F: \mathbb{R}^2 \to \mathbb{R}^2$ is an orientation-preserving homeomorphism and P is a periodic orbit of F, then the braid type $bt(P, F)$ is the isotopy class of F relative to P, up to topological conjugacy. In other words, the braid type of P is determined by fixing the action of F on P but allowing it to be deformed by isotopy in the complement of P, and also allowing a global change of coordinates.

The periodic orbit structure of maps in the quadratic family—or, indeed, of any unimodal map f—is easily understood using techniques of kneading theory. The critical point c is used to divide the line into left and right halves, and the kneading sequence of f is the itinerary of the critical value $f(c)$: the sequence of lefts and rights along the orbit $(f^n(c))_{n \geq 1}$. There is then a simple recipe for generating the set of all itineraries of points $x \in \mathbb{R}$ from this kneading sequence. Permutations associated to periodic orbits of unimodal maps—which are called *unimodal permutations*—are determined by the itineraries of the points on the orbit. The set of permutations of periodic orbits of a unimodal map f is therefore determined by the kneading sequence of the map, and can be enumerated by a straightforward algorithm.

The situation for the Hénon family is quite different. We have very little idea, to this day, of the way in which braid types of periodic orbits are built up in the family, going from none to a full horseshoe's worth, as the parameter a increases. In fact, by the result of Kan, Koçak and Yorke [11], periodic orbits of the Hénon family are both created and destroyed near every homoclinic tangency, and it is not even known whether or not all periodic orbits which appear in the Hénon family have the same braid type as periodic orbits of the horseshoe.

The diagrams in Fig. 1 show regions in the parameter plane for the Hénon family where attracting periodic orbits of periods 8, 9, 10, and 11 were found. (Similar loci, for various periods, were first considered by El Hamouly and Mira [7].) These plots have a very rich structure, and understanding how the dynamics varies in the Hénon family includes explaining this structure. In this paper we are particularly interested in the hook-like structures, also called *swallow configurations* by Milnor [13] in the one-dimensional cubic case. These structures are open sets consisting of a main body and four limbs, two of which intersect the a-axis in two distinct (small) intervals.

(a)   **(b)**

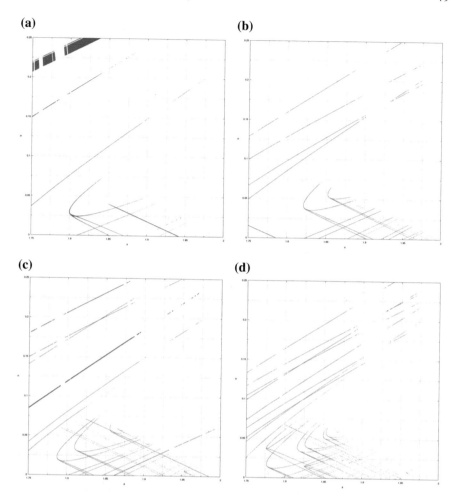

(c)   **(d)**

Fig. 1 Scatterplots of parameters $(a, b) \in [1.75, 2.00] \times [0.0, 0.25]$ for which the Hénon map $F_{a,b}$ possesses an attracting periodic orbit of period $p = 8, 9, 10, 11$. One of the "hook-like" structures for period 8 can be seen intersecting the a-axis in two intervals at approximately $a = 1.8517$ and $a = 1.87$ respectively

Each of these hooks indicates that there is one attracting periodic orbit in the Hénon family which can be deformed into two different attracting periodic orbits of the quadratic family. That is, we expect each of the hooks to be associated to attracting periodic orbits of the corresponding Hénon maps whose braid type is constant in the region $b > 0$ and which degenerate into periodic orbits of the quadratic family with two different permutations as $b \downarrow 0$ along each of the two ends of the hook. Viewing this in the opposite direction, the hooks indicate that there are certain pairs of unimodal permutations which coalesce into a single Hénon periodic orbit.

In this paper we identify mechanisms which lead to this coalescence on the level of unimodal permutations. More precisely, we describe two mechanisms which associate to unimodal permutations pairs of equivalent braids, and provide numerical evidence showing that the hooks in Fig. 1 can all be explained in terms of these mechanisms.

There are at least two distinct reasons for which it is useful to be able to relate unimodal permutations to braid equivalences in the Hénon family. First, as mentioned above, our understanding of the periodic orbit structure of Hénon maps is very limited, and being able to connect Hénon braid types to unimodal permutations provides a means of deducing information about the former from the latter (about which we know everything); one of the consequences of our results is that we predict the existence of infinitely many hooks in the Hénon family, each of which associates a braid type with a pair of intervals in the parameter space of the quadratic family. Second, the more general problem of deciding whether or not two braid types are equal is a difficult one, and mechanisms for constructing equivalent braids can be useful. For example, there are conjectural constraints [4] on the possible orders in which horseshoe braid types can be built up in families which pass from trivial dynamics to a full horseshoe; and if this conjecture holds, then each pair of equivalent braids generates an infinite family of pairs of equivalent braids.

We next review some definitions and terminology which will make it possible to give rough descriptions of the two mechanisms mentioned above. A geometric braid on n strands (see Fig. 3) is a diagram with n arcs (strands) connecting two ordered sets of n points lined up vertically, so that only double intersections are allowed and at each of them it is specified which strand goes above and which goes below. The points at the top are called *initial endpoints* and those at the bottom are called *terminal endpoints*. To each geometric braid is associated a braid type, and braid types determine geometric braids up to conjugacy (these facts will be discussed further in the text). Geometric braids also induce permutations on n elements in the obvious way: associate to the initial endpoint the terminal endpoint along the same strand. Since this association forgets all information about crossings of strands, it is far from being one-to-one. It is possible, however, to associate a unique braid to a unimodal permutation by requiring that each pair of strands crosses at most once and that, when two strands cross, the one which started to the left goes above the one which started to the right. In this way we can talk about the unimodal braid associated with a unimodal permutation.

Given a unimodal permutation υ, let f be a unimodal map realising υ as its critical orbit. The *dynamical* preimage of a point in the critical orbit is the preimage under f which is also contained in the critical orbit (corresponding to the unique preimage at the level of the permutation). The other preimage of a point of the critical orbit is called the *non-dynamical* preimage.

The first mechanism is as follows. First we 'break' f at the dynamical preimage of the critical point: that is, we perturb f in a neighbourhood of this point. Assuming that the break is small, we get a new point which is a closest return to the critical point. Make this new point follow the sequence of forward iterates of the critical point: we may do this for any finite time by making the break sufficiently small. If we arrive near the non-dynamical preimage of the critical point, we can 'reconnect' this iterate of the new point to the critical point. Again, this means that we perturb f in a neighborhood of this iterate so that its image is the critical point. Depending on which side of the critical point the new point lies, we get a pair of distinct critical orbit types. (In terms of braids this construction is a generalisation of the cabling construction for a braid, followed by a half-twist between strands.)

The second mechanism takes a pair of equivalent unimodal permutations υ_- and υ_+, such as those generated by the first mechanism, and produces another equivalent pair υ_-^1 and υ_+^1 of unimodal permutations. Let f_- and f_+ denote unimodal maps whose critical orbits realise υ_- and υ_+ respectively. As before, we break f_- and f_+ at the dynamical preimage, but in both cases the break is large so that the new point is a second closest return to the critical point. This is done to ensure that the image of the new point is close to the image of the first closest return. We make this image of the new point follow the forward iterates of the first closest return until it arrives close to the dynamical preimage of the critical point. It is then reconnected to the critical point as before. By repeating this process, a chain of pairs of equivalent braids can be generated.

In Sect. 2 we present some background on braids and unimodal maps; in Sects. 3 and 4 the first and second constructions of braid equivalences are described; and in Sect. 5 we discuss applications to the Hénon family and numerical results.

2 Notation and Terminology

2.1 Braids and Braid Equivalence

Let $\Delta \subset \mathbb{R}^2$ denote the closed unit disk. Let $\mathrm{Homeo}_+(\Delta)$ denote the group of orientation-preserving homeomorphisms of Δ. Given a subset $A \subset \Delta$ let $\mathrm{Homeo}_+(\Delta, A)$ denote the group of orientation-preserving homeomorphisms f of Δ satisfying $f(A) = A$. Where necessary we endow these groups with the uniform topology.

2.1.1 Braid Equivalence

Let $p \in \mathbb{N}$. Let $P_{\mathrm{can}} \subset \overset{\circ}{\Delta}$ denote the unique set of p points contained in the horizontal axis, whose complement in this axis has connected components all of equal length. Let $F \in \mathrm{Homeo}_+(\Delta, P_{\mathrm{can}})$. Denote by $[F]_{P_{\mathrm{can}}}$ the isotopy class of F rel P_{can}. When it is clear from the context, we will also use the notation $[F]$.

The group of such isotopy classes under composition is called the *mapping class group* of $(\Delta, P_{\mathrm{can}})$ and is denoted by MCG_p.

Let $F \in \mathrm{Homeo}_+(\Delta)$ possess a periodic orbit P of smallest period p. We assume P is in the interior of Δ: if not, we extend F arbitrarily over a collaring of Δ. Let $H: (\Delta, P_{\mathrm{can}}) \to (\Delta, P)$ be any homeomorphism. Then the *braid type* of (P, F)

is the conjugacy class $\langle [H^{-1} \circ F \circ H] \rangle$ of $[H^{-1} \circ F \circ H]$ in MCG_p. Denote the braid type of (P, F) by $\mathrm{bt}(P, F)$. Let BT_p denote the set of all braid types of a fixed period p.

Remark 2.1 The braid type is independent of the collaring and the choice of homeomorphism H (see [8] for more details).

Let $F_0, F_1 \in \mathrm{Homeo}_+(\Delta)$ possess periodic orbits P_0 and P_1 respectively. We say that the pair (P_0, F_0) and (P_1, F_1) are *braid equivalent* if $\mathrm{bt}(P_0, F_0) = \mathrm{bt}(P_1, F_1)$. Denote this equivalence by $(P_0, F_0) \sim_{BE} (P_1, F_1)$. Equivalently, $(P_0, F_0) \sim_{BE} (P_1, F_1)$ if there exists a homeomorphism $H : (\Delta, P_0) \to (\Delta, P_1)$ such that $F_0 \simeq H^{-1} \circ F_1 \circ H$ rel P_0 in Δ.

2.1.2 Braids

We now relate the notion of braid equivalence to that of a braid. Let B_p denote the braid group on p strands (see [1]). Denote the composition of braids $\alpha, \beta \in B_p$ by $\alpha \cdot \beta$. If α and β are *conjugate* we write $\alpha \sim \beta$. If α and β are *reverse-conjugate*, i.e. $\alpha \cdot \gamma = \gamma^{-1} \cdot \beta$ for some $\gamma \in B_p$, we write $\alpha \sim_r \beta$.

Remark 2.2 We will consider braids, geometric braids and braid diagrams. Typically we will not make the distinction. However, where necessary we will use the notation $a \simeq b$ to denote that the geometric braids a and b are isotopic. We will also denote their product by $a \cdot b$, whenever the set of terminal endpoints of b coincides with the set of initial endpoints of a.

Let $Z(B_p)$ denote the centre of B_p. It is known that $B_p / Z(B_p)$ is naturally isomorphic to MCG_p (see [9]). Hence BT_p is in one-to-one correspondence with the set of conjugacy classes of $B_p / Z(B_p)$ whose underlying permutation is a cycle (and hence close up to give a knot, rather than just a link). Therefore, given (P_F, F) we denote by $\beta(P_F, F)$ the conjugacy class in $B_p / Z(B_p)$ corresponding to $\mathrm{bt}(P_F, F)$.

Remark 2.3 The following is well-known. Let (P_0, F_0) and (P_1, F_1) have associated braids β_0 and β_1. Then $(P_0, F_0) \sim_{BE} (P_1, F_1)$ if and only if there exists a braid σ and $m \in \mathbb{Z}$ such that $\sigma^{-1} \cdot \beta_0 \cdot \sigma$ can be deformed into $\beta_1 \cdot \tau^m$ by a sequence of isotopies, 2nd and 3rd Reidemeister moves and their inverses, where τ is a braid associated to a generator of $Z(B_p)$.

Finally, we observe that if $F \in \overline{\mathrm{Homeo}_+(\Delta)}$ possesses a periodic orbit P of smallest period p which corresponds to an isolated fixed point of non-zero index for the iterate F^p, then any small perturbation $F' \in \mathrm{Homeo}_+(\Delta)$ will also possess a periodic orbit P' which is a continuation of P and of the same period. Therefore we can define $\mathrm{bt}(P, F) = \mathrm{bt}(P', F')$. Since having an isolated fixed point of non-zero index is an open property, this is well-defined.

2.1.3 Braid Equivalence in Families

Definition 2.4 (*Braid equivalence in parametrised families*) Let $k \in \mathbb{N}$. Let $B \subset \mathbb{R}^k$ be a contractible bounded open set. Let $F \in C(B, \overline{\mathrm{Homeo}_+(\Delta)})$ be a k-parameter family of continuous self-maps. We will use the notation F_b for the map $F(b)$. Let $B^0 = \{b \in B : F_b \in \mathrm{Homeo}_+(\Delta)\}$. Let $b_0, b_1 \in B$ satisfy the property that F_{b_0} and F_{b_1} have periodic orbits P_0 and P_1 respectively, both of smallest period p. Then (P_0, F_{b_0}) and (P_1, F_{b_1}) are *braid equivalent in the family* F if

1. there exists a collection of p pairwise distinct points

$$P(t) = \{P^1(t), P^2(t), \ldots, P^p(t)\} \subset \Delta \qquad (2.1)$$

   where $P^i(t)$ varies continuously with $t \in [0, 1]$ for each $i = 1, 2, \ldots, p$, such that $P_0 = P(0)$ and $P_1 = P(1)$;
2. there exists a path $\gamma : [0, 1] \to B$ such that $\gamma(0) = b_0$, $\gamma(1) = b_1$ and $P(t)$ is a periodic orbit for $F_{\gamma(t)}$, for each $t \in [0, 1]$.

(Observe, necessarily $P(t)$ must have smallest period p for $F_{\gamma(t)}$ as the $P^i(t)$ are distinct.) In other words, the one-parameter sub-family $F_{\gamma(t)}$ realises a strong Nielsen equivalence between (P_0, F_{b_0}) and (P_1, F_{b_1}).

Remark 2.5 Braid equivalence in a parametrised family implies braid equivalence.

We will be specifically interested in the case when $b_0, b_1 \in B \setminus B^0$ and $\gamma(b_0, b_1) \subset \mathrm{Homeo}_+(\Delta)$. (The motivating example will be that F denotes the family of Hénon maps and F_{b_0}, F_{b_1} will correspond to quadratic maps.)

2.1.4 Braids and Braid Diagrams

Recall that braids can be represented by braid diagrams. Braid diagrams will be normalised in the following way: They lie in the unit square $[-1, 1] \times [-1, 1]$. Each strand has an *initial endpoint* lying in $[-1, 1]^{\mathrm{init}} = [-1, 1] \times \{1\}$. Each strand has a *terminal endpoint* lying in $[-1, 1]^{\mathrm{term}} = [-1, 1] \times \{-1\}$. The vertical line through any initial endpoint contains a terminal endpoint (and vice versa). All crossings are transverse. Only double points are allowed. The strands are directed downwards. (The last condition ensures no strand can 'backtrack'.) Consequently, if α and β are braids with associated diagrams, then the braid diagram corresponding to the product $\alpha \cdot \beta$ is the braid formed by placing the diagram corresponding to β directly above the diagram corresponding to α and rescaling.

Remark 2.6 We adopt this convention as it coincides with the convention of composing maps or isotopies from the right. For example, if the isotopy F_t^α realises the braid α and the isotopy F_t^β realises the braid β then $F_t^\alpha \cdot F_t^\beta$ realises the braid $\alpha \cdot \beta$.

Let β be a braid diagram. Denote by S_β the set of strands. Denote by E_β^{init} the set of initial endpoints. We will denote the points in E_β^{init}, ordered from left to right, by

0^{init}, 1^{init}, ..., $(p-1)^{\text{init}}$. Similarly, denote by E_β^{term} the set of terminal endpoints. We will denote the points in E_β^{term}, ordered from left to right, by 0^{term}, 1^{term}, ..., $(p-1)^{\text{term}}$. We denote the strand emanating from i^{init} by $s_\beta(i)$. Let $\pi: [-1, 1]^2 \to [-1, 1]$ denote the projection to the first coordinate. Then by assumption $\pi(i^{\text{init}}) = \pi(i^{\text{term}})$ for all $i = 0, 1, ..., p-1$.

Given a strand $s \in S_\beta$ let s^{init} and s^{term} denote its initial and terminal endpoints respectively. When it is clear which braid β is being considered we will drop β from our notation, so that $s_\beta(i)$ becomes $s(i)$, $s_\beta(i)^{\text{init}}$ becomes $s(i)^{\text{init}}$, and so on.

2.2 Unimodal Dynamics

2.2.1 Unimodal Permutations

Remark 2.7 In this section we consider only intervals in $\{0, 1, ..., p-1\}$, e.g. for $i, j \in \{0, 1, ..., p-1\}$ satisfying $i < j$ we let

$$[i, j] = \{k \in \{0, 1, ..., p-1\} : i \le k \le j\} \tag{2.2}$$

Define (i, j), $[i, j)$ and $(i, j]$ similarly. In later sections we may also assume they are embedded in $\mathbb{R}$ but it will be clear from the context what is meant.

Definition 2.8 Let $p \in \mathbb{N}$. Endow the set $\{0, 1, ..., p-1\}$ with its natural ordering. A permutation υ of the set $\{0, 1, ..., p-1\}$ is *unimodal* if there exists $m \in (0, p-1)$, such that

1. υ is order-preserving on the interval $[0, m]$;
2. υ is order-reversing on the interval $[m, p-1]$.

We call m the *folding point* of υ. We denote the set of unimodal permutations on $\{0, 1, ..., p-1\}$ by U_p. Let $U = \bigcup_{p \in \mathbb{N}} U_p$.

If a unimodal permutation υ is cyclic then we can introduce the following, which we call *cyclic notation*: Define o: $\{0, 1, ..., p-1\} \to \{0, 1, ..., p-1\}$ by $o(\upsilon^i(m)) = i$ for $i = 0, 1, ..., p-1$. Then we may uniquely represent υ by $(o(0), o(1), o(2), ..., o(p-1))$. Observe that o is a bijection. Hence we can recover υ by setting $\upsilon(o^{-1}(i)) = o^{-1}(i+1 \mod p)$.

Example 2.9 The cyclic unimodal permutation υ of $\{0, 1, 2, 3, 4\}$ given by

$$\upsilon(0) = 1, \ \upsilon(1) = 3, \ \upsilon(2) = 4, \ \upsilon(3) = 2, \ \upsilon(4) = 0 \tag{2.3}$$

has folding point $m = 2$ and therefore $o(2) = 0$. Applying υ iteratively gives

$$o(4) = 1, \ o(0) = 2, \ o(1) = 3, \ o(3) = 4. \tag{2.4}$$

In cyclic notation[2] we write $(o(0), o(1), o(2), o(3), o(4)) = (\underline{2}, 3, \underline{0}, 4, \underline{1})$. (In other words, cyclic notation is just shorthand for the collection of inequalities $\upsilon^2(m) < \upsilon^3(m) < m < \upsilon^4(m) < \upsilon(m)$.)

Definition 2.10 Let $\upsilon \in U_p$ be cyclic. Let $q \in \{0, 1, \ldots, p-1\}$. We call q a *right closest return time* to the folding point if $\upsilon^q(m) < m$ and the interval $(\upsilon^q(m), m)$ does not contain $\upsilon^r(m)$ for any integer r. Similarly, call q a *left closest return time* to the folding point if $m > \upsilon^q(m)$ and the interval $(m, \upsilon^q(m))$ does not contain $\upsilon^r(m)$ for any integer r. Finally, we call $q \in \{0, 1, \ldots, p-1\}$ a *closest return time* to the folding point if the interval $(\upsilon^{q+1}(m), \upsilon(m))$ does not contain any point $\upsilon^r(m)$ for any integer r.

Note that a closest return time will either be a left or right closest return time.

Definition 2.11 Let $\upsilon \in U_p$. Let $i, j, k \in \{0, 1, \ldots, p-1\}$, $i < j$. We say the closed interval $[i, j]$ *maps over* k if $k \in \upsilon[i, j]$ and *maps strictly over* k if $k \in \upsilon(i, j)$.

Remark 2.12 For each $k \in \{0, 1, \ldots, p-1\}$, $k \neq 0$ or $p-1$, it is clear that there exists an interval mapping strictly over k. It is also clear that each interval mapping strictly over the folding point m contains at least one subinterval of shortest length which also maps strictly over m. (The interval $[\upsilon^{-1}(k) - 1, \upsilon^{-1}(k) + 1]$ maps strictly over k and no strict subinterval also satisfies this property.) In fact, there are at most two intervals mapping over k of shortest length.

Definition 2.13 Let $\upsilon \in U_p$. For each $k \in \{0, 1, \ldots, p-1\}$ we call $\upsilon^{-1}(k)$ the *dynamical preimage* of k. If $k \neq 0, p-1$, the *interval containing the dynamical preimage of k* is the shortest closed interval mapping strictly over k which contains the dynamical preimage $\upsilon^{-1}(k)$ of k. If $k = 0$ or $p-1$, the *interval containing the dynamical preimage of k* is the shortest closed interval mapping over k which contains $\upsilon^{-1}(k)$.

The other shortest closed interval strictly mapping over k, when it exists, is called the *interval containing the non-dynamical preimage of k*.

Example 2.14 The unimodal cyclic permutation υ of $\{0, 1, 2, 3\}$ given by

$$\upsilon(0) = 2, \ \upsilon(1) = 3, \ \upsilon(2) = 1, \ \upsilon(3) = 0 \tag{2.5}$$

(which in cyclic notation is given by $(\underline{2}, \underline{0}, 3, \underline{1})$) does not have an interval containing the non-dynamical preimage of the folding point $m = 1$. To see this, observe that 1 has dynamical preimage 2 lying in the right interval. The only non-empty subinterval of the left interval is the left interval itself, $[0, 1]$. This maps to the interval $[2, 3]$. As $[2, 3]$ doesn't contain the folding point, the unimodal permutation υ does not have an interval containing the non-dynamical preimage of the folding point.

---

[2]Note that we will use underlinings to distinguish representations, so $\underline{3}$ denotes the third point in the orbit of $\underline{1} = p - 1$, but 3 denotes the third point from the left.

Definition 2.15 Let $\upsilon \in U_p$ be cyclic. Given $i, j \in \{0, 1, \ldots, p - 1\}$, denote by $\kappa = \kappa(i, j) > 0$ the smallest positive integer such that $j = \upsilon^\kappa(i)$. If $\upsilon^l(i) \neq m$ for all $0 \leq l < \kappa$, let

$$\rho(i, j) = \text{card}\{0 \leq l < \kappa : \upsilon^l(i) > m\} \qquad (2.6)$$

Definition 2.16 Let $\upsilon \in U_p$ be cyclic. We say that υ is *reconnectable at the dynamical preimage* if the following property holds: let $D = [d^-, d^+]$ denote the interval containing the dynamical preimage d of $\underline{0}$. Then either $\rho(\underline{1}, d^-)$ is odd or $\rho(\underline{1}, d^+)$ is even, or both.

We say that υ is *reconnectable at the non-dynamical preimage* if the following properties hold:

1. [*Preimage condition*] The interval containing the non-dynamical preimage of m exists. Denote it by $E = [e^-, e^+]$.
2. [*Parity condition*] Either $\rho(\underline{1}, e^-)$ is odd or $\rho(\underline{1}, e^+)$ is even, or both.

Remark 2.17 It will become clear in what follows that this notion also makes sense for other points $k \in \{0, 1, \ldots, p - 1\}, k \neq m$. However, for simplicity we will only consider reconnections at the folding point.

Recall that we aim to construct braid-equivalent pairs of unimodal combinatorial types starting from a given unimodal combinatorial type. In what follows it will become clear that the construction we propose works precisely when the initial combinatorial type is reconnectable, either at the dynamical preimage or non-dynamical preimage.

Example 2.18 Given a cyclic unimodal permutation υ, even if the interval containing the non-dynamical preimage of m exists it may not be reconnectable. For example consider, in cyclic notation, $(\underline{2}, \underline{7}, \underline{3}, \underline{8}, \underline{0}, \underline{5}, \underline{4}, \underline{9}, \underline{6}, \underline{1})$. The dynamical preimage $\underline{9}$ of the folding point $\underline{0}$ lies to the right of $\underline{0}$. The interval containing the non-dynamical preimage of $\underline{0}$ is $[\underline{7}, \underline{3}]$. To calculate $\rho(\underline{1}, \underline{3})$ we count the number of elements of $\{\underline{1}, \underline{2}\}$ lying to the right of $\underline{0}$, of which there is 1. Similarly there are 4 elements of $\{\underline{1}, \underline{2}, \underline{3}, \underline{4}, \underline{5}, \underline{6}\}$ lying to the right of $\underline{0}$. Hence $\rho(\underline{1}, \underline{3}) = 1$ and $\rho(\underline{1}, \underline{7}) = 4$. Since $\underline{7} < \underline{3}$ the parity condition isn't satisfied.

2.2.2   Unimodal Braids

We call a braid *positive* if each crossing is positive (see Fig. 2). A braid is *direct* if each strand crosses any other strand at most once.[3] It is known that for any permutation $\upsilon \in U_p$ there is a unique positive, direct braid β which induces υ. We call such braids *unimodal*. Denote the set of unimodal braids on p strands by UB_p. Let $UB = \bigcup_{p \in \mathbb{N}} UB_p$.

---

[3]In the literature such braids are called permutation braids—however, it will be useful to have an adjective to describe this property, as in [8].

Fig. 2 The two types of crossings used to construct braid diagrams

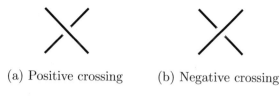

(a) Positive crossing   (b) Negative crossing

Fig. 3 The braid β for the cyclic unimodal permutation $(\underline{2}, \underline{3}, \underline{0}, \underline{4}, \underline{1})$

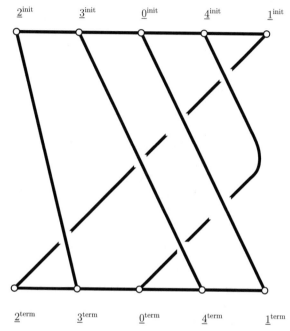

A unimodal braid β possesses a canonical braid diagram satisfying:

1. All strands initiated at m or to the left of m travel rightwardly.
2. All strands initiated at a point to the right of m travel rightwardly, touch the *folded line* $\pi^{-1}(p-1)$, then travel leftwardly.

We call strands satisfying Properties 1 and 2 *unimodal strands*. Henceforth we will identify a unimodal braid β with the corresponding canonical unimodal braid diagram. For example, the canonical unimodal braid diagram for the permutation $(\underline{2}, \underline{3}, \underline{0}, \underline{4}, \underline{1})$, is shown in Fig. 3.

Let S be a collection of unimodal strands, not necessarily forming a braid, which are positive, direct and which contains a folding strand $s(\underline{0})$. Let s be a strand in S. The strand r *unimodally follows* s in S if

1. r is unimodal, r^{init} neighbours s^{init}, and r^{term} neighbours s^{term},
2. the collection $S \cup \{r\}$ is positive and direct,
3. r has crossings of the following types

- if t is a strand crossing s: then it also crosses r and the crossing is of the same type. All such crossings occur in the same order for r as they do for s.
- if s hits the folded line: then r hits the folded line and they cross once, when s is moving rightwardly but r is still moving leftwardly, or vice versa. Otherwise there is no crossing.

Let s and s' be strands in S such that $s^{\text{init}} < s'^{\text{init}}$ are neighbours lying on the same side of $\underline{0}$. A strand r *unimodally follows* the pair s, s' in S if

1. r is unimodal, r^{init} lies between s^{init} and s'^{init}, and r^{term} lies between s^{term} and s'^{term},
2. the collection $S \cup \{r\}$ is positive and direct,
3. r has crossings of three types

   - if t is a strand crossing both s and s': then t makes the same type of crossing with r between its crossings with s' and s;
   - if t is a strand crossing s' but not s: if r^{term} lies between s'^{term} and t^{term}, then r makes the same type of crossing with t after s' crosses t. Otherwise there is no crossing. Similarly, if t crosses s but not s';
   - if s and s' hit the folded line, then r makes a single crossing both with s and with s', either side of the single crossing between s and s'. Otherwise there are no crossings.

Before proceeding, let us make the following trivial observation concerning the action of half-twists on a cabled pair of strands.

Observation 2.19 (Fundamental Observation) Let β be a braid. Let s be a strand of β. Form a new strand s' which follows the strand s (i.e. makes the same crossings with all other strands and in the same order). Allow an arbitrary number of crossings between s and s'. Let β' denote the resulting braid.[4] Let τ_{init} denote a positive half-twist between s^{init} and s'^{init}. Let τ_{term} denote a positive half-twist between s^{term} and s'^{term}. Then $\beta' \cdot \tau_{\text{init}} = \tau_{\text{term}} \cdot \beta'$.

3 Breaking Braids via First Closest Returns

3.1 Cabling

Our first construction extends that by Holmes [10] which generalised the cabling[5] construction for iterated torus knots to horseshoe braids. Our description is in terms of braids rather than templates, but they are equivalent. Note that Holmes did not

---

[4] Formally, these are not braids as they do not have the same set of initial and terminal endpoints. To avoid confusion, let us call these objects *almost-braids*.

[5] Although Holmes only used the term cabling for iterated torus knots (not iterated horseshoe knots) we will use the term for both cases.

consider braid equivalence of the pairs of braids created by this construction. He considered only the single unimodal braid that resulted.

We now give an informal description of the cabling procedure. Given a cyclic braid $\beta \in B_p$, take the folding strand s_0. 'Break' the braid by disconnecting s_0 from s_0^{term} and 'glue' it to a neighbouring point r_0^{term} which is not already the endpoint of some strand. Let r_1^{init} and s_1^{init} denote the vertical projection of r_0^{term} and s_0^{term} respectively. Let s_1 denote the strand emanating from s_1^{init}.

Add a strand r_1 with initial point r_1^{init} and which unimodally follows s_1. Then r_1 has a terminal point r_1^{term} neighbouring s_1^{term}. Now repeat the process: Let r_2^{init} and s_2^{init} denote the vertical projection of r_1^{term} and s_1^{term} respectively. Add a strand r_2 unimodally following s_2, etc.

The final strand r_{p-1} has initial point r_{p-1}^{init} and terminal point r_0^{term}. However, it may not be possible for r_{p-1} to follows s_{p-1} unimodally, as r_0^{term} may lie on the wrong side of s_0^{term}. Therefore r_{p-1} unimodally follows s_{p-1} until all other crossing in have been made, then makes a negative crossing between s_{p-1} and r_{p-1} before connecting to r_0^{term}.

Given a unimodal braid β we can cable it in two distinct ways: to the left or right of the folding strand. It can be shown that this pair of braids are conjugate or reverse-conjugate. However, they are not both unimodal.

Observe that the cabling can be closed-up to form a braid whenever we land inside the interval containing the dynamical preimage of the folding point. More precisely, choose $q \in \mathbb{N}$ so that r_q^{term} lies in an interval $(s_q^{\text{term}}, s_{q'}^{\text{term}})$, not containing any terminal endpoints of β, and such that it maps over s_0^{term}. Let r_{q+1}^{init} denote the vertical projection of r_q^{term}. At this last step, construct a strand r_q emanating from r_q^{term} which, rather than completely following s_q, only follows s_q until we can reconnect r_q to s_0^{term} without creating any further crossings. We call this a *generalised cabling at the dynamical preimage*. Again there are two distinct cablings: to the left or right of the folding strand. As in the cabling case, the pair of braids are conjugate or reverse-conjugate. However, they are never both unimodal.

In Construction 3.1 below, we observe that the cabling can be closed-up at any point in time at which we lie in the interval containing the non-dynamical preimage of the folding point. We call this *generalised cabling at the non-dynamical preimage*. There are two preferred cablings, either side of the folding strand. Theorem 3.3 says this pair is conjugate or reverse-conjugate. What is important is that, unlike the previous two constructions, the braids produced by this process are both unimodal.

Construction 3.1 (Generalised Cabling—At the Non-Dynamical Preimage) Let $\beta \in UB_p$ be cyclic and reconnectable at the non-dynamical preimage. Let $\upsilon \in U_p$ denote the corresponding unimodal permutation. Let $E = [\underline{e}^-, \underline{e}^+]$ denote the interval containing the non-dynamical preimage of the folding point. Then either $\rho(\underline{1}, \underline{e}^-)$ is odd or $\rho(\underline{1}, \underline{e}^+)$ is even, or both. Let $q - 1$ denote either of the points $\underline{e}^-$ or $\underline{e}^+$ satisfying this parity condition.

Choose $\epsilon > 0$ small. Let $r_- \in (\underline{0} - \epsilon, \underline{0})$ and $r_+ \in (\underline{0}, \underline{0} + \epsilon)$. Let $r_1 \in (\underline{1} - \epsilon, \underline{1})$, $r_2 \in (\underline{2}, \underline{2} + \epsilon)$ and choose points $r_i \in (\underline{i} - \epsilon, \underline{i} + \epsilon), i = 3, \ldots, q - 1$, which satisfy

$$r_i \in \begin{cases} (\underline{i} - \epsilon, \underline{i}) & \text{if } \rho(\underline{1}, \underline{i}) \text{ is even} \\ (\underline{i}, \underline{i} + \epsilon) & \text{if } \rho(\underline{1}, \underline{i}) \text{ is odd} \end{cases} \tag{3.1}$$

For $i = \pm, 1, 2 \ldots, q-1$, let r_i^{init} and r_i^{term} denote the projection of r_i to $[-1, 1]^{\text{init}}$ and $[-1, 1]^{\text{term}}$ respectively. These will be endpoints for the strands constructed below.

First, let $r_\pm(0)$ denote the strand with initial endpoint $r_\pm^{\text{init}}$ and terminal endpoint r_1^{term}, which follows unimodally the strand $s(0)$.

Secondly, assume $r_\pm(i)$ have been constructed for $i = 0, \ldots, j-1$, where $0 < j \le q-2$. Let $r_+(j) = r_-(j)$ denote the strand with initial endpoint r_i^{init} and terminal endpoint r_{i+1}^{term} which follows unimodally the strand $s(i)$.

Thirdly, assume that the strands $r_\pm(i)$ have been constructed for all $i = 0, 1, \ldots, q-2$. Construct the strand $r_\pm(q-1)$ as follows. As $[\underline{e}^-, \underline{e}^+]$ maps over the folding point, one of the strands $s(\underline{e}^-)$ or $s(\underline{e}^+)$ has a terminal endpoint which lies on the same side of the folding point as its initial endpoint. Call it s and the other strand s'. Let $r_\pm(q-1)$ denote the strand with initial endpoint r_{q-1}^{init} and terminal endpoint $r_\pm^{\text{term}}$ which unimodally follows s and s'. We make modifications to $r_\pm(q-1)$ as follows:

(I) [*If* $\underline{0} < \underline{p} - 1$.] Since $r_+^{\text{term}} > s(\underline{p} - 1)^{\text{term}}$ and $r_{q-1}^{\text{init}} < \underline{0}^{\text{init}} < s(\underline{p} - 1)^{\text{init}}$ we need an additional crossing between $r_+(q-1)$ and $s(\underline{p} - 1)$. Make a single *negative* crossing between $r_+(q-1)$ and $s(\underline{p} - 1)$ after all other crossings, resulting from $r_+(q-1)$ following s and s', have been made.

Since $r_-^{\text{term}} < s(\underline{p} - 1)^{\text{term}}$ and $r_{q-1}^{\text{init}} < \underline{0}^{\text{init}} < s(\underline{p} - 1)^{\text{init}}$ no additional crossings between $r_-(q-1)$ and $s(\underline{p} - 1)$ are necessary.

(II) [*If* $\underline{p} - 1 < \underline{0}$.] Since $r_-^{\text{term}} < s(\underline{p} - 1)^{\text{term}}$ and $s(\underline{p} - 1)^{\text{init}} < \underline{0}^{\text{init}} < r_{q-1}^{\text{init}}$ we need an additional crossing between $r_+(q-1)$ and $s(\underline{p} - 1)$. Make a single *negative* crossing between $r_+(q-1)$ and $s(\underline{p} - 1)$ after all other crossings, resulting from $r_+(q-1)$ following s, have been made.

Since $r_+^{\text{term}} > s(\underline{p} - 1)^{\text{term}}$ and $s(\underline{p} - 1)^{\text{init}} < \underline{0}^{\text{init}} < r_{q-1}^{\text{init}}$ no additional crossings between $r_+(q-1)$ and any other strand are necessary.

Let $\alpha_\pm$ denote the braid consisting of the strands of β, together with the strands $r_\pm(j), j = 0, 1, \ldots, q-1$ formed above. Note that $\alpha_\pm$ is not cyclic. In fact, it is not necessarily unimodal. Let $\tau_\pm$ denote the positive half-twist between $r_\pm$ and $\underline{0}$. (That is, $\tau_\pm$ is the Artin generator exchanging $\underline{0}$ and $r_\pm$ via a single positive crossing.) Let $\beta_\pm = \tau_\pm \cdot \alpha_\pm$. Note that $\beta_\pm$ is a cyclic braid. Moreover, after cancelling appropriate positive and negative crossings between the strands $r_\pm(q-1)$ and $s(\underline{p} - 1)$ via Reidemeister moves and isotopy, we see that, in both cases (I) and (II), that β_- and β_+ are unimodal. See Fig. 4.

Example 3.2 Consider the cyclic unimodal permutation υ from Example 2.9. In cyclic notation it is given by $(\underline{2}, \underline{3}, \underline{0}, \underline{4}, \underline{1})$. Then the interval containing the non-dynamical preimage is $E = [\underline{2}, \underline{3}]$. Since $\rho(\underline{1}, \underline{2}) = 1$ is odd, υ is reconnectable at the nondynamical preimage. Let β denote the corresponding unimodal braid and $\beta_\pm$

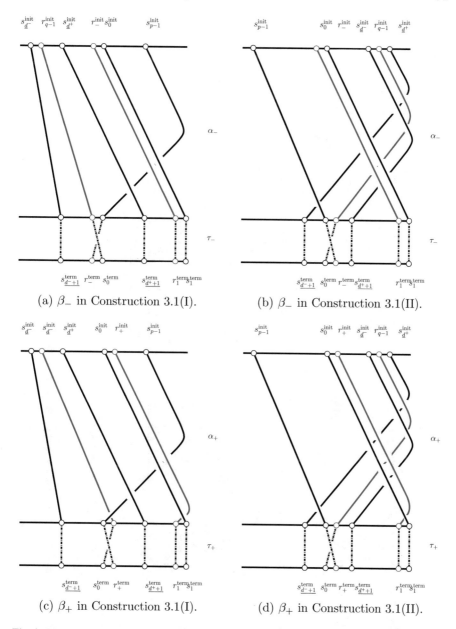

(a) β_- in Construction 3.1(I).

(b) β_- in Construction 3.1(II).

(c) β_+ in Construction 3.1(I).

(d) β_+ in Construction 3.1(II).

Fig. 4 The braids β_- and β_+ from Construction 3.1. Only the necessary strands are shown. Strands from the original braid β are black, while new strands are red. (Only the first and last new strands are depicted)

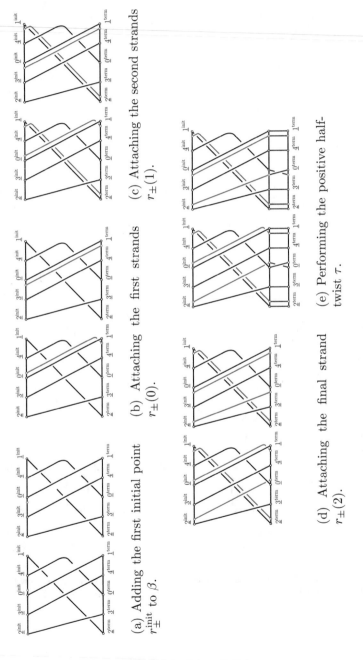

Fig. 5 Adding strands to two copies of β from Example 3.2 resulting in the new braids β_- and β_+

the pair of braids from Construction 3.1. If $\upsilon_\pm$ denotes the permutation corresponding to $\beta_\pm$ then

$$\upsilon_- = (\underline{2}, \underline{7}, \underline{3}, \underline{5}, \underline{0}, \underline{4}, \underline{6}, \underline{1}) \qquad \upsilon_+ = (\underline{2}, \underline{7}, \underline{3}, \underline{0}, \underline{5}, \underline{4}, \underline{6}, \underline{1}) \qquad (3.2)$$

See Fig. 5 for the construction of $\beta_\pm$ in this case.

More generally, consider the case when the dynamical preimage of the folding point lies to the right of m. Then β_- and β_+ are depicted in Fig. 4a, c respectively. Identifying the sets of initial and terminal endpoints of β_- and β_+ in an order-preserving manner, then precomposing β_- with a positive half-twist between the strand $s(\underline{0})$ and its neighbour r_-, and post-composing by the inverse of this twist, yields the braid β_+. A similar argument can be given when the dynamical preimage of the folding point lies to the left of m. Then β_- and β_+ are shown in Fig. 4b, d. However, in this case we precompose and post-compose by the same positive half-twist. Hence we have the following Theorem (See Fig. 5).

Theorem 3.3 *Let $p \in \mathbb{N}$. Let $\beta \in U B_p$ be cyclic. Assume β is reconnectable at the non-dynamical preimage. Let β_- and β_+ denote the braids from Construction 3.1.*

1. *If the non-dynamical preimage lies to the left of m then $\beta_+ \sim \beta_-$.*
2. *If the non-dynamical preimage lies to the right of m then $\beta_+ \sim_r \beta_-$.*

Theorem 3.3 and Remark 2.3 imply the following important corollary.

Corollary 3.4 *Let $p \in \mathbb{N}$. Let $\upsilon \in U_p$ be reconnectable at the non-dynamical preim-age. Let υ_- and υ_+ denote the unimodal permutations from Construction 3.1.*

If $f_-, f_+ \in C([-1, 1], [-1, 1])$ are unimodal maps with periodic critical orbits C_- and C_+ of type υ_- and υ_+ respectively, then $(C_-, f_-) \sim_{BE} (C_+, f_+)$.

4 Breaking Braids via Second Closest Returns

4.1 Generalised Cabling Process

We now describe a second construction that, given a braid-equivalent pair of unimodal permutations υ_- and υ_+ satisfying conditions given below, generates another pair of braid-equivalent unimodal permutations. Moreover, the new pair satisfy the same conditions and hence we can apply the construction once more. The idea is the following. Take $\upsilon_\pm$ as given by Construction 3.1. Break the connection at $\underline{0}^{\text{term}}$ and glue the free strand to a point just outside the interval $[\underline{p}^{\text{term}}_-, \underline{p}^{\text{term}}_+]$, making a *second closest return*. Then consecutively add strands, starting from this point, that unimodally follow the strands from initial endpoints $\underline{p}^{\text{init}}_\pm$, $\underline{p+1}^{\text{init}}_\pm$, etc. Stop once a terminal endpoint lands inside D, the interval containing the dynamical preimage of the folding point. Finally, to close-up the braids, glue the last strand back to $\underline{0}^{\text{term}}$.

However, since we will define the process inductively we need to enlarge the space of such pairs (not just those coming from Construction 3.1). Before giving the general construction, however, let us consider the following example.

Example 4.1 Let $\beta_\pm$ denote the braids from Example 3.2. For notational clarity, let us initially denote the underlying permutations by

$$\upsilon_- = (\underline{2}_-, \underline{7}_-, \underline{3}_-, \underline{5}_-, \underline{0}_-, \underline{4}_-, \underline{6}_-, \underline{1}_-) \tag{4.1}$$

and

$$\upsilon_+ = (\underline{2}_+, \underline{7}_+, \underline{3}_+, \underline{0}_+, \underline{5}_+, \underline{4}_+, \underline{6}_+, \underline{1}_+). \tag{4.2}$$

The braids $\beta_\pm$ are shown in Fig. 7a. The initial and terminal endpoints of $\beta_\pm$ are denoted by $\underline{0}_\pm^{\text{init}}, \underline{1}_\pm^{\text{init}}, \ldots, \underline{7}_\pm^{\text{init}}$ and $\underline{0}_\pm^{\text{term}}, \underline{1}_\pm^{\text{term}}, \ldots, \underline{7}_\pm^{\text{term}}$ respectively. Denote the corresponding strands by $s(\underline{0}_\pm), \ldots, s(\underline{7}_\pm)$. We may assume that the endpoints of the strands in $\beta_\pm$ have been moved, in an order-preserving manner, so that for all $k \neq 5$, $\underline{k}_-^{\text{init}} = \underline{k}_+^{\text{init}}$ and $\underline{k}_-^{\text{term}} = \underline{k}_+^{\text{term}}$. We may also assume that the strands have been deformed so that $s(\underline{k}_-) = s(\underline{k}_+)$ for all $k \neq 4$ or 5 (i.e., any strand whose set of initial or terminal endpoints does not contain $\underline{5}_\pm^{\text{init}}$ or $\underline{5}_\pm^{\text{term}}$). Therefore, for all $k \neq 5$, denote the points $\underline{k}_\pm^{\text{init}}$ and $\underline{k}_\pm^{\text{term}}$ by $\underline{k}^{\text{init}}$ and $\underline{k}^{\text{term}}$ respectively. Similarly, for all $k \neq 4$ or 5, denote the strands $s(\underline{k}_+)$ simply by $s(\underline{k})$. See Fig. 6. Let us do the following to the braids β_- and β_+. As shown in Fig. 7a, add an initial endpoint, shown in green, to the left of $\underline{5}^{\text{init}}$ in both diagrams. Add a strand, also shown in green, from this new initial endpoint which unimodally follows the strand $s(\underline{5}_-)$. See Fig. 7b. Next, add a new initial endpoint directly above the terminal endpoint of this last strand, which neighbours $\underline{6}^{\text{init}}$. From this initial endpoint add a new strand which unimodally follows the next strand $s(\underline{6})$. See Fig. 7c. Observe that the new terminal endpoint lies inside the interval containing the dynamical preimage. Add a final initial endpoint directly above this terminal endpoint, necessarily neighbouring $\underline{7}^{\text{init}}$. From this initial endpoint add a strand which unimodally follows the strand $s(\underline{7})$ until $s(\underline{7})$ has made all its crossings *except* possibly for a single crossing with $s(\underline{4})$. Make the new strand form a negative crossing with the strand $s(\underline{7})$ before ending at a terminal endpoint directly below the very first initial endpoint we started with. See Fig. 7d. This gives a pair of unimodal braids, which are non-cyclic (there are exactly two cycles). To form a pair of cyclic braids, perform a negative half-twist between this final terminal endpoint and $s(\underline{0})^{\text{term}}$. There is a complication in that for one of these braids, there is another terminal endpoint, namely $\underline{5}_+^{\text{term}}$, lying in between: we take the strand from this endpoint going *under* the half-twist. See Fig. 7e. Denote the resulting braids by β_-^1 and β_+^1. Conjugating β_+^1 by a positive half-twist between $\underline{0}$ and $\underline{5}_+$, the resulting braid can be deformed by a series of isotopies and Reidemeister moves into the braid β_+^1.

Next, we need the following notions to simplify exposition. Given a unimodal permutation $\upsilon \in U_p$ and $\imath \in \{0, 1, \ldots, p-1\}$, $\imath \neq \underline{0}$, we add a point $\bar{\imath}$ to the linearly ordered set $\{0, 1, \ldots, p-1\}$, which we call the *opposite* of $\imath$ with respect to υ, which satisfies

(i) $\bar{\imath} < \underline{0}$ if and only if $\imath > \underline{0}$,

Fig. 6 The braids $\beta_\pm$ with strands in the same diagram, from Example 4.1. The strands with long dashes come from β_-, while the strands with short dashes come from β_+

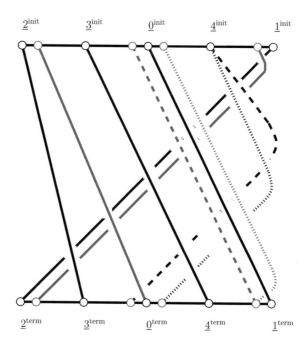

$$2^{\text{init}} \qquad 3^{\text{init}} \qquad 0^{\text{init}} \qquad 4^{\text{init}} \qquad 1^{\text{init}}$$

$$2^{\text{term}} \qquad 3^{\text{term}} \qquad 0^{\text{term}} \qquad 4^{\text{term}} \qquad 1^{\text{term}}$$

(ii) $\underline{i} \in (\underline{j}, \underline{k})$ if and only if $\upsilon(\underline{i}) \in (\upsilon(\underline{j}), \upsilon(\underline{k}))$.

Remark 4.2 Observe that, if the set $\{1, \dots, p-1\}$ is embedded in the line in an order-preserving manner and υ is realised by a unimodal endomorphism f, then the opposite of a point $\underline{i}$ just corresponds to the preimage of $f(\underline{i})$ under f which is not $\underline{i}$.

In what follows we only need to consider closest return times to the folding point, so we will simply say that p is the closest return time. (See Sect. 2.2 for the definition.)

For positive integers p and q, let $U_{p,q}$ denote the set of cyclic unimodal permutations of length $p + q$ with closest return time p. Given $\upsilon \in U_{p,q}$, let $C = [p, \bar{p}]$, where $\bar{p}$ denotes the opposite point of p with respect to υ. Let $D = [d^-, d^+]$ denote the interval containing the dynamical preimage $\underline{d} = p + q - 1$ of $\underline{0}$. Let $D^- = [d^-, \underline{d}]$ and $D^+ = [\underline{d}, d^+]$. Let $E = [e^-, e^+]$ denote the interval containing the non-dynamical preimage.

Example 4.3 Observe that the unimodal permutations constructed in Example 3.2 and also considered in Example 4.1,

$$\upsilon_- = (\underline{2}, \underline{7}, \underline{3}, \underline{5}_-, \underline{0}, \underline{4}, \underline{6}, \underline{1}) \text{ and } \upsilon_+ = (\underline{2}, \underline{7}, \underline{3}, \underline{0}, \underline{5}_+, \underline{4}, \underline{6}, \underline{1}) \qquad (4.3)$$

both lie in $U_{5,3}$, where we have kept the notation convention of Example 4.1. If we denote the intervals C, D, $D^\pm$, E for the permutation $\upsilon_\pm$ by $C_\pm$, $D_\pm$, $D_\pm^\pm$, $E_\pm$, then we find that $C_- = [\underline{5}_-, \underline{5}_+] = C_+$, $D_- = [\underline{2}, \underline{3}] = D_+$, $D_-^- = [\underline{2}, \underline{7}] = D_+^-$, $D_-^+ = [\underline{7}, \underline{3}] = D_+^+$, $E_- = [\underline{0}, \underline{4}]$, and $E_+ = [\underline{4}, \underline{6}]$.

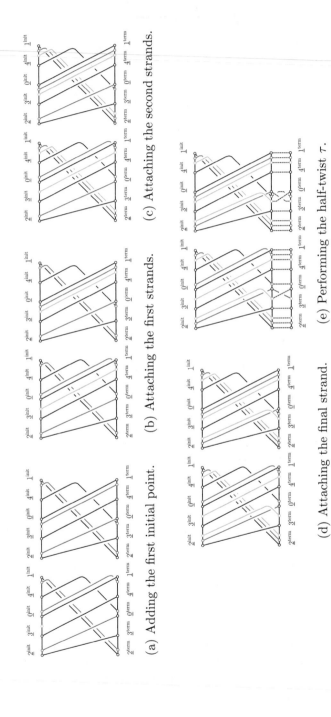

(a) Adding the first initial point.

(b) Attaching the first strands.

(c) Attaching the second strands.

(d) Attaching the final strand.

(e) Performing the half-twist τ.

Fig. 7 Adding strands to $\beta_\pm$ from Example 3.2 resulting in the new braids β_-^1 and β_+^1

Now consider the general situation. Let $v_\pm \in U_{p,q}$. In cyclic notation denote $v_\pm$ by $(\underline{2}_\pm, \ldots, \underline{0}_\pm, \ldots, \underline{1}_\pm)$. Assume the following:

1. (*closest returns lie on opposite sides of the folding point*) $\underline{p}_- < \underline{0}_-$ and $\underline{0}_+ < \underline{p}_+$.
2. (*all remaining points are in order-preserving bijection*) for all $k, l \neq p, \underline{k}_- < \underline{l}_-$ if and only if $\underline{k}_+ < \underline{l}_+$. Hence, if $\underline{k}_\pm$ are embedded in the line in an order-preserving manner, as in the preceding example, we may assume that $\underline{k}_- = \underline{k}_+$ for all $k \neq p$. Consequently, for each $k \neq p$, we may denote the point $\underline{k}_\pm$ by $\underline{k}$.
3. (*the closest return is not contained in the interval containing the dynamical preimage of the folding point*) $\underline{p}_\pm \notin \partial D_\pm$.
4. (*the dynamical preimage of the folding point and the dynamical preimage of the closest return lie on opposite sides of the folding point*) either $\underline{p+q-1}_\pm < \underline{0}_\pm < \underline{p-1}_\pm$, or $\underline{p-1}_\pm < \underline{0}_\pm < \underline{p+q-1}_\pm$.

Let $C_\pm, D_\pm, D_\pm^\pm, E_\pm$ denote the corresponding intervals for $v_\pm$. Then by (1) we may assume that $\underline{p}_- = \overline{p}_+$ and hence $C_- = C_+$. Denote this interval by C. Properties (2) and (3) then imply that $D_- = D_+$. Denote this interval by D. Similarly $D_-^\pm = D_+^\pm$ and we may denote this interval by $D^\pm$. Note that, as we saw in the previous Example 4.3, E_- and E_+ do not necessarily coincide.

By the discussion in Sect. 3, Condition (2) implies the following property, that will be used in the proof of Theorem 4.8 below.

2'. Let $\beta_\pm$ denote the canonical unimodal braid of $v_\pm$. Then there exists a braid γ with $p + q$ strands, with the property that outside of $C = [\underline{p}_-, \overline{p}_+]$, γ is trivial, and such that one of the following holds:

$2'_+$. $\beta_- \cdot \gamma = \gamma \cdot \beta_+$
$2'_-$. $\beta_- \cdot \gamma = \gamma^{-1} \cdot \beta_+$

Finally we will need the following trivial observation:

5. (*existence of the transit time from the closest return to the interval containing the dynamical preimage of the folding point*) The orbit segment $p + 1, p + 2, \ldots, p + q - 1$, of the image of the closest return point to the interval D does not intersect the interval C.

Call $t = q - 2$ the *transit time*. We say it is a *left transit time* if the parity, given by $\rho(p+1, p+t+1)$, is even and a *right transit time* if this parity is odd.

Example 4.4 Let us continue considering $v_\pm$ from Example 4.3. Then the transit time is $t = 1$, since $q = 3$. It is a right transit time. If we imagine a point neighbouring $p + 1$ to the left, then the image of that point lies in the interval containing the dynamical preimage, i.e., we land inside the interval containing the dynamical preimage after a single iterate. Moreover, this image lies to the right of the dynamical preimage of the folding point, therefore it is a right transit time.

Remark 4.5 Any pair (v_-, v_+) of equivalent unimodal permutations from Construction 3.1 satisfies Properties (1)–(5). In particular, the unimodal permutations from Example 4.3 satisfy these conditions.

Construction 4.6 (Generalised Cabling Process.) Let (υ_-, υ_+) denote a pair of cyclic unimodal permutations satisfying properties (1)–(4). Recall that $C_\pm$, $D_\pm$, etc., denote the corresponding intervals for $\upsilon_\pm$. Recall that we may assume $C_- = C_+$ and $D_- = D_+$. Denote these intervals by C and D respectively. Observe that $\upsilon_-(D) = \upsilon_+(D)$ and this interval contains $\underline{0}$ and $\underline{p}_-$ in its interior.

Let $\beta_\pm$ denote the canonical braid corresponding to $\upsilon_\pm$. Denote the set of strands for $\beta_\pm$ by $S_\pm$. For $k = 0, 1, \ldots, p + q - 1$, denote the k-th strand by $s_\pm(\underline{k})$. The strand $s_\pm(\underline{k})$ lies between initial endpoint $s_\pm(\underline{k})^{\text{init}} = \underline{k}_\pm^{\text{init}}$ and terminal endpoint $s_\pm(\underline{k})^{\text{term}} = \underline{k+1}_\pm^{\text{term}}$, where addition is taken modulo $p + q$. By property (4) above we may assume, after applying an isotopy if necessary, that $s_-(\underline{k}) = s_+(\underline{k})$ for all $k \neq p - 1$ or p.

Remark 4.7 We will also use the notation s_k for the projection to $[-1, 1]$ of the initial endpoint of $s(\underline{k})$ for $k = 0, 1, \ldots, p_-, p_+, \ldots, p + q - 1$.

Let t be the transit time. Construct a pair $(\upsilon_-^1, \upsilon_+^1)$ of unimodal permutations as follows:

To begin, we define points r_i, r_i^{init} and r_i^{term} for $i = 0, 1, \ldots, t + 1$. Choose $\epsilon > 0$ small. Let $r_1 \in (p + 1 - \epsilon, p + 1)$, and choose points $r_i \in (p + i - 1, p + i + 1)$, $i = 2, \ldots, t + 1$, which satisfy

$$r_i \in \begin{cases} (\underline{p+i} - \epsilon, \underline{p+i}) & \text{if } \rho(\underline{p+1}, \underline{p+i}) \text{ even} \\ (\underline{p+i}, \underline{p+i} + \epsilon) & \text{if } \rho(\underline{p+1}, \underline{p+i}) \text{ odd} \end{cases} \tag{4.4}$$

Finally, choose $r_0 \in (\underline{p}_- - 1, \underline{p}_+ + 1)$ which satisfies

$$r_0 \in \begin{cases} (\underline{p}_- - \epsilon, \underline{p}_-) & \text{if } D < \underline{0} \text{ and } t \text{ a right time, or } \underline{0} < D \text{ and } t \text{ a left time.} \\ (\underline{p}_+, \underline{p}_+ + \epsilon) & \text{if } D < \underline{0} \text{ and } t \text{ a left time, or } \underline{0} < D \text{ and } t \text{ a right time.} \end{cases} \tag{4.5}$$

For $i = 0, 1, 2, \ldots, t + 1$ let r_i^{init} and r_i^{term} denote the projections of r_i onto $[-1, 1]^{\text{init}}$ and $[-1, 1]^{\text{term}}$ respectively. As before, these will be endpoints for the strands constructed inductively below.

First, let $r_\pm(\underline{1})$ denote the strand with initial endpoint r_1^{init} and terminal endpoint r_2^{term} which follows unimodally the strand $s_\pm(\underline{p+1})$.

Secondly, assume that the strands $r_\pm(\underline{i})$ have been constructed for $i = 1, \ldots, j - 1$, for some $0 < j < t + 1$. Let $r_\pm(\underline{j})$ denote the strand with initial endpoint r_j^{init} and terminal endpoint r_{j+1}^{term} which follows unimodally the strand $s_\pm(\underline{p+j})$.

Thirdly, assume that the strands $r_\pm(\underline{i})$ have been constructed for all $i = 1, 2, \ldots, t$. Then $r_\pm(\underline{t+1})$ denotes the strand with initial endpoint r_{t+1}^{init} and terminal endpoint r_0^{term} which, in each of the four cases below, satisfies the following:

(Ii) [*If $p + q - 1 < \underline{0}$ and t is a left transit time.*]
Let $r_-(\underline{t+1})$ unimodally follow the strand $s_-(\underline{p+t+1})$ rightwardly until $s_-(\underline{p+t+1})$ has only one crossing to make, which is necessarily with $s_-(\underline{p-1})$, before reconnecting to its terminal endpoint. Then $r_-(\underline{t+1})$ forms

a *negative* crossing with $s_-(p+t+1)$, after which it makes a *positive* crossing with $s_-(p-1)$ before connecting to its terminal endpoint. See Fig. 8a.

Let $r_+(t+1)$ unimodally follow the strand $s_+(p+t+1)$ rightwardly until $s_+(p+t+1)$ has no more crossings to make before reconnecting to its terminal endpoint. Then $r_+(t+1)$ forms a *negative* crossing with $s_+(p+t+1)$, after which it makes a *positive* crossing with $s_+(p-1)$ before connecting to its terminal endpoint. See Fig. 8c.

(Ii) [*If* $p+q-1 < 0$ *and* t *a right transit time.*]

Let $r_-(t+1)$ unimodally follow the strand $s_-(p+t+1)$ rightwardly until $s_-(p+t+1)$ has only one crossing to make, which is necessarily with $s_-(p-1)$, before reconnecting to its terminal endpoint. Then $r_-(t+1)$ forms a *negative* crossing with $s_-(p+t+1)$ before connecting to its terminal endpoint. See Fig. 8b.

Let $r_+(t+1)$ unimodally follow the strand $s_+(p+t+1)$ rightwardly until $s_+(p+t+1)$ has no further crossings to make before reconnecting to its terminal endpoint. Then $r_+(t+1)$ forms a *negative* crossing with $s_+(p+t+1)$ before connecting to its terminal endpoint. See Fig. 8d.

(IIi) [*If* $0 < p+q-1$ *and* t *a left transit time.*]

Let $r_-(t+1)$ unimodally follow $s_-(p+t+1)$ until $s_-(p+t+1)$ has no more crossings to make before reconnecting to its terminal endpoint. Then $r_-(t+1)$ forms a *negative* crossing with $s_-(p+t+1)$, after which it makes a *positive* crossing with $s_-(p-1)$ before connecting to its terminal endpoint. See Fig. 9a.

Let $r_+(t+1)$ unimodally follow $s_+(p+t+1)$ until $s_+(p+t+1)$ has only one crossing to make, which is necessarily with $s_+(p-1)$, before reconnecting to its terminal endpoint. Then $r_+(t+1)$ forms a *positive* crossing with $s_+(p-1)$, after which it makes a *negative* crossing with $s_+(p+t+1)$ before connecting to its terminal endpoint. See Fig. 9c.

(IIii) [*If* $0 < p+q-1$ *and* $t+1$ *a right transit time.*]

Let $r_-(t+1)$ unimodally follow $s_-(p+t+1)$ until $s_-(p+t+1)$ has no further crossings to make before connecting to its terminal endpoint. Then $r_-(t+1)$ forms a single *negative* crossing with $s_-(p+t+1)$ before connecting to its terminal endpoint. See Fig. 9b.

Let $r_+(t+1)$ unimodally follow $s_+(p+t+1)$ until $s_+(p+t+1)$ has only one crossing to make, which is necessarily with $s_+(p-1)$, before connecting to its terminal endpoint. Then $r_+(t+1)$ forms a single *negative* crossing with $s_+(p+t+1)$ before connecting to its terminal endpoint. See Fig. 9d.

Finally, let $r_\pm(0)$ denote the strand with initial endpoint r_0^{init} and terminal endpoint r_1^{term} which, in cases (Ii) and (IIii), follows unimodally the strand $s(p_+)$ and, in cases (Iii) and (IIi), follows unimodally the strand $s(p_-)$.

Let $\alpha_\pm^1 \in B_{p+q+t+2}$ denote the braid consisting of the strands of $\beta_\pm$ together with the strands $r_\pm(i)$, $i = 0, 1, \ldots, t+1$, formed above. (As before, after relabelling the endpoints this gives a well-defined braid diagram normalised as in Sect. 2.2.2. However, for the moment we will keep the labelling as above.) Note

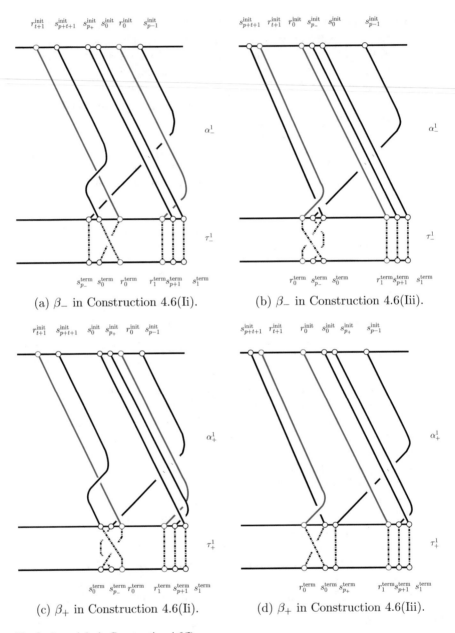

(a) β_- in Construction 4.6(Ii). (b) β_- in Construction 4.6(Iii).

(c) β_+ in Construction 4.6(Ii). (d) β_+ in Construction 4.6(Iii).

Fig. 8 β_- and β_+ in Construction 4.6(I)

that $\alpha_\pm^1$ is not a cyclic braid. Note also that $\alpha_\pm^1$ is not necessarily a unimodal braid. In fact, in all cases $\alpha_\pm^1$ is direct but neither positive nor unimodal. See Figs. 8a–9d.

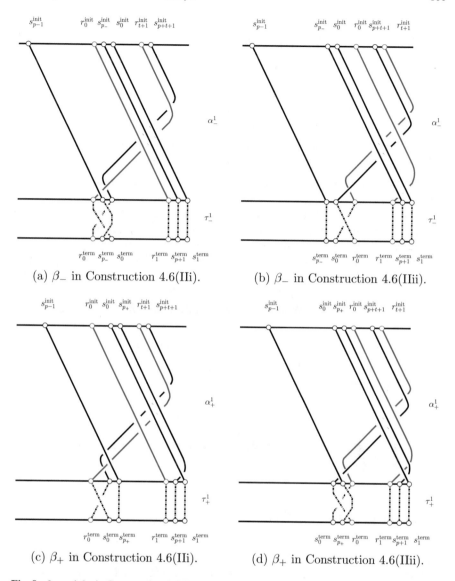

(a) β_- in Construction 4.6(IIi).   (b) β_- in Construction 4.6(IIii).

(c) β_+ in Construction 4.6(IIi).   (d) β_+ in Construction 4.6(IIii).

Fig. 9 β_- and β_+ in Construction 4.6(II)

As in the previous construction, we now compose $\alpha_\pm^1$ with a positive half-twist $\tau_\pm^1$. However, in terms of the diagram there is an extra complication due to either s_{p_-} or s_{p_+} lying between s_0 and r_0. (Recall that, for each k, s_k denotes the projection to $[-1, 1]$ of the initial endpoint of $s(\underline{k})$.) Therefore let $\tau(i, i + 1)$ denote the Artin generator positively interchanging i and $i + 1$. Then

(Ii) τ_-^1 denotes a single positive half-twist $\tau(s_0, r_0)$ between s_0 and r_0.

τ_+^1 denotes $\tau(s_0, s_{p_+})^{-1} \cdot \tau(s_{p_+}, r_0) \cdot \tau(s_0, s_{p_+})$.

(Ii) τ_-^1 denotes $\tau(s_{p_-}, s_0) \cdot \tau(r_0, s_{p_-}) \cdot \tau(s_{p_-}, s_0)^{-1}$.

τ_+^1 denotes a single positive half-twist $\tau(r_0, s_0)$ between r_0 and s_0.

(IIi) τ_-^1 denotes $\tau(s_{p_-}, s_0)^{-1} \cdot \tau(r_0, s_{p_-}) \cdot \tau(s_{p_-}, s_0)$.

τ_+^1 denotes a single positive half-twist $\tau(r_0, s_0)$ between r_0 and s_0.

(IIii) τ_-^1 denotes a single positive half-twist $\tau(s_0, r_0)$ between s_0 and r_0.

τ_+^1 denotes $\tau(s_{p_+}, r_0)^{-1} \cdot \tau(s_0, s_{p_+}) \cdot \tau(s_{p_+}, r_0)$.

Let $\beta_\pm^1 = \tau_\pm^1 \cdot \alpha_\pm^1$. Note that $\beta_\pm^1$ is a cyclic braid. Moreover, after cancelling appropriate crossings between strands $r_\pm(t + 1)$, $s_\pm(p + t + 1)$ and $s_\pm(p - 1)$ via Reidemeister moves and isotopy, we find that β_-^1 and β_+^1 are both unimodal. Again, see Figs. 8a–9d.

Let use investigate what is going on in more detail. Consider the case (Ii). Then β_- and β_+ are depicted in Fig. 8a, c respectively. Identifying the sets of initial and terminal endpoints of β_- and β_+ in an order-preserving manner, then precomposing β_- with a positive half-twist between the strand $s(\underline{0})$ and its neighbour $s(p_-)$, and post-composing by the inverse of this twist, yields the braid β_+. A similar argument can be given in cases (Iii)–(IIii). See Figs. 8b–9d. However, in cases (IIi) and (IIii) we pre-compose and post-compose by the same positive half-twist. Hence we have the following.

Theorem 4.8 *Given* (υ_-, υ_+), *satisfying Properties (1)–(4). Let* β_- *and* β_+ *denote the corresponding unimodal braids. Let* (β_-^1, β_+^1) *denote the pair of braids produced from Construction 4.6.*

1. *If* $\beta_- \sim \beta_+$ *then* $\beta_-^1 \sim \beta_+^1$.
2. *If* $\beta_- \sim_r \beta_+$ *then* $\beta_-^1 \sim_r \beta_+^1$.

Moreover, the corresponding unimodal permutations $(\upsilon_-^1, \upsilon_+^1)$ *also satisfy the Properties (1)–(4).*

5   Applications to the Hénon Family

We have constructed two mechanisms for constructing braid equivalences. However, we would like to restrict ourselves to equivalences realised in the Hénon family.

Before proceeding, let us give a brief description of the parameter space of the quadratic family and the Hénon family. Recall from the introduction that $f_a(x) = a - x^2$ denotes the quadratic family and $F_{a,b}(x, y) = (a - x^2 - by, x)$ denotes the Hénon family. In the (a, b)-plane, for b positive the map $F_{a,b}$ is an orientation-preserving diffeomorphism. For $b \in (0, 1)$ the map $F_{a,b}$ is area-contracting. In fact, $\mathrm{Jac}(F_{a,b})(x, y) = b$ for all $(a, b) \in \mathbb{R}^2$ and $(x, y) \in \mathbb{R}^2$.

The parabola

$$(1 + b)^2 + 4a = 0 \tag{5.1}$$

is the saddle-node bifurcation locus. For all parameters (a, b) on this curve $F_{a,b}$ possesses a unique fixed point. The parabola passes through the a-axis at $a = -1/4$. All parameters to the left of this curve possess no fixed point and the iterates of all points escape to infinity. All parameters to the right of the curve possess two fixed points, one saddle and one sink, and hence the non-wandering set is non-trivial.

The curve

$$a = (5 + 2\sqrt{5})(1 + b)^2 \tag{5.2}$$

lies to the right of the saddle-node bifurcation curve above. It was shown by Devaney-Nitecki [6] that for all parameters (a, b) lying to the right of this curve the map $F_{a,b}$ possesses a full horseshoe, and all points either escape to infinity or converge to this invariant set under iteration.

For a fixed b sufficiently small, increasing a from the saddle-node bifurcation locus to the horseshoe locus the map $F_{a,b}$ undergoes a period-doubling cascade. Each period-doubling bifurcation curve is algebraic and they accumulate upon an analytic curve which intersects the a-axis at the Feigenbaum-Collet-Tresser parameter $a = 1.401....$ For parameters to the left of the accumulation of period-doubling, the map $F_{a,b}$ has simple dynamics: there are finitely many periodic orbits each of period 2^n for some non-negative integer n.

After this accumulation of period-doubling less is known. However, restricting to the a-axis we know more. Here between the accumulation of period-doubling and the horseshoe locus, uncountably many bifurcations occur. For each periodic kneading sequence there corresponds a hyperbolic component, i.e., an interval such that for each parameter in this interval $F_{a,b}$ has a periodic attractor whose itinerary is determined by the given periodic kneading sequence. For example, there is an interval around the point $a = 1.7549..$, for which every parameter has an attractive cycle of period three. (This parameter is actually the critically-periodic parameter, or centre, of the unique period-three hyperbolic component.)

These hyperbolic intervals extend to open subsets of the (a, b)-plane, where the periodic attractor persists. The loci of all such parameters, for fixed periods, were first considered by El-Hamouly and Mira [7]. Many components of this locus have the following structure: there exists a main 'body' out of which four 'limbs' emanate. Then limbs do not intersect; the union of the limbs and body is simply connected; two of the limbs intersect $\{b = 1\}$; the two remaining limbs intersect $\{b = 0\}$. (Similar configurations have been observed for the one-dimensional cubic family. These configurations have been called *swallow configurations* by Milnor [13].) Numerical investigations into the braid-equivalences exhibited in the (a, b)-plane were carried out by Holmes [10]. (See also Sannami [15].) However, currently very little is understood about these configurations. Their apparent prevalence, in the chaotic parameter region for the Hénon family as well as in other families, also requires explanation.

Remark 5.1 The braid equivalences we constructed were based on the initial unimodal permutation υ possessing a non-dynamical preimage. This can only be satisfied if the corresponding kneading sequence satisfies $\kappa \succ 10C$ or, equivalently, has hyperbolic parameter interval lying to the right of the period-three hyperbolic param-

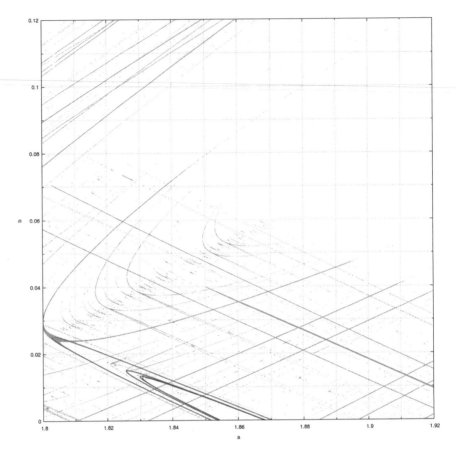

Fig. 10 The isotracal curves of periods 8, 11, 14 and 17 generated from head $1001C$ are shown in red. The grey region show the scatterplots for periods 8, 11, 14 and 17. The darker the colour, the higher the period

eter interval. Consequently, all numerical example given below intersect the a-axis in the interval [1.7549..., 2]. However, in [5] we will describe a generalisation of the construction of braid equivalences given here which do not have this restriction.

Following the numerical evidence given below, we ask the following questions. Let υ_- and υ_+ be an arbitrary pair of combinatorial types from Construction 3.1.

Question A. Let $a_-, a_+ \in [-1/4, 2]$ be such that f_{a_-} and f_{a_+} have critical orbits c_- and c_+ of types υ_- and υ_+ respectively. Let C_- and C_+ denote the corresponding periodic orbits for $F_{a_-,0}$ and $F_{a_+,0}$ respectively. Does there exist a braid equivalence in the family $F_{a,b}$ connecting $(C_-, F_{a_-,0})$ and $(C_+, F_{a_+,0})$?

The above question only deal with braid equivalences coming from Construction 3.1. Now, given an initial braid equivalent pair $(\upsilon_-^0, \upsilon_+^0)$, let us consider a sequence of equivalent pairs $(\upsilon_-^i, \upsilon_+^i)$ coming from Construction 4.6.

(a)   **(b)**

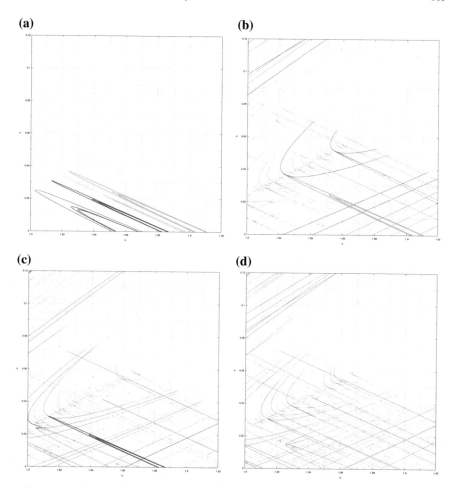

(c)   **(d)**

Fig. 11 For the Hénon family $F_{a,b}$, in Fig. 11a zero isotracal paths are plotted for heads $1001C$, $10011C$, $100111C$ and $1001111C$. In Figs. 10, 11b, c the same paths are plotted, for a fixed head, together with scatterplots of parameters with attracting points of the same period

Question B. For each positive integer i, let a^i_-, $a^i_+ \in [-1/4, 2]$ be parameters such that $f_{a^i_-}$ and $f_{a^i_+}$ have critical orbits c^i_- and c^i_+ of types υ^i_- and υ^i_+ respectively. Let C^i_- and C^i_+ denote the corresponding periodic orbits for $F_{a^i_-,0}$ and $F_{a^i_+,0}$ respectively. Does there exist, for each i, a braid equivalence in the family $F_{a,b}$ connecting $(C^i_-, F_{a^i_-,0})$ and $(C^i_+, F_{a^i_+,0})$? Are the paths γ^i in the (a, b)-plane which realise these braid equivalences pairwise disjoint?

We now give numerical evidence suggesting A and B are true. For simplicity we represent unimodal combinatorial types by the itinerary of the critical point (with respect to the standard partition $I_0 = [0, m)$, $I_C = m$, $I_1 = (m, p - 1]$.) Fig. 1a–d showed plots of parameters for which $F_{a,b}$ possessed an attracting periodic orbit of a fixed period p. These regions connected distinct degenerate Hénon parameters.

Numerically we construct a curve between such parameters as follows. First, take a unimodal permutation υ satisfying the hypotheses of Construction 3.1, i.e., it is reconnectable at the non-dynamical preimage, and the non-dynamical preimage lies to the left of the folding point. Call υ the *head*. Apply Construction 3.1, giving a braid-equivalent pair υ_-^0 and υ_+^0. Then apply Construction 4.6 inductively, giving braid-equivalent pairs υ_-^i and υ_+^i for $i = 1, 2, 3$. This was repeated for various υ. See Table 1 for orbits listed by itinerary and grouped by head, with the associated permutation given in cyclic notation. They are listed in pairs obtained as just described: the first pair is obtained from the head by Construction 3.1 and subsequent ones from Construction 4.6, inductively.

For each pair υ_-^i and υ_+^i we computed the superattracting parameters a_- and a_+ in the quadratic family with critical orbit of type υ_- and υ_+ respectively. Then the parameter locus of

$$F_{a,b}^p(x_{in}, y_{in}) = (x_{in}, y_{in}), \qquad \text{tr} DF_{a,b}^p(x_{in}, y_{in}) = 0. \tag{5.3}$$

passing through the parameters $(a_-, 0)$ and $(a_+, 0)$ was computed. We call a parameter curve lying in the locus (5.3) a *zero isotracal path*. (More generally an *isotracal path* satisfies (5.3), but with the trace set to some fixed constant instead of zero.) We compute an isotracal curve iteratively by starting from the initial data given by

$$a = a_\pm, \quad b = 0, \quad x_{in} = a_\pm, \quad y_{in} = 0. \tag{5.4}$$

On each slice $\{b = b_0\}$, Newton's method was used to find $a = a(b_0), x_{in} = x_{in}(b_0)$ and $y_{in} = y_{in}(b_0)$ satisfying Eq. (5.3). The value of b was then incremented and the values of a, x_{in} and y_{in} from the previous step were used as initial data. The algorithm terminated once $a_-(b)$ and $a_+(b)$ were sufficiently close.

Table 2 shows some of the braid equivalences which were realised in the Hénon family. The paths in the parameter region of $(a, b) \in [1.8, 1.9] \times [0.0, 0.1]$ are shown in Fig. 11a. The red curves show the period 8, 11, 14 and 17 curves associated with head $1001C$ given in Table 2. They are shaded so that the darker the curve is, the higher the period. Similarly, the green curves show the period 9, 12, 15 and 18 curves associated with the head $10011C$ from Table 2. The blue curves show the period 10, 13, 16 and 19 curves associated with the head $100111C$ from Table 2. The yellow curves show the period 11, 14, 17 and 20 curves associated with the head $1001111C$ from Table 2. In each of these cases, the darker the curve is, the higher the period.

These plots were then superimposed with the scatterplots in Figs. 10–11d. The grey region denoting the data from the first algorithm. So, for example, Fig. 10 shows the curves of period 8, 11, 14 and 17 in Fig. 11a together with the scatterplot data from the introduction for periods 8, 11, 14, and 17, where the darker the grey is the lower the period. Figure 11b, c are similar.

Table 2 also gives the prefix and decoration to compare the current mechanism with that given by the first and second authors (see [4] for more details on prefixes and decorations). For example consider the kneading sequences $10011110C$ and $10011010C$ (see Table 2). These have the same prefix 1001 but different decorations,

Table 1 Braid equivalences in cyclic notation

Critical itinerary	Cyclic notation
1001010C	2,7,3,5,0,4,6,1
1001110C	2,7,3,0,5,4,6,1
1001010010C	2,7,10,3,8,5,0,4,9,6,1
1001110010C	2,7,10,3,8,0,5,4,9,6,1
1001010010110C	2,7,13,10,3,8,5,0,11,4,9,12,6,1
1001110010110C	2,7,13,10,3,8,0,5,11,4,9,12,6,1
1001010010110110C	2,7,16,13,10,3,8,5,0,14,11,4,9,12,15,6,1
1001110010110110C	2,7,16,13,10,3,8,0,5,14,11,4,9,12,15,6,1
10011110C	2,8,3,0,6,4,5,7,1
10011010C	2,8,3,6,0,4,5,7,1
10011110010C	2,8,11,3,9,0,6,4,5,10,7,1
10011010010C	2,8,11,3,9,6,0,4,5,10,7,1
10011110010110C	2,8,14,11,3,9,0,6,12,4,5,10,13,7,1
10011010010110C	2,8,14,11,3,9,6,0,12,4,5,10,13,7,1
10011010010110110C	2,8,17,14,11,3,9,6,0,15,12,4,5,10,13,16,7,1
10011110010110110C	2,8,17,14,11,3,9,0,6,15,12,4,5,10,13,16,7,1
100111010C	2,9,3,7,0,5,4,6,8,1
100111110C	2,9,3,0,7,5,4,6,8,1
100111010010C	2,9,12,3,10,7,0,5,4,6,11,8,1
100111110010C	2,9,12,3,10,0,7,5,4,6,11,8,1
100111010010110C	2,9,15,12,3,10,7,0,13,5,4,6,11,14,8,1
100111110010110C	2,9,15,12,3,10,0,7,13,5,4,6,11,14,8,1
100111010010110110C	2,9,18,15,12,3,10,7,0,16,13,5,4,6,11,14,17,8,1
100111110010110110C	2,9,18,15,12,3,10,0,7,16,13,5,4,6,11,14,17,8,1
1001111110C	2,10,3,8,0,6,4,5,7,9,1
1001111010C	2,10,3,8,6,0,4,5,7,9,1
1001111110010C	2,10,13,3,11,0,8,6,4,5,7,12,9,1
1001111010010C	2,10,13,3,11,8,0,6,4,5,7,12,9,1
1001111110010110C	2,10,16,13,3,11,0,8,14,6,4,5,7,12,15,9,1
1001111010010110C	2,10,16,13,3,11,8,0,14,6,4,5,7,12,15,9,1
1001111110010110110C	2,10,19,16,13,3,11,0,8,17,14,6,4,5,7,12,15,18,9,1
1001111010010110110C	2,10,19,16,13,3,11,8,0,17,14,6,4,5,7,12,15,18,9,1
100110010100110C	2,11,6,15,3,12,7,9,0,4,13,8,14,5,10,1
100110011100110C	2,11,6,15,3,12,7,0,9,4,13,8,14,5,10,1
100110010100110010C	2,11,18,6,15,22,3,12,19,7,16,9,0,4,13,20,8,21,14,5,17,10,1
100110011100110010C	2,11,18,6,15,22,3,12,19,7,16,0,9,4,13,20,8,21,14,5,17,10,1
100110010100110010110100110C	2,11,25,18,6,15,29,22,3,12,26,19,7,16,9,0,23,4,13,27,20,8, 21,28,14,5,17,24,10,1
100110011100110010110100110C	2,11,25,18,6,15,29,22,3,12,26,19,7,16,0,9,23,4,13,27,20,8, 21,28,14,5,17,24,10,1
100110010100110010110010110100110C	2,11,32,25,18,6,15,36,29,22,3,12,33,26,19,7,16,9,0,30,23,4, 13,34,27,20,8,21,28,35,14,5,17,24,31,10,1
100110011100110010110010110100110C	2,11,32,25,18,6,15,36,29,22,3,12,33,26,19,7,16,0,9,30,23,4,13, 34,27,20,8,21,28,35,14,5,17,24,31,10,1

Table 2 Braid equivalences and associated data

Head	Period	Unimodal a	Critical itinerary	Prefix	Decoration
1001C	8	1.85173004941	1001010C	1001	10
	8	1.87000388083	1001110C	1001	10
	11	1.85376146047	1001010010C	1001	10010
	11	1.86808014899	1001110010C	1001	10010
	14	1.85420779956	1001010010110C	1001	10010110
	14	1.86768154112	1001110010110C	1001	10010110
	17	1.85429374549	1001010010110110C	1001	10010110110
	17	1.86760846133	1001110010110110C	1001	10010110110
10011C	9	1.90311677305	10011110C	1001	110
	9	1.91144463147	10011010C	1001	010
	12	1.90383983136	10011110010C	1001	110010
	12	1.91074213574	10011010010C	1001	010010
	15	1.90396328656	10011110010110C	1001	110010110
	15	1.91062464072	10011010010110C	1001	010010110
	18	1.90398296779	10011010010110110C	1001	010010110110
	18	1.91060625301	10011110010110110C	1001	110010110110
100111C	10	1.88240793375	1001110010C	1001	1010
	10	1.88717220640	1001111110C	1001	1110
	13	1.88286957759	1001110100010C	1001	1010010
	13	1.88672860609	1001111110010C	1001	1110010
	16	1.88295631360	1001110100010110C	1001	1010010110
	16	1.88664609851	1001111110010110C	1001	1110010110
	19	1.88297117918	100111010010110110C	1001	1010010110110
	19	1.88663211208	100111110010110110C	1001	1110010110110

(continued)

Table 2 (continued)

Head	Period	Unimodal a	Critical itinerary	Prefix	Decoration
1001111C	11	1.8957542001	1001111110C	1001	11110
	11	1.8980890902792	1001111010C	1001	11010
	14	1.8959632393	1001111110010C	1001	11110010
	14	1.89787945132	1001111010010C	1001	11010010
	17	1.89600029216	1001111110010110C	1001	11110010110
	17	1.89784256570	1001111010010110C	1001	11010010110
	20	1.89600636282	1001111110010110110C	1001	11110010110110
	20	1.89783655715	1001111010010110110C	1001	11010010110110
10011001C	16	1.93213488504	100110010100110C	10011001	100110
	16	1.93235574679	100110011100110C	10011001	100110
	23	1.93213896428	1001100101001100100110C	10011001	100110010110
	23	1.93235156026	1001100111001100100110C	10011001	100110010110
	30	1.93213911573	100110010100100100110011001100110C	10011001	1001100100110110011001100110
	30	1.93235140816	100110011100100100110011001100110C	10011001	1001100100110110011001100110
	37	1.93213912112	100110010100100100110110110011001100110C	10011001	1001100100110110110011001100110
	37	1.93235140286	100110011100100100110110110011001100110C	10011001	1001100100110110110011001100110

110 and 010 respectively, showing that braid equivalent sequences constructed by the mechanism in this article may have different decorations.

We can weaken Questions A and B by asking if these braid equivalences are realisable in a more general class of maps containing the Hénon family. The typical generalisation of the Hénon family is that of *Hénon-like maps*. These are maps of the form

$$F(x, y) = (f(x) - \varepsilon(x, y), x) \tag{5.5}$$

where f is unimodal on some interval J and $\epsilon : J \times J \to \mathbb{R}$ satisfies $\partial_y \epsilon > 0$. These are diffeomorphisms onto their images which appear, after a suitable change of variables, when considering maps in the neighbourhood of a homoclinic bifurcation [14].

> *Question A'.* Given unimodal maps f_- and f_+ of type υ_- and υ_+ respectively, does there exist a family F_t, $t \in [-1, 1]$, of Hénon-like diffeomorphisms such that $F_{-1} = \iota(f_-)$ and $F_{+1} = \iota(f_+)$, where ι is some embedding of the set of unimodal maps into the boundary of the space of Hénon-like diffeomorphisms?

and

> *Question B'.* For each positive integer i, let $f_{-,i}$ and $f_{+,i}$ have critical orbits c_-^i and c_+^i of types υ_-^i and υ_+^i respectively. Let C_-^i and C_+^i denote the corresponding periodic orbits for the degenerate Hénon-like maps $F_{-,i,0}$ and $F_{+,i,0}$ respectively. Does there exist, for each i, a one-parameter family of Hénon-like maps $F_{i,t}$ realising a braid equivalence connecting $(C_-^i, F_{-,i,0})$ and $(C_+^i, F_{+,i,0})$?

Acknowledgements A. de Carvalho was partially supported by FAPESP grant 2011/16265-8. T. Hall was partially supported by FAPESP grant 2011/17581-0. P. Hazard was supported by FAPESP grant 2008/10659-1 and Leverhulme Trust grant RPG-279. We would like to thank the institutions IME- USP, IMPA and the IMS at Stony Brook for their hospitality during the time in which parts of this work was done. Thanks also to A. Hammerlindl for his help with numerical computations. Finally, we would like to thank C. Tresser for useful discussions on braids and the Hénon family.

References

1. J.S. Birman, *Braids, Links, and Mapping Class Groups*, in Annals of Mathematics Studies (Princeton University Press, Princeton, N.J., 1974) (No. 82. MR 0375281 (51 #11477))
2. P. Boyland, *Braid Types and a Topological Method of Proving Positive Entropy* (Boston University, 1984)
3. P. Boyland, Topological methods in surface dynamics. Topol. Appl. **58**(3), 223–298 (1994) (MR 1288300 (95h:57016))
4. A. de Carvalho, T. Hall, The forcing relation for horseshoe braid types. Exp. Math. **11**(2), 271–288 (2002)
5. A. de Carvalho, T. Hall, P. Hazard, *Braid Equivalence in the Hénon Family II* (2015) (in preparation)
6. R. Devaney, Z. Nitecki, Shift automorphisms in the Hénon mapping. Commun. Math. Phys. **67**(2), 137–146 (1979) (MR 539548 (80f:58035))
7. H. El Hamouly, C. Mira, Singularités dues au feuilletage du plan des bifurcations d'un difféomorphisme bi-dimensionnel, C. R. Acad. Sci. Paris Sér. I Math. **294**(12), 387–390 (1982) (MR 659728 (83m:58056))

8. T. Hall, The creation of horseshoes. Nonlinearity **7**, 861–924 (1994)
9. V.L. Hansen, in *Braids and Coverings: Selected Topics*, ed. by L. Gæde, H.R. Morton. London Mathematical Society Student Texts, vol. 18 (Cambridge University Press, Cambridge, 1989, With appendices. MR 1247697 (94g:57004))
10. P. Holmes, Knotted periodic orbits in suspensions of Smale's horseshoe: period multiplying and cabled knots. Phys. D **21**(1), 7–41 (1986) (MR 860006 (88b:58112))
11. I. Kan, H. Koçak, J.A. Yorke, Antimonotonicity: concurrent creation and annihilation of periodic orbits. Ann. Math. (2) **136**(2), 219–252 (1992) (MR 1185119 (94c:58135))
12. M. Lyubich, The quadratic family as a qualitatively solvable model of chaos. Not. Am. Math. Soc. **47**(9), 1042–1052 (2000) (MR 1777885 (2001g:37063))
13. J. Milnor, Remarks on iterated cubic maps. Exp. Math. **1**(1), 5–24 (1992) (MR 1181083 (94c:58096))
14. J. Palis Jr., F. Taken, *Hyperbolicity & Sensitive Chaotic Dynamics at Homoclinic Bifurcations* Cambridge Studies in Advanced Mathematics, vol. 35 (Cambridge University Press, 1993)
15. A. Sannami, A topological classification of the periodic orbits of the Henon family. Technical Report (Hokkaido University, 1988)

Empiric Stochastic Stability of Physical and Pseudo-physical Measures

Eleonora Catsigeras

Dedicated to the memory of Prof. Welington de Melo

Abstract We define the empiric stochastic stability of an invariant measure in the finite-time scenario, adapting the classical definition of stochastic stability. We prove that an invariant measure of a continuous system is empirically stochastically stable if and only if it is physical. We also define the empiric stochastic stability of a weak*-compact set of invariant measures instead of a single measure. Even when the system has not physical measures it still has minimal empirically stochastically stable sets of measures. We prove that such sets are necessarily composed by pseudo-physical measures. Finally, we apply the results to the one-dimensional C1-expanding case to conclude that the measures of empirically stochastically sets satisfy Pesin Entropy Formula.

Keywords Empiric stochastic stability · Physical measures · Pseudo physical measures · Pesin Entropy Formula

1   Introduction

The purpose of this paper is to study a type of stochastic stability of invariant measures, which we call "empiric stochastic stability" for continuous maps $f : M \mapsto M$ on a compact Riemannian manifold M of finite dimension, with or without boundary. In particular, we are interested on the empirically stochastically stable measures of one-dimensional continuous dynamical systems, and among them, the C^1-expanding maps on the circle.

E. Catsigeras (✉)
Instituto de Matemática y Estadística "Rafael Laguardia", Universidad de la República, Av. J. Herrera y Reissig 565, Montevideo, Uruguay
e-mail: eleonora@fing.edu.uy

© Springer Nature Switzerland AG 2019
M. J. Pacifico and P. Guarino (eds.), *New Trends in One-Dimensional Dynamics*, Springer Proceedings in Mathematics & Statistics 285,
https://doi.org/10.1007/978-3-030-16833-9_7

Let us denote by (M, f) the deterministic (zero-noise) dynamical system obtained by iteration of f, and by (M, f, P_ε) the randomly perturbed system whose noise amplitude is ε. Even if we will work on a wide scenario which includes any continuous dynamical system (M, f), we restrict the stochastic system (M, f, P_ε) by assuming that the noise probability distribution is uniform (i.e. it has constant density) on all the balls of radius $\varepsilon > 0$ of M (for a precise statement of this assumption see formula (1) below). We call ε the *noise level*, or also the *amplitude of the random perturbation*. To define the empiric stochastic stability we will take $\varepsilon \to 0^+$.

In the stochastic system (M, f, P_ε), the symbol P_ε denotes the family of probability distributions, which are called *transition probabilities*, according to which the noise is added to $f(x)$ for each $x \in M$. Precisely, each transition probability is, for all $n \in \mathbb{N}$, the distribution of the state x_{n+1} of the noisy orbit conditioned to $x_n = x$, for each $x \in M$. As said above, the transition probability is supported on the ball with center at $f(x)$ and radius $\varepsilon > 0$. So, the zero-noise system (M, f) is recovered by taking $\varepsilon = 0$; namely, $(M, f) = (M, f, P_0)$. The observer naturally expects that if the amplitude $\varepsilon > 0$ of the random perturbation were small enough, then the ergodic properties of the stochastic system "remembered" those of the zero-noise system.

The foundation and tools to study the random perturbations of dynamical systems were early provided in [4, 19, 28]. The stochastic stability appears in the literature mostly defined through the stationary meaures μ_ε of the stochastic system (M, f, P_ε). Classically, the authors prove and describe, under particular conditions, the existence and properties of the f-invariant measures that are the weak*-limit of ergodic stationary measures as $\varepsilon \to 0^+$. See for instance the early results of [8, 20–22, 30]), and the later works of [1–3, 25]. For a review on stochastic and statistical stability of randomly perturbed dynamical systems, see for instance [29] and Appendix D of [7].

The stationary measures of the ramdom perturbations provide the probabilistic behaviour of the noisy system asymptotically in the future. Nevertheless, from a rather practical or experimental point of view the concept of stochastic stability should not require the knowledge a priori of the limit measures of the perturbed system as $n \to +\infty$. For instance [15] presents numerical experiments on the stability of one-dimensional noisy systems in a finite time. The ergodic stationary measure is in fact substituted by an empirical (i.e. obtained after a finite-time observation of the system) probability. Also in other applications of the theory of random systems (see for instance [16, 18]), the stationary measures are usually unkown, are not directly obtained from the experiments, but substituted by the finite-time empiric probabilities which approximate the stationary measures if the observations last enough.

Summarizing, for a certain type of stochastically stable properties, one should not need the infinite-time noisy orbits. Instead, one may take the noisy orbits *up to a large finite time n, which* are indeed those that the experimenter observes and predicts. The statistics of the observations and predictions of the noisy orbits still reflect, for the experimenter and the predictor, the behaviour of the stochastic system, *but only up to some finite horizon.*

Motivated by the above arguments, in Sect. 2 we will define *the empiric stochastic stability.* Roughly speaking, an f-invariant probability for the zero-noise system

(M, f) is empirically stochastically stable if it approximates, up to an arbitrarily small error $\rho > 0$, the statistics of sufficiently large pieces of the noisy orbits, for some fixed time n, provided that the noise-level $\varepsilon > 0$ is small enough (see Definition 4). This concept is a reformulation in a *finite-time* scenario of one of the usual definition of infinite-time stochastic stability (see for instance [1, 8, 30]).

1.1 Setting the Problem

Let $\varepsilon > 0$ and $x \in M$. Denote by $B_\varepsilon(x) \subset M$ the open ball of radius ε centered at x. Consider the Lebesgue measure m, i.e. the finite measure obtained from the volume form induced by the Riemannian structure of the manifold. For each point $x \in M$, we take the restriction of m to the ball $B_\varepsilon(f(x))$. Precisely, we define the probability measure $p_\varepsilon(x, \cdot)$ by the following equality:

$$p_\varepsilon(x, A) := \frac{m\big(A \cap B_\varepsilon(f(x))\big)}{m\big(B_\varepsilon(f(x))\big)} \quad \forall\, A \in \mathscr{A}, \tag{1}$$

where $\mathscr{A}$ is the Borel sigma-algebra in M.

Definition 1 (*Stochastic system with noise-level ε.*) For each value of $\varepsilon > 0$, consider the stochastic process or Markov chain $\{x_n\}_{n \in \mathbb{N}} \subset M^{\mathbb{N}}$ in the measurable space $(M, \mathscr{A})$ such that, for all $A \in \mathscr{A}$:

$$\mathrm{prob}(x_0 \in A) = m(A), \quad \mathrm{prob}(x_{n+1} \in A | x_n = x) = p_\varepsilon(x, A),$$

where $p_\varepsilon(x, \cdot)$ is defined by equality (1).

The system whose stochastic orbits are the Markov chains as above is called *stochastic system with noise-level ε*. We denote it by (M, f, P_ε), where

$$P_\varepsilon := \{p_\varepsilon(x, \cdot)\}_{x \in M}.$$

The stochastic systems with noise-level $\varepsilon > 0$ are usually studied by assuming certain regularity of the zero-noise systems (M, f), and by taking the ergodic stationary measures μ_ε of the stochastic system (M, f, P_ε) (see for instance [30]). When assuming that the transition probabilities satisfy equality (1), all the stationary probability measures become absolutely continuous with respect to the Lebesgue measure m (see for instance [6]). Therefore, if a property holds for the noisy orbits for μ_ε-a.e initial state $x \in M$, it also holds for a Lebesgue-positive set of states.

When looking at the noisy system, the experimenter usually obtains the values of several bounded measurable functions φ, which are called *observables*, along the stochastic orbits $\{x_n\}_{n \in \mathbb{N}}$. From Definition 1, the expected value of φ at instant 0 is $E(\varphi)_0 = \int \varphi(x_0)\, dm(x_0)$. Besides, from the definition of the transition probabilities by equality (1), for any given state $x \in M$ the expected value of $\varphi(x_{n+1})$ conditioned to $x_n = x$ is $\int \varphi(y)\, p_\varepsilon(x, dy)$. So, in particular at instant 1 the expected value of φ is

$$E(\varphi)_1 = \iint \varphi(x_1)\, p_\varepsilon(x_0, dx_1)\, dm(x_0),$$

and its expected value at instant 2 is

$$E(\varphi)_2 = \iiint \varphi(x_2)\, p_\varepsilon(x_1, dx_2)\, p_\varepsilon(x_0, dx_1)\, dm(x_0).$$

Analogously, by induction on n we obtain that for all $n \geq 1$, the expected value $E(\varphi)_n$ of the observable φ is

$$E(\varphi)_n = \iiint \ldots \int \varphi(x_n) p_\varepsilon(x_{n-1}, dx_n)\ldots p_\varepsilon(x_1, dx_2)\, p_\varepsilon(x_0, dx_1)\, dm(x_0). \quad (2)$$

Since the Lebesgue measure m is not necessarily stationary for the system (M, f, P_ε), the expected value of the same function φ at each instant n, if the initial distribution is m, may change with n.

As said at the beginning, we assume that the experimenter only sees the values of the observable functions along finite pieces of the noisy orbits because his experiment and his empiric observations can not last forever. When analyzing the statistics of the observed data, he considers for instance the time average of the collected observations along those finitely elapsed pieces of randomly perturbed orbits. These time averages can be computed by the integrals of the observable functions with respect to certain probability measures, which are called *empiric stochastic probabilities for finite time* n (see Definition 3). Precisely, for any any fixed time $n \geq 1$ and for any initial state $x_0 \in M$, the empiric stochastic probability $\sigma_{\varepsilon,n,x_0}$ is defined such that the time average of the expected values of any observable φ at instants $1, 2, \ldots, n$ along the noisy orbit initiating at x_0, can be computed by the following equality:

$$\frac{1}{n} \sum_{j=1}^{n} E(\varphi(x_j)|x_0) = \int \varphi(y) d\sigma_{\varepsilon,n,x_0}(y),$$

where

$$E(\varphi(x_j)|x_0) = \iint \ldots \int \varphi(x_j)\, p_\varepsilon(x_{j-1}, dx_j)\ldots p_\varepsilon(x_1, dx_2) p_\varepsilon(x_0, dx_1). \quad (3)$$

We also assume that the experimenter only sees Lebesgue-positive sets in the phase space M. So, when analyzing the statistics of the observed data in the noisy system, he will not observe all the empiric stochastic distributions $\sigma_{\varepsilon,n,x}$, but only those for Lebesgue-positive sets of initial states $x \in M$. If besides he can only manage a finite set of continuous observable functions, then he will not see the exact probability distributions, but some weak* approximations to them up to an error $\rho > 0$, in the metric space $\mathcal{M}$ of probability measures.

For some classes of mappings on the manifold M, even with high regularity (for instance Morse-Smale C^∞ diffeomorphisms with two or more hyperbolic sinks), one single measure μ is not enough to approximate the empiric stochastic probabilities of the noisy orbits for Lebesgue-a.e. $x \in M$. The experimenter may need a set $\mathscr{K}$ composed by several probability measures instead of a single measure. Motivated by this phenomenon, we define the *empiric stochastic stability of a weak*-compact set* $\mathscr{K}$ of f-invariant probability measures (see Definition 8). This concept is similar to the empiric stochastic stability of a single measure, with two main changes: first, it substitutes the measure μ by a weak*-compact set $\mathscr{K}$ of probabilities; and second, it requires $\mathscr{K}$ be minimal with the property of empiric stochastic stability, when restricting the stochastic system to a fixed Lebesgue-positive set of noisy orbits. In particular, a *globally* empirically stochastically stable set $\mathscr{K}$ of invariant measures minimally approximates the statistics of Lebesgue-a.e. noisy orbits. We will prove that it exists and is unique.

1.2 Main Results

A classical concept in the ergodic theory of zero-noise dynamical systems is that of *physical measures* [14]. In brief, a physical measure is an f-invariant measure μ whose basin of statistical attraction has positive Lebesgue measure. This basin is composed by the zero-noise orbits such that the time average probability up to time n converges to μ in the weak*-topology as $n \to +\infty$ (see Definitions 11 and 12).

One of the main purposes of this paper is to answer the following question:

Question 1. Is there some relation between the empirically stochastically stable measures and the physical measures? If yes, how are they related?

We will give an answer to this question in Theorem 1 and Corollary 1 (see Sect. 2.1 for their precise statements). In particular, we will prove the following result:

Theorem. *An f-invariant measure is empirically stochastically stable if and only if it is physical.*

A generalization of physical measures, is the concept of *pseudo-physical probability measures,* which are sometimes also called SRB-like measures [10–12]. They are defined such that, for all $\rho > 0$, their weak* ρ-neighborhood, has a (weak) basin of statistical attraction with positive Lebesgue measure (see Definitions 11 and 12).

To study this more general scenario of pseudo-physics, our second main purpose is to answer the following question:

Question 2. Do empirically stochastically stable sets of measures relate with pseudo-physical measures? If yes, how do they relate?

We will give an answer to this question in Theorem 2 and its corollaries, whose precise statements are in Sect. 2.1. In particular, we will prove the following result:

Theorem. *A weak*-compact set of invariant probability measures is empirically stochastically stable only if all its measures are pseudo-physical. Conversely, any pseudo-physical measure belongs to the unique globally empirically stochastically stable set of measures.*

2   Definitions and Statements

We denote by $\mathcal{M}$ the space of Borel probability measures on the manifold M, endowed with the weak*-topology; and by $\mathcal{M}_f$ the subspace of f-invariant probabilities, where (M, f) is the zero-noise dynamical system. Since the weak* topology in $\mathcal{M}$ is metrizable, we can choose and fix a metric dist* that endows that topology.

To make formula (2) and other computations concise, it is convenient to introduce the following definition:

Definition 2 (*The transfer operators $\mathcal{L}_\varepsilon$ and $\mathcal{L}_\varepsilon^*$*). Denote by $C^0(M, \mathbb{C})$ the space of complex continuous functions defined in M. For the stochastic system (M, f, P_ε), we define the *transfer operator* $\mathcal{L}_\varepsilon : C^0(M, \mathbb{C}) \mapsto C^0(M, \mathbb{C})$ as follows:

$$(\mathcal{L}_\varepsilon \varphi)(x) := \int \varphi(y)\, p_\varepsilon(x, dy) \ \forall\, x \in M, \ \forall\, \varphi \in C^0(M, \mathbb{C}). \tag{4}$$

From equality (1) it is easy to prove that $p_\varepsilon(x, \cdot)$ depends continuously on $x \in M$ in the weak* topology. So, $\mathcal{L}_\varepsilon \varphi$ is a continuous function for any $\varphi \in C^0(M, \mathbb{C})$.

Through Riesz representation theorem, for any measure $\mu \in \mathcal{M}$ there exists a unique measure, which we denote by $\mathcal{L}_\varepsilon^* \mu$, such that

$$\int \varphi\, d(\mathcal{L}_\varepsilon^* \mu) := \int (\mathcal{L}_\varepsilon \varphi)\, d\mu \ \forall\, \varphi \in C^0(M, \mathbb{C}). \tag{5}$$

We call $\mathcal{L}_\varepsilon^* : \mathcal{M} \mapsto \mathcal{M}$ the dual transfer operator or also, *the transfer operator in the space of measures*.

From the above definition, we obtain the following property for any *observable function* $\varphi \in C^0(M, \mathbb{C})$: its expected value at the instant n along the stochastic orbits with noise level ε is

$$E(\varphi)_n = \int (\mathcal{L}_\varepsilon{}^n \varphi)\, dm = \int \varphi\, d(\mathcal{L}_\varepsilon^{*n} m).$$

We are not only interested in the expected values of the observables φ, but also *in the statistics* (i.e time averages of the observables) along the individual noisy orbits. With such a purpose, we first consider the following equality:

$$(\mathcal{L}_\varepsilon{}^n \varphi)(x) = \int \varphi\, d(\mathcal{L}_\varepsilon^{*n} \delta_x) \ \ \forall\, x \in M, \tag{6}$$

where δ_x denotes the Dirac probability measure supported on $\{x\}$. Second, we introduce the following concept of empiric probabilities for the stochastic system:

Definition 3 (*Empiric stochastic probabilities*) For any fixed instant $n \geq 1$, and for any initial state $x \in M$, we define the *empiric stochastic probability* $\sigma_{\varepsilon, n, x}$ of the noisy orbit with noise-level $\varepsilon > 0$, with initial state x, and up to time n, as follows:

$$\sigma_{\varepsilon,n,x} := \frac{1}{n} \sum_{j=1}^{n} \mathscr{L}_{\varepsilon}^{*j} \delta_x.$$ (7)

Note that the empiric stochastic probabilities for Lebesgue almost $x \in M$ allow the computation of the time averages of the observable φ along the noisy orbits. Precisely,

$$\frac{1}{n} \sum_{j=1}^{n} (\mathscr{L}_{\varepsilon}^{j} \varphi)(x) = \int \varphi(y)\, d\sigma_{\varepsilon,n,x}(y) \quad \forall\, \varphi \in C^0(M, \mathbb{C}).$$ (8)

Definition 4 (*Empiric stochastic stability of a measure*) We call a probability measure $\mu \in \mathcal{M}_f$ *empirically stochastically stable* if there exists a measurable set $\widehat{A} \subset M$ with positive Lebesgue measure such that:

For all $\rho > 0$ and for all $n \in \mathbb{N}^+$ large enough there exists $\varepsilon_0 > 0$ (which may depend on ρ and on n but not on x) satisfying

$$\text{dist}^*(\sigma_{\varepsilon,n,x}, \mu) < \rho \quad \forall\, 0 < \varepsilon \le \varepsilon_0, \text{ for Lebesgue a.e. } x \in \widehat{A}.$$

Definition 5 (*Basin of empiric stochastic stability of a measure*) For any probability measure μ, we construct the following (maybe empty) set in the ambient manifold M:

$$\widehat{A}_\mu := \Big\{ x \in M : \ \forall \rho > 0 \ \exists\, N = N(\rho) \text{ such that } \forall\, n \ge N \ \exists\, \varepsilon_0 = \varepsilon_0(\rho, n) > 0 \text{ satisfying}$$

$$\text{dist}^*(\sigma_{\varepsilon,n,x}, \mu) < \rho \quad \forall\, 0 < \varepsilon \le \varepsilon_0 \Big\}.$$ (9)

We call the set $\widehat{A}_\mu \subset M$ the *basin of empiric stochastic stability of μ*. Note that it is defined for any probability measure $\mu \in \mathcal{M}$, but it may be empty, or even if nonempty, it may have zero Lebesgue-measure when μ is not empirically stochastically stable.

The set $\widehat{A}_\mu$ is measurable (see Lemma 2). According to Definition 4, a probability measure μ is empirically stochastically stable if and only if the set $\widehat{A}_\mu$ has positive Lebesgue measure (see Lemma 3).

Definition 6 (*Global empiric stochastic stability of a measure*) We say that $\mu \in \mathcal{M}_f$ is *globally empirically stochastically stable* if it is empirically stochastically stable, and besides its basin $\widehat{A}_\mu$ of empiric stability has full Lebesgue measure.

Definition 7 (*Basin of empiric stochastic stability of a set of measures*) For any nonempty weak*-compact set $\mathcal{K} \subset \mathcal{M}$, we construct the following (maybe empty) set in the space manifold M:

$$\widehat{A}_\mathcal{K} := \{ x \in M : \ \forall \rho > 0 \ \exists\, N = N(\rho) \text{ such that } \forall\, n \ge N \ \exists\, \varepsilon_0 = \varepsilon_0(\rho, n) > 0 \text{ satisfying}$$

$$\text{dist}^*(\sigma_{\varepsilon,n,x}, \mathcal{K}) < \rho \quad \forall\, 0 < \varepsilon \le \varepsilon_0 \}.$$ (10)

We call $\widehat{A}_{\mathcal{K}} \subset M$ *the basin of empiric stochastic stability of* $\mathcal{K}$.

Note that $\widehat{A}_{\mathcal{K}}$ is defined for any nonempty weak*-compact set $\mathcal{K} \subset \mathcal{M}$. But it may be empty, or even if nonempty, it may have zero Lebesgue measure when $\mathcal{K}$ is not empirically stochastically stable, according to the following definition:

Definition 8 (*Empiric stochastic stability of a set of measures*) We call a nonempty weak*-compact set $\mathcal{K} \subset \mathcal{M}_f$ of f-invariant probability measures *empirically stochastically stable* if :

(a) There exists a measurable set $\widehat{A} \subset M$ with positive Lebesgue measure, such that:
For all $\rho > 0$ and for all $n \in \mathbb{N}^+$ large enough, there exists $\varepsilon_0 > 0$ (which may depend on ρ and n, but not on x), satisfying:

$$\text{dist}^*(\sigma_{\varepsilon,n,x}, \ \mathcal{K}) < \rho \quad \forall \, 0 < \varepsilon \leq \varepsilon_0, \quad \forall \, x \in \widehat{A}.$$

(b) $\mathcal{K}$ is minimal in the following sense: if $\mathcal{K}' \subset \mathcal{M}_f$ is nonempty and weak*-compact, and if $\widehat{A}_{\mathcal{K}} \subset \widehat{A}_{\mathcal{K}'}$ Lebesgue-a.e., then $\mathcal{K} \subset \mathcal{K}'$.

By definition, if $\mathcal{K}$ is empirically stochastically stable, then the set $\widehat{A} \subset M$ satisfying condition (a), has positive Lebesgue measure and is contained in $\widehat{A}_{\mathcal{K}}$. Since $\widehat{A}_{\mathcal{K}}$ is measurable (see Lemma 4), we conclude that it has positive Lebesgue measure.

Nevertheless, for a nonempty weak*-compact set $\mathcal{K}$ be empirically stochastically stable, it is not enough that $\widehat{A}_{\mathcal{K}}$ has positive Lebesgue measure. In fact, to avoid the whole set $\mathcal{M}_f$ of f-invariant measures be always an empirically stochastically stable set, we ask $\mathcal{K}$ to satisfy condition (b). In brief, we require a property of minimality of $\mathcal{K}$ with respect to Lebesgue-a.e. point of its basin $\widehat{A}_{\mathcal{K}}$ of empiric stochastic stability.

Definition 9 (*Global empiric stochastic stability of a set of measures*) We say that a nonempty weak*-compact set $\mathcal{K} \in \mathcal{M}_f$ is *globally empirically stochastically stable* if it is empirically stochastically stable, and besides its basin $\widehat{A}_{\mathcal{K}}$ of empiric stability has full Lebesgue measure.

We recall the following definitions from [11]:

Definition 10 (*Empiric zero-noise probabilities and $p\omega$-limit sets*) For any fixed natural number $n \geq 1$, the *empiric probability* $\sigma_{n,x}$ of the orbit with initial state $x \in M$ and up to time n of the zero-noise system (M, f), is defined by the following equality:

$$\sigma_{n,x} := \frac{1}{n} \sum_{j=1}^{n} \delta_{f^j(x)}.$$

It is standard to check, from the construction of the empiric stochastic probabilities in Definition 3, that $\sigma_{\varepsilon,n,x}$ is absolutely continuous with respect to the Lebesgue measure

m. In contrast, the empiric probability $\sigma_{n,x}$ for the zero-noise orbits is *atomic*, since it is supported on a finite number of points.

The *p-omega limit set* $p\omega_x$ in the space $\mathcal{M}$ of probability measures, corresponding to the orbit of $x \in M$, is defined by:

$$p\omega_x := \{\mu \in \mathcal{M} : \exists\, n_i \to +\infty \text{ such that } \lim\nolimits^*_{i \to +\infty} \sigma_{n_i,x} = \mu\},$$

where $\lim^*$ is taken in the weak*-topology of $\mathcal{M}$. It is standard to check that $p\omega_x \subset \mathcal{M}_f$ for all $x \in M$.

Definition 11 (*Strong and ρ-weak basin of statistical attraction*) For any f-invariant probability measure $\mu \in \mathcal{M}_f$, the (strong) *basin of statistical attraction of μ* is the (maybe empty) set

$$A_\mu : = \{x \in M : p\omega_x = \{\mu\}\}. \tag{11}$$

For any f-invariant probability measure $\mu \in \mathcal{M}_f$, and for any $\rho > 0$, the *ρ-weak basin of statistical attraction of μ* is the (maybe empty) set

$$A_\mu^\rho : = \{x \in M : \operatorname{dist}^*(p\omega_x, \{\mu\}) < \rho\}.$$

Definition 12 (*Physical and pseudo-physical measures*) For the zero-noise dynamical system (M, f), an f-invariant probability measure μ is *physical* if its strong basin of statistical attraction A_μ has positive Lebesgue measure.

An f-invariant probability measure μ is *pseudo-physical* if for all $\rho > 0$, its ρ-weak basin of statistical attraction A_μ^ρ has positive Lebesgue measure.

It is standard to check that, even if the ρ-weak basin of statistical attraction A_μ^ρ depends on the chosen weak*-metric in the space $\mathcal{M}$ of probabilities, the set of pseudo-physical measures remains the same when changing this metric (provided that the new metric also induces the weak*-topology).

Note that the strong basin of statistical attraction of any measure is always contained in the ρ-weak basin of the same measure. Hence, any physical measure (if there exists some) is pseudo-physical. But not all the pseudo-physical measures are necessarily physical (see for instance example 5 of [10]).

We remark that we *do not require the ergodicity of μ* to be physical or pseudo-physical. In fact, in [17] it is proved that the C^∞ diffeomorphism, popularly known as the Bowen Eye, exhibits a segment of pseudo-physical measures whose extremes, and so all the measures in the segement, are non ergodic. Also, for some C^0-version of Bowen Eye (see example 5 B of [10]) there is a unique pseudo-physical measure, it is physical and non-ergodic.

2.1 Statement of the Results

Theorem 1 (Characterization of empirically stochastically stable measures) *Let $f : M \mapsto M$ be a continuous map on a compact Riemannian manifold M. Let μ be an f-invariant probability measure. Then, μ is empirically stochastically stable if and only if it is physical.*

Besides, if μ is physical, then its basin $\widehat{A}_\mu \subset M$ of empiric stochastic stability equals Lebesgue-a.e. its strong basin $A_\mu \subset M$ of statistical attraction.

We will prove Theorem 1 and the following corollaries in Sect. 3.

Corollary 1 *Let $f : M \mapsto M$ be a continuous map on a compact Riemannian manifold M. Then, the following conditions are equivalent:*

(i)   There exists an f-invariant probability measure μ_1 that is globally empirically stochastically stable.

(ii)   There exists an f-invariant probability measure μ_2 that is physical and such that its strong basin of statistical attraction has full Lebesgue measure.

(iii)   There exists a unique f-invariant probability measure μ_3 that is pseudophysical.

Besides, if (i), (ii) or (iii) holds, then $\mu_1 = \mu_2 = \mu_3$, this measure is the unique empirically stochastically stable, and the set $\{\mu_1\}$ is the unique weak*-compact set in the space of probability measures that is empirically stochastically stable.

Before stating the next corollary, we fix the following definition: we say that a property of the maps on M is C^1-*generic* if it holds for a countable intersection of open and dense sets of maps in the C^1- topology.

Corollary 2 *For C^1-generic and for all C^2 expanding maps of the circle, there exists a unique ergodic measure μ that is empirically stochastically stable. Besides μ is globally empirically stochastically stable and it is the unique measure that satisfies the following Pesin Entropy Formula [23, 24]:*

$$h_\mu(f) = \int \log |f'| \, d\mu. \tag{12}$$

Theorem 1 is a particular case of the following result.

Theorem 2 (Empirically stochastically stable sets and pseudo-physics)

Let $f : M \mapsto M$ be a continuous map on a compact Riemannian manifold M.
(a) If $\mathcal{K}$ is a nonempty weak*-compact set of f-invariant measures that is empirically stochastically stable, then any $\mu \in \mathcal{K}$ is pseudo-physical.
(b) A set $\mathcal{K}$ of f-invariant measures is globally empirically stochastically stable if and only if it coincides with the set of all the pseudo-physical measures.

We will prove Theorem 2 and the following corollaries in Sect. 4.

Corollary 3 *For any continuous map $f : M \mapsto M$ on a compact Riemannian manifold M, there exists and is unique the nonempty weak*-compact set $\mathcal{K}$ of f-invariant measures that is globally stochastically stable. Besides, $\mu \in \mathcal{K}$ if and only if μ is pseudo-physical.*

Corollary 4 *If a pseudo-physical measure μ is isolated in the set of pseudo-physical measures, then it is empirically stochastically stable; hence physical.*

Corollary 5 *Let $f : M \mapsto M$ be a continuous map on a compact Riemannian manifold M. Then, the following conditions are equivalent:*

(i)   The set of pseudo-physical measures is finite.
(ii)   There exists a finite number of (individually) empirically stochastically stable measures, hence physical measures, and the union of their strong basins of statistical attraction covers Lebesgue a.e.

Corollary 6 *If the set of pseudo-physical measures is countable, then there exists countably many empirically stochastically stable measures, hence physical, and the union of their strong basins of statistical attractions covers Lebesgue a.e.*

Corollary 7 *For all C^1-expanding maps of the circle, all the measures of any empirically stochastically stable set $\mathcal{K}$ satisfy Pesin Entropy Formula (12).*

Corollary 8 *For C^0-generic maps of the interval, the globally empirically stochastically stable set $\mathcal{K}$ of invariant measures includes all the ergodic measures but is meager in the whole space of invariant measures.*

3   Proof of Theorem 1 and its Corollaries

We decompose the proof of Theorem 1 into several lemmas:

Lemma 1 *For $\varepsilon > 0$ small enough:*
(a) The transformation $x \in M \mapsto p_\varepsilon(x, \cdot) \in \mathcal{M}$ is continuous.
(b) The transfer operator $\mathcal{L}_\varepsilon^* : \mathcal{M} \mapsto \mathcal{M}$ is continuous.
(c) The transformation $x \in M \mapsto \sigma_{\varepsilon,n,x} \in \mathcal{M}$ is continuous.
(d) $\lim_{\varepsilon \to 0^+}^* p_\varepsilon(x, \cdot) = \delta_{f(x)}$ uniformly on M.
(e) $\lim_{\varepsilon \to 0^+}^* \mathcal{L}_\varepsilon^{*n} \delta_x = \delta_{f^n(x)}$ uniformly on M.
(f) $\lim_{\varepsilon \to 0^+}^* \sigma_{\varepsilon,n,x} = \sigma_{n,x}$ uniformly on M.

Proof (a): It is immediate from the construction of the probability measure $p_\varepsilon(x, \cdot)$ by equality (1), and taking into account that the Lebesgue measure restricted to a ball of radius ε depends continuously on the center of the ball.
(b): Take a convergent sequence $\{\mu_i\}_{i \in \mathbb{N}} \subset \mathcal{M}$ and denote $\mu = \lim_i^* \mu_i$. For any continuous function $\varphi : M \mapsto M$, we have

$$\int \varphi \, d\mathcal{L}_\varepsilon^* \mu_i = \int \mathcal{L}_\varepsilon \varphi \, d\mu_i. \tag{13}$$

Since $(\mathcal{L}_\varepsilon \varphi)(x) = \int \varphi(y) p_\varepsilon(x, dy)$ and $p_\varepsilon(x, \cdot)$ depends continuously on x, we deduce that $\mathcal{L}_\varepsilon \varphi$ is a continuous function. So, from (13) and the definition of the weak* topology in $\mathcal{M}$, we obtain:

$$\lim_{i \to +\infty} \int \varphi \, d\mathcal{L}_\varepsilon^* \mu_i = \lim_{i \to +\infty} \int \mathcal{L}_\varepsilon \varphi \, d\mu_i = \int \mathcal{L}_\varepsilon \varphi \, d\mu = \int \varphi \, d\mathcal{L}_\varepsilon^* \mu.$$

We conclude that $\lim_i^* \mathcal{L}_\varepsilon^* \mu_i = \mathcal{L}_\varepsilon^* \mu$, hence $\mathcal{L}_\varepsilon^*$ is a continuous operator on $\mathcal{M}$.

(c): Since the composition of continuous operators is continuous, we have that $\mathcal{L}_\varepsilon^{*j} : \mathcal{M} \mapsto \mathcal{M}$ is continuous for each fixed $j \in \mathbb{N}^+$. Besides, it is immediate to check that the transformation $x \in M \mapsto \delta_x \in \mathcal{M}$ is continuous. Thus, also the transformation $x \in M \mapsto \mathcal{L}_\varepsilon^{*j} \delta_x \in \mathcal{M}$ is continuous. We conclude that, for fixed $\varepsilon > 0$ and fixed $n \in \mathbb{N}^+$, the transformation

$$x \in M \mapsto \sigma_{\varepsilon,n,x} = \frac{1}{n} \sum_{j=1}^n \mathcal{L}_\varepsilon^{*j} \delta_x \in \mathcal{M}$$

is continuous.

(d): For any given $\rho > 0$ we shall find $\varepsilon_0 > 0$ (independent on $x \in M$) such that, $\text{dist}^*(p_\varepsilon(x, \cdot), \, \delta_{f(x)}) < \rho$ for all $0 < \varepsilon < \varepsilon_0$ and for all $x \in M$. For any metric dist^* that endows the weak* topology in $\mathcal{M}$, the inequality $\text{dist}^*(p_\varepsilon(x, \cdot), \, \delta_{f(x)}) < \rho$ holds, if and only if, for a *finite* number (which depends on ρ and on the metric) of continuous functions $\varphi : M \mapsto \mathbb{C}$, the difference $|\int \varphi(y) \, p_\varepsilon(x, dy) - \varphi(f(x))|$ is smaller than a certain $\varepsilon' > 0$ (which depends on ρ and on the metric). Let us fix such a continuous function φ. Since M is compact, φ is uniformly continuous on M. Thus, for any $\varepsilon' > 0$ there exists ε_0 such that, if $\text{dist}(y_1, y_2) < \varepsilon \leq \varepsilon_0$, then $|\varphi(y_1) - \varphi(y_2)| < \varepsilon'$. Since $p_\varepsilon(x, \cdot)$ is supported on the ball $B_\varepsilon(f(x))$, we deduce:

$$\left| \int \varphi(y) p_\varepsilon(x, dy) - \varphi(f(x)) \right| \leq \int |\varphi(y) - \varphi(f(x))| \, p_\varepsilon(x, dy) \leq \varepsilon',$$

because $\text{dist}(y, f(x)) < \varepsilon \leq \varepsilon_0$ for $p_\varepsilon(x, \cdot)$- a.e. $y \in M$.

Since ε_0 does not depend on x, we have proved that $\lim_{\varepsilon \to 0^+}^* p_\varepsilon(x, \cdot) = \delta_{f(x)}$ uniformly for all $x \in M$.

(e): Let us prove that $\lim_{\varepsilon \to 0^+} \mathcal{L}_\varepsilon^{*n} \delta_x = \delta_{f^n(x)}$ uniformly on $x \in M$. By induction on $n \in \mathbb{N}^+$:

If $n = 1$, for any continuous function $\varphi : M \mapsto \mathbb{C}$ we compute the following integral

$$\int \varphi \, d\mathcal{L}_\varepsilon^* \delta_x = \int (\mathcal{L}_\varepsilon \varphi) \, d\delta_x = (\mathcal{L}_\varepsilon \varphi)(x) = \int \varphi(y) \, p_\varepsilon(x, dy).$$

From the unicity of the probability measure of Riesz Representation Theorem, we obtain $\mathscr{L}_\varepsilon^* \delta_x = p_\varepsilon(x, \cdot)$. Applying part d), we conclude

$$\lim\nolimits_{\varepsilon \to 0^+}^* \mathscr{L}_\varepsilon^* \delta_x = \lim\nolimits_{\varepsilon \to 0^+}^* p_\varepsilon(x, \cdot) = \delta_{f(x)}, \text{ uniformly on } x \in M.$$

Now, assume that, for some $n \in \mathbb{N}^+$, the following assertion holds:

$$\lim\nolimits_{\varepsilon \to 0^+}^* \mathscr{L}_\varepsilon^{*n} \delta_x = \delta_{f^n(x)}, \text{ uniformly on } x \in M. \tag{14}$$

Let us prove the same assertion for $n + 1$, instead of n: Fix a continuous function $\varphi : M \mapsto \mathbb{C}$. As proved in part (d), for any $\varepsilon' > 0$, there exists $\varepsilon_0 > 0$ (independent on $x \in M$) such that

$$|(\mathscr{L}_\varepsilon \varphi)(x) - \varphi(f(x))| = \left| \int \varphi(y) p_\varepsilon(x, dy) - \varphi(f(x)) \right| < \frac{\varepsilon'}{2} \quad \forall\, 0 < \varepsilon \le \varepsilon_0, \ \forall\, x \in M.$$

Thus

$$\left| \int \varphi\, d\mathscr{L}_\varepsilon^{*n+1} \delta_x - \int (\varphi \circ f)\, d\mathscr{L}_\varepsilon^{*n} \delta_x \right| = \left| \int (\mathscr{L}_\varepsilon \varphi)\, d\mathscr{L}_\varepsilon^{*n} \delta_x - \int (\varphi \circ f)\, d\mathscr{L}_\varepsilon^{*n} \delta_x \right|$$

$$\le \int \left| (\mathscr{L}_\varepsilon \varphi) - \varphi \circ f \right| d\mathscr{L}_\varepsilon^{*n} \delta_x < \frac{\varepsilon'}{2} \quad \forall\, 0 < \varepsilon \le \varepsilon_0, \ \forall\, x \in M. \tag{15}$$

Besides, the induction assumption (14) implies that, if ε_0 is chosen small enough, then for the continuous function $\varphi \circ f$ the following inequality holds:

$$\left| \int (\varphi \circ f)\, d\mathscr{L}_\varepsilon^{*n} \delta_x - \varphi(f^{n+1}(x)) \right| =$$

$$= \left| \int (\varphi \circ f)\, d\mathscr{L}_\varepsilon^{*n} \delta_x - \int (\varphi \circ f)\, d\delta_{f^n(x)} \right| < \frac{\varepsilon'}{2} \quad \forall\, 0 < \varepsilon \le \varepsilon_0, \ \forall\, x \in M. \tag{16}$$

Joining inequalities (15) and (16) we deduce that for all $\varepsilon' > 0$, there exists $\varepsilon_0 > 0$ (independent of x) such that

$$\left| \int \varphi\, d\mathscr{L}_\varepsilon^{*n+1} \delta_x - \int \varphi\, d\delta_{f^{n+1}(x)} \right| < \varepsilon' \quad \forall\, 0 < \varepsilon \le \varepsilon_0, \ \forall\, x \in M.$$

In other words:

$$\lim\nolimits_{\varepsilon \to 0^+}^* \mathscr{L}_\varepsilon^{*n+1} \delta_x = \delta_{f^{n+1}(x)} \text{ uniformly on } x \in M,$$

ending the proof of part (e).

(f): Since $\sigma_{\varepsilon,n,x} = \frac{1}{n}\sum_{j=1}^{n}\mathscr{L}_{\varepsilon}^{*j}\delta_x$, applying part (e) to each probability measure $\mathscr{L}_{\varepsilon}^{*j}\delta_x$, we deduce that

$$\lim\nolimits_{\varepsilon\to 0^+}^{*}\mathscr{L}_{\varepsilon}^{*n+1}\delta_x = \frac{1}{n}\sum_{j=1}^{n}\delta_{f^j(x)} = \sigma_{n,x} \quad \text{uniformly on } x \in M,$$

ending the proof of Lemma 1.   □

Lemma 2 *For any probability measure μ consider the (maybe empty) basin of stochastic stability $\widehat{A}_\mu$ defined by equality (9), and the (maybe empty) strong basin of statistical attraction A_μ defined by equality (11).*

Then, $\widehat{A}_\mu$ and A_μ are measurable sets and coincide. Besides, they satisfy the following equality:

$$\widehat{A}_\mu = A_\mu = \bigcap_{k\in\mathbb{N}^+}\bigcup_{N\in\mathbb{N}^+}\bigcap_{n\geq N} C_{n,\,1/k}(\mu), \tag{17}$$

where, for any real number $\rho > 0$ and any natural number $n \geq 1$, the set $C_{n,\,\rho}(\mu)$ is defined by

$$C_{n,\,\rho}(\mu) := \{x \in M: \ dist^*(\sigma_{n,x},\ \mu) < \rho\}.$$

Proof From equality (11), we re-write the strong basin of statistical attraction of μ as follows:

$$A_\mu = \left\{x \in M: \ \lim\nolimits_{n\to+\infty}^{*}\sigma_{n,x} = \mu\right\} = \bigcap_{\rho>0}\bigcup_{N\in\mathbb{N}^+}\bigcap_{n\geq N} C_{n,\rho}(\mu). \tag{18}$$

From equality (9) we have:

$$\widehat{A}_\mu = \bigcap_{\rho>0}\bigcup_{N\in\mathbb{N}^+}\bigcap_{n\geq N} D_{n,\rho}(\mu), \tag{19}$$

where $D_{n,\rho}(\mu)$ is defined by

$$D_{n,\rho}(\mu) := \bigcup_{\varepsilon_0>0}\bigcap_{0<\varepsilon\leq\varepsilon_0} \{x \in M: \ dist^*(\sigma_{\varepsilon,n,x},\ \mu) < \rho\}.$$

The assertion $dist^*(\sigma_{\varepsilon,n,x},\ \mu) < \rho$ for all $0 < \varepsilon \leq \varepsilon_0$ implies

$$\lim_{\varepsilon\to 0^+} dist^*(\sigma_{\varepsilon,n,x},\ \mu) \leq \rho < 2\rho.$$

Thus, applying part (f) of Lemma 1, we deduce that $dist^*(\sigma_{n,x},\mu) < 2\rho$ for all $x \in D_{n,\rho}(\mu)$. In other words,

$$D_{n,\rho}(\mu) \subset C_{n,2\rho}(\mu),$$

which, joint with equalities (18) and (19), implies:

$$\widehat{A}_\mu \subset A_\mu.$$

To prove the converse inclusion, we apply again part (f) of Lemma 1 to write:

$$C_{n,\rho}(\mu) = \{x \in X : \ \mathrm{dist}^*(\lim^*_{\varepsilon \to 0^+} \sigma_{\varepsilon,n,x}, \mu) < \rho\}$$

Therefore

$$\lim_{\varepsilon \to 0^+} \mathrm{dist}^*(\sigma_{\varepsilon,n,x}, \mu) < \rho \ \ \forall x \in C_{n,\rho}(\mu).$$

Thus,

$$C_{n,\rho}(\mu) \subset \bigcup_{\varepsilon_0 > 0} \bigcup_{0 < \varepsilon \le \varepsilon_0} \{x \in M : \ \mathrm{dist}^*(\sigma_{\varepsilon,n,x}, \mu) < \rho\} = D_{n,\rho}(\mu).$$

The above inclusion, joint with equalities (18) and (19), implies

$$A_\mu \subset \widehat{A}_\mu.$$

We have proved that

$$\widehat{A}_\mu = A_\mu = \bigcap_{\rho>0} \bigcup_{N\in\mathbb{N}^+} \bigcap_{n\ge N} C_{n,\rho}(\mu).$$

Since the set $C_{n,\rho}(\mu)$ decreases when ρ decreases (with n and μ fixed), the family

$$\left\{ \bigcup_{N\in\mathbb{N}^+} \bigcap_{n\ge N} C_{n,\rho}(\mu). \right\}_{\rho>0},$$

whose intersection is A_μ, is decreasing when ρ decreases. Therefore, its intersection is equal to the intersection of its countable subfamily

$$\left\{ \bigcup_{N\in\mathbb{N}^+} \bigcap_{n\ge N} C_{n,\, 1/k}(\mu). \right\}_{k\in\mathbb{N}^+}.$$

We have proved equality (13) of Lemma 2.

Finally, note that the set $C_{n,\, 1/k}(\mu) \subset M$ is open, because $\sigma_{n,x} = (1/n)\sum_{j=1}^n \delta_{f^j(x)}$ (with fixed n) depends continuously on x. Since equality (13) states that $\widehat{A}_\mu = A_\mu$ is the countable intersection of a countable union of a countable intersection of open sets, we conclude that it is a measurable set, ending the proof of Lemma 2. $\square$

Lemma 3 *A probability measure μ is empirically stochastically stable, according to Definition 4, if and only if its basin $\widehat{A}_\mu$ of empiric stability, defined by equality (9), has positive Lebesgue measure.*

Proof If μ is empirically stochastically stable, then from Definition 4, there exists a Lebesgue-positive set $\widehat{A} \subset M$ such that $\widehat{A} \subset \widehat{A}_\mu$. Hence $m(\widehat{A}_\mu) > 0$.

To prove the converse assertion, assume that $m(\widehat{A}_\mu) = \alpha > 0$. Let us construct a positive Lebesgue set $\widehat{A} \subset \widehat{A}_\mu$ such that for any $\rho > 0$, there exists $N \in \mathbb{N}^+$ (uniform on $x \in \widehat{A}$), such that for all $n \geq N$ there exists $\varepsilon_0 > 0$ (uniform on $x \in \widehat{A}$) satisfying

$$\mathrm{dist}^*(\sigma_{\varepsilon,n,x}, \mu) < \rho \quad \forall \, 0 < \varepsilon \leq \varepsilon_0, \quad \forall \, x \in \widehat{A} \quad \text{(to be proved)}. \tag{20}$$

Applying Lemma 2 we have

$$\widehat{A}_\mu = \bigcap_{k \in \mathbb{N}^+} \bigcup_{N \in \mathbb{N}^+} E_{N,1/k}, \quad \text{where} \quad E_{N,1/k} := \bigcap_{n \geq N} C_{n,1/k}(\mu).$$

For fixed $k \in \mathbb{N}^+$ we have $E_{N+1,1/k} \subset E_{N,1/k}$ for all $N \geq 1$, and

$$\widehat{A}_\mu = \bigcup_{N \in \mathbb{N}^+} (E_{N,1/k} \cap \widehat{A}_\mu). \quad \text{Then} \quad \lim_{N \to +\infty} m(E_{N,1/k} \cap \widehat{A}_\mu) = m(\widehat{A}_\mu) = \alpha.$$

Therefore, for each $k \geq 1$ there exists $N(k) \geq 1$ such that

$$\alpha(1 - 1/3^k) \leq m(E_{N(k),1/k} \cap \widehat{A}_\mu) \leq \alpha.$$

We construct

$$\widehat{A} := \bigcap_{k \in \mathbb{N}^+} (E_{N(k),1/k} \cap \widehat{A}_\mu).$$

We will prove that $\widehat{A}$ has positive Lebesgue measure and that assertion (20) is satisfied uniformly for all $x \in \widehat{A}$. First,

$$m(\widehat{A}_\mu \setminus \widehat{A}) = m\left(\bigcup_{k \geq 1} (\widehat{A}_\mu \setminus E_{N(k),1/k})\right) \leq \sum_{k=1}^{+\infty} (\alpha - m(E_{N(k),1/k} \cap \widehat{A}_\mu)) \leq \sum_{k=1}^{+\infty} \frac{\alpha}{3^k} = \frac{\alpha}{2},$$

from where

$$m(\widehat{A}) = m(\widehat{A}_\mu) - m(\widehat{A}_\mu \setminus \widehat{A}) \geq \alpha - \frac{\alpha}{2} = \frac{\alpha}{2} > 0.$$

Second, for all $\rho > 0$, there exists a natural number $k \geq 2/\rho$, and a set $B_{N(k),1/k} \supset \widehat{A}$ such that

$$x \in C_{n,1/k}(\mu) \quad \forall n \geq N(k), \quad \forall \, x \in B_{N(k),1/k}.$$

Therefore, for all $n \geq N(k)$ (which is independent on x) we obtain:

$$\mathrm{dist}^*(\sigma_{n,x}, \mu) < \frac{1}{k} \le \frac{\rho}{2} \quad \forall x \in \widehat{A}. \tag{21}$$

Finally, applying part (f) of Lemma 1, for each fixed $n \ge N(k)$ there exists $\varepsilon_0 > 0$ (independent of x), such that

$$\mathrm{dist}^*(\sigma_{\varepsilon,n,x}, \sigma_{n,x}) < \frac{\rho}{2} \quad \forall 0 < \varepsilon \le \varepsilon_0, \ \forall x \in \widehat{M}. \tag{22}$$

Inequalities (21) and (22) end the proof of inequality (20); hence Lemma 3 is proved. $\qquad\square$

End of the proof of Theorem 1.

Proof From Lemma 3, μ is empirically stochastically stable if and only if $m(\widehat{A}_\mu) > 0$. From Definition 12, μ is physical if and only if $m(A_\mu) > 0$. Applying Lemma 2 we have $\widehat{A}_\mu = A_\mu$. We conclude that μ is empirically stochastically stable if and only if μ is physical. $\qquad\square$

Before proving Corollary 1, we recall the following theorem taken from [11]:

Theorem 3 *Let $f : M \mapsto M$ be a continuous map on a compact Riemannian manifold M. Then, the set $\mathcal{O}_f$ of pseudo-physical measures for f is nonempty and weak*-compact, and contains $p\omega_x$ for Lebesgue-a.e. $x \in M$.*

Moreover, $\mathcal{O}_f$ is the minimal nonempty weak*-compact set of probability measures that contains $p\omega_x$ for Lebesgue-a.e. $x \in M$.

Proof See [11, Theorem 1.5].

Proof of Corollary 1.

Proof (i) implies (ii): If μ_1 is globally empirically stable, then by Definition 6 $m(\widehat{A}_{\mu_1}) = m(M)$. Applying Theorem 1, μ_1 is physical. Besides, from Lemma 2, we know $\widehat{A}_{\mu_1} = A_{\mu_1}$. Then $m(A_{\mu_1}) = m(M)$. So, there exists $\mu_2 = \mu_1$ that is physical and whose strong basin of statistical attraction has full Lebesgue measure, as wanted.

(ii) implies (iii): If μ_2 is physical and $m(A_{\mu_2}) = m(M)$, then from Definitions 10 and 11, we deduce that the set $\{\mu_2\}$ contains $p\omega_x$ for Lebesgue-a.e. $x \in M$. Besides $\{\mu_2\}$ is nonempty and weak*-compact. Hence, applying the last assertion of Theorem 3, we deduce that $\{\mu_2\}$ is the whole set $\mathcal{O}_f$ of pseudo physical measures for f. In other words, there exists a unique measure $\mu_3 = \mu_2$ that is pseudo-physical, as wanted.

(iii) implies (i): If there exists a unique measure μ_3 that is pseudo-physical for f, then, applying Theorem 3 we know that that the set $\{\mu_3\}$ contains $p\omega_x$ for Lebesgue-a.e. $x \in M$. From Definitions 10 and 11, we deduce that the strong basin A_{μ_3} of statistical attraction of μ_3 has full Lebesgue measure. Then, μ_3 is physical, and applying Theorem 1 μ_3 is empirically stochastically stable. Besides, from Lemma 2, we obtain that the basin $\widehat{A}_{\mu_3}$ of empiric stochastic stability of μ_3 coincides with A_{μ_3};

hence it has full Lebesgue measure. From Definition 6 we conclude that there exists a measure $\mu_1 = \mu_3$ that is globally empirically stochastically stable, as wanted.

We have proved that (i), (ii) and (iii) are equivalent conditions. Besides, we have proved that if these conditions holds, the three measures μ_1, μ_2 and μ_3 coincide. This ends the proof of Corollary 1.   □

Proof of Corollary 2.

Proof On the one hand, a classical theorem by Ruelle states that any C^2 expanding map f of the circle S^1 has a unique invariant measure μ that is ergodic and absolutely continuous with respect to the Lebesgue measure. Thus, from Pesin's Theory [26, 27], it is the unique invariant measure that satisfies Pesin Entropy Formula (12).

On the other hand, Campbell and Quas [9] have proved that C^1-generic expanding maps in the circle have a unique invariant measure μ that satisfies Pesin Entropy Formula, but nevertheles μ is mutually singular with the Lebesgue measure (see also [5]).

Applying the above known results, to prove this corollary we will first show that for any C^1 expanding map f, if it exhibits a unique invariant measure μ that satisfies (12), then μ is the unique empirically stochastically stable measure. In fact, in [12] it is proved that any pseudo-physical measure of any C^1 expanding map of S^1 satisfies Pesin Entropy Formula (12). Hence, we deduce that, for our map f, μ is the unique pseudo-physical measure. Besides in [11], it is proved that if the set of pseudo-physical or SRB-like measures is finite, then all the pseudo-physical measures are physical. We deduce that our map f has a unique physical measure μ. Applying Theorem 1, μ is the unique empirically stochastically stable measure, as wanted.

Now, to end the proof of this corollary, let us show that the measure μ that was considered above, is globally empirically stochastically stable. From Theorem 3, the set $\mathscr{O}_f$ of all the pseudo-physical measures is the minimal weak*-compact set of invariant measures such that $p\omega(x) \subset \mathscr{O}_f$ for Lebesgue-a.e. $x \in S^1$. But, in our case, we have $\mathscr{O}_f = \{\mu\}$; hence $p\omega(x) = \{\mu\}$ for Lebesgue-a.e. $x \in S^1$. Applying Definition 11, we conclude that the strong basin of statistical attraction A_μ has full Lebesgue measure; and so, by Theorem 1 the basin $\widehat{A}_\mu$ of empirically stochastic stability of μ covers Lebesgue-a.e. the space; hence μ is globally empirically stochastically stable.   □

4   Proof of Theorem 2 and its Corollaries

For any nonempty weak*-compact set $\mathscr{K}$ of f-invariant measures, recall Definition 7 of the (maybe empty) basin $\widehat{A}_\mathscr{K} \subset M$ of empiric stochastic stability of $\mathscr{K}$ constructed by equality (10).

Similarly to Definition 11, in which the strong basin A_μ of statistical attraction of a single measure μ is constructed, we define now *the (maybe empty) strong basin of statistical attraction $A_\mathscr{K} \subset M$ of the set $\mathscr{K} \subset \mathscr{M}$*, as follows:

$$A_{\mathscr{K}} := \{x \in M, \ p\omega_x \subset \mathscr{K}\}, \tag{23}$$

where $p\omega_x$ is the p-omega limit set (limit set in the space $\mathscr{M}$ of probabilities) for the empiric probabilities along the orbit with initial state in $x \in M$ (recall Definition 10).

We will prove the following property of the basins $\widehat{A}_{\mathscr{K}}$ and $A_{\mathscr{K}}$:

Lemma 4 *For any nonempty weak*-compact set $\mathscr{K}$ in the space $\mathscr{M}$ of probability measures, the basins $\widehat{A}_{\mathscr{K}} \subset M$ and $A_{\mathscr{K}} \subset M$, defined by equalities (10) and (23) respectively, are measurable sets and coincide. Moreover*

$$\widehat{A}_{\mathscr{K}} = \widehat{A}_{\mu} = \bigcap_{k \in \mathbb{N}^+} \bigcup_{N \in \mathbb{N}^+} \bigcap_{n \geq N} C_{n,1/k}(\mathscr{K}),$$

where, for all $\rho > 0$ the set $C_{n,\rho}(\mathscr{K}) \subset M$ is defined by

$$C_{n,\rho}(\mathscr{K}) = \{x \in M : \ dist^*(\sigma_{n,x}, \ \mathscr{K}) < \rho\}.$$

Proof Repeat the proof of Lemma 2, with the set $\mathscr{K}$ instead of the single measure μ, and using equalities (10) and (23), instead of (9) and (11) respectively. $\qquad\square$

Lemma 5 *The set $\mathscr{O}_f$ of all pseudo-physical measures is globally empirically stochastically stable.*

Proof From Theorem 3, $p\omega_x \subset \mathscr{O}_f$ for Lebesgue-a.e. $x \in M$. Thus, the strong basin of statistical attraction $A_{\mathscr{O}_f}$ of $\mathscr{O}_f$, defined by equality (23), has full Lebesegue measure. After Lemma 4, the basin $\widehat{A}_{\mathscr{O}_f}$ of empiric stochastic stability of $\mathscr{O}_f$, has full Lebesgue measure. Therefore, if we prove that $\mathscr{O}_f$ is empirically stochastically stable, it must be globally so.

We now repeat the proof of Lemma 3, using $\mathscr{O}_f$ instead of a single measure μ, to construct a Lebesgue-positive set $\widehat{A} \subset M$ such that, for all $\rho > 0$ and for all n large enough, there exists $\varepsilon_0 > 0$ (independenly of $x \in \widehat{A}$) such that

$$dist^*(\sigma_{\varepsilon,n,x}, \ \mathscr{O}_f) < \rho \quad \forall \, 0 < \varepsilon \leq \varepsilon_0, \ \forall \, x \in \widehat{A}.$$

Thus, $\mathscr{O}_f$ satisfies condition (a) of Definition 8, to be empirically stochastically stable. Let us prove that $\mathscr{O}_f$ also satisfies condition (b):

Assume that $\mathscr{K} \subset \mathscr{M}_f$ is nonempty and weak*-compact and $\widehat{A}_{\mathscr{O}_f} \subset \widehat{A}_{\mathscr{K}}$ Lebesgue-a.e. We shall prove that $\mathscr{O}_f \subset \mathscr{K}$. Arguing by contradiction, assume that there exists a probability measure $\nu \in \mathscr{O}_f \setminus \mathscr{K}$. Choose

$$0 < \rho < \frac{dist^*(\nu, \ \mathscr{K})}{2} \tag{24}$$

On the one hand, since ν is pseudo-physical, applying Definitions 11 and 12, the ρ-weak basin A_ν^ρ of statistical attraction of ν has positive Lebesgue measure. In brief:

$$m(\{x \in M : \liminf_{n \to +\infty} \mathrm{dist}^*(\sigma_{n,x}, \ \nu) < \rho\}) > 0. \tag{25}$$

From inequalities (24) and (25), and applying equality (23), we deduce that

$$m(\{x \in M: \ p\omega_x \not\subset \mathscr{K}\}) > 0, \quad m(A_{\mathscr{K}}) < m(M). \tag{26}$$

On the other hand, applying Lemma 4 and the hypothesis $\widehat{A}_{\mathscr{O}_f} \subset \widehat{A}_{\mathscr{K}}$ Lebesgue-a.e., we deduce

$$A_{\mathscr{O}_f} \subset A_{\mathscr{K}} \text{ Lebesgue a.e..}$$

Applying Theorem 3 and equality (23), we have

$$m(A_{\mathscr{O}_f}) = m(M), \quad \text{from where we deduce } m(A_{\mathscr{K}}) = m(M),$$

contradicting the inequality at right in (26).

We have proved that $\mathscr{O}_f \subset \mathscr{K}$. Thus $\mathscr{O}_f$ satisfies condition (b) of Definition 8, ending the proof of Lemma 5. $\qquad\square$

End of the proof of Theorem 2.

Proof We denote by $\mathscr{O}_f$ the set of all pseudo-physical measures.

(a) Let $\mathscr{K} \subset \mathscr{M}_f$ be empirically stochastically stable, according to Definition 8. We shall prove that $\mathscr{K} \subset \mathscr{O}_f$. Assume by contradiction that there exists $\nu \in \mathscr{K} \setminus \mathscr{O}_f$. So, ν is not pseudo-physical, and applying Definition 12, there exists $\rho > 0$ such that the ρ-weak basin A_ν^ρ of statistical attraction of ν has zero Lebesgue measure. In brief, after Definition 11, we have

$$m(\{x \in M: \ \mathrm{dist}^*(p\omega_x, \ \nu) < \rho\}) = 0,$$

from where we deduce that

$$p\omega_x \ \subset \ \mathscr{M}_f \setminus \mathscr{B}_\rho(\nu) \quad \text{Lebesgue-a.e. } x \in M, \tag{27}$$

where $\mathscr{B}_\rho(\nu)$ is the open ball in the space $\mathscr{M}$ of probability measures, with center at ν and radius ρ.

Applying Lemma 4 and equality (23) we have

$$\widehat{A}_{\mathscr{K}} = A_{\mathscr{K}} = \{x \in X: \ p\omega_x \subset \mathscr{K}\}.$$

Joining with assertion (27), we deduce that $A_{\mathscr{K}} \subset A_{\mathscr{K} \setminus \mathscr{B}_\rho(\nu)}$ Lebesgue-a.e.; and applying again Lemma 4 we deduce:

$$\widehat{A}_{\mathscr{K}} \subset \widehat{A}_{\mathscr{K} \setminus \mathscr{B}_\rho(\nu)} \quad \text{Lebesgue-a.e.}$$

But, by hypothesis $\mathscr{K}$ is empirically stochastically stable. Thus, it satisfies condition (b) of Definition 8. We conclude that $\mathscr{K} \subset \mathscr{K} \setminus \mathscr{B}_\rho(\nu)$, which is a contradiction, ending the proof of part (a) of Theorem 2.

(b) According to Lemma 5, if $\mathscr{K} = \mathscr{O}_f$, then $\mathscr{K}$ is globally empirically stochastically stable. Now, let us prove the converse assertion. Assume that $\mathscr{K}$ is globally empirically stochastically stable. We shall prove that $\mathscr{K} = \mathscr{O}_f$. Applying part (a) of Theorem 2, we know that $\mathscr{K} \subset \mathscr{O}_f$. So, it is enough to prove now that $\mathscr{O}_f \subset \mathscr{K}$.

By hypothesis $m(\widehat{A}_{\mathscr{K}}) = m(M)$. From Lemma 4 we have $\widehat{A}_{\mathscr{K}} = A_{\mathscr{K}}$). We deduce that $m(A_{\mathscr{K}}) = m(M)$. From this latter assertion and equality (23), we obtain

$$p\omega_x \subset \mathscr{K} \text{ for Lebesgue-a.e. } x \in M.$$

Finally, we apply the last assertion of Theorem 3 to conclude that $\mathscr{O}_f \subset \mathscr{K}$, as wanted. This ends the proof of Theorem 2. □

Proof of Corollary 3.

Proof This corollary is immediate after Theorem 2 and Lemma 5. In fact, Lemma 5 states that the set $\mathscr{O}_f$, which is composed by all the pseudo-physical measures, is globally empirically stochastically stable. And part (b) of Theorem 2, states that $\mathscr{O}_f$ is the unique set of f-invariant measures that is globally empirically stochastically stable. □

Before proving Corollaries 4, 5 and 6, we recall the following known result:

Theorem 4 *For all $x \in M$ the p-omega limit set $p\omega_x$ has the following property:*
For any pair of measures $\mu_0, \mu_1 \in p\omega_x$ and for every real number $0 \leq \lambda \leq 1$ there exists a measure μ_λ such that $dist^*(\mu_0, \mu_\lambda) = \lambda dist^*(\mu_0, \mu_1)$.

Proof See [11, Theorem 2.1]. □

Proof of Corollary 4.

Proof Assume that μ is pseudo-physical and isolated in the set $\mathscr{O}_f$ of all pseudo-physical measures. Then, there exists $\rho > 0$ such that:

$$\text{if } \nu \in \mathscr{O}_f \text{ and } dist^*(\nu, \mu) < \rho, \quad \text{then} \quad \nu = \mu. \tag{28}$$

Since μ is pseudo-physical, from Definition 12 we know that the ρ-weak basin A_μ^ρ of statistical attraction of μ has positive Lebesgue measure. From Definition 11 we deduce that

$$m(A_\mu^\rho = m(\{x \in M : \ dist^*(p\omega_x, \mu) < \rho\}) > 0. \tag{29}$$

Applying Theorem 3, we know that $p\omega_x \subset \mathscr{O}_f$ for Lebesgue-a.e. $x \in M$. Joining the latter assertion with (28) and (29) we deduce that

$$\{\mu\} = p\omega_x \bigcap \mathscr{B}_\rho \mu \text{ for Lebesgue-a.e. } x \in A_\mu^\rho,$$

where $\mathscr{B}_\rho\mu$ is the ball in the space of probability measures, with center at μ and radius ρ.

Besides, from Theorem 4 we deduce that $p\omega_x = \{\mu\}$ for Lebesgue-a.e. $x \in A^\rho_\mu$, hence for a Lebesgue-positive set of points $x \in M$. Applying Definition 12, we conclude that the given pseudo-physical measure μ is physical; hence, from Theorem 1, μ is empirically stochastically stable. □

Proof of Corollary 5.

Proof (i) implies (ii): If the set $\mathscr{O}_f$ of pseudo-physical measures is finite, then all the pseudo-physical are physical due to Corollary 4. Then, applying Theorem 1, all of them are (individually) empirically stochastically stable. Besides the union of their strong basins of statistical attraction has full Lebesgue measure: In fact, applying Definition 11 and equality (23), that union is the set $A_{\mathscr{O}_f}$; and, due to Theorem 3, the set $A_{\mathscr{O}_f}$ has full Lebesgue measure. So, assertion (ii) is proved.

(ii) implies (i): Assume that there exists a finite number $r \geq 1$ of empirically stochastically stable measures $\mu_1, \mu_2, \ldots, \mu_r$ (hence, physical measures, due to Theorem 1). Assume also that the strong basins A_{μ_i} of statistical attraction have an union $\bigcup_{i=1}^r A_{\mu_i}$ that covers Lebesgue-a.e.. Applying Definition 11 and equality (23), we deduce that $A_{\{\mu_1,\ldots,\mu_r\}} = \bigcup_{i=1}^r A_{\mu_i}$ has full Lebesgue measure. So, from the last assertion of Theorem 3, $\mathscr{O}_f \subset \{\mu_1, \ldots, \mu_r\}$. In other words, the set $\mathscr{O}_f$ of pseudo-physical measures is finite, proving assertion (i). □

Proof of Corollary 6.

Proof If the set $\mathscr{O}_f$ is finite, then we apply Corollary (5) to deduce that there exists a finite number of empirically stochastically stable measures, hence physical, and that the union of their strong basins of statistical attraction has full Lebesgue measure.

Now let us consider the case for which, by hypothesis, the set $\mathscr{O}_f$ of pseudo-physical measures is countably infinite. In brief: $\mathscr{O}_f = \{\mu_i\}_{i \in \mathbb{N}}$.

Applying Theorem 3, the p-omega limit sets $p\omega_x$ are contained in $\mathscr{O}_f$ for Lebesgue-a.e. $x \in M$. But, from Theorem 4 we know that $p\omega_x$ is either a single measure or uncountably infinite. Since it is contained in the countable set $\mathscr{O}_f$, we deduce the $p\omega_x$ is composed by a single measure of $\mathscr{O}_f$ for Lebesgue-a.e. $x \in M$. Now, recalling Definition 11 and equality (23), we deduce that

$$A_{\mathscr{O}_f} = \bigcup_{i=1}^{+\infty} A_{\mu_i}, \qquad \sum_{i=1}^{+\infty} m(A_{\mu_i}) = m(M).$$

Therefore, there exists finitely many or countable infinitely many pseudo-physical measures $\mu_{i_n} : 1 \leq n \leq r \in \mathbb{N}^+ \cup \{+\infty\}$ such that

$$\mu(A_{\mu_{i_n}}) > 0 \ \forall \ 1 \leq n \leq r, \qquad \sum_{n=1}^{r} m(A_{\mu_n}) = m(M). \tag{30}$$

From Definition 12, each measure μ_{i_n} is physical; hence empirically stochastically stable due to Theorem 1. Besides, from equality at right in (30), we deduce that the union $\bigcup_{n=1}^{r} A_{\mu_{i_n}}$ has full Lebesgue measure, as wanted.

Finally, to end the proof of Corollary 6, let us show that the set $\{\mu_{i_n} : 1 \le n \le r$ of physical measures above constructed, can not be finite. In brief, let us prove that $r = +\infty$. In fact, if there existed a finite number $r \in \mathcal{N}^+$ of physical measures whose basins of statistical attraction have an union with full Lebesgue measure, then, we would apply Corollary 5 and deduce that the set $\mathcal{O}_f$ of pseudo-physical measures is finite. But in our case, by hypothesis, $\mathcal{O}_f$ is countably infinite, ending the proof of Corollary 6. □

Proof of Corollary 7.

Proof From part (a) of Theorem 2 we know that all the measures of any empirically stochastically stable set $\mathcal{K} \subset \mathcal{M}_f$ is pseudo-physical. Besides, in [12] it is proved that, for any C^1 expanding map f of the circle, any pseudo-physical or SRB-like measure satisfies Pesin Entropy Formula (12). We conclude that all the measures of $\mathcal{K}$ satisfy this formula. □

Proof of Corollary 8.

Proof From part (b) of Theorem 2 we know that the globally empirically stochastically stable set $\mathcal{K}$ coincides with the set $\mathcal{O}_f$ of pseudo-physical measures. Besides, in [13] it is proved that, for C^0-generic maps f of the interval, any ergodic measure belongs to $\mathcal{O}_f$ but, nevertheless $\mathcal{O}_f$ is a weak*-closed with empty interior in the space $\mathcal{M}_f$ of invariant measures. We conclude that all ergodic measures belong to the globally empirically stochastically stable set $\mathcal{K}$ and that this set of invariant measures is meager in $\mathcal{M}_f$, as wanted. □

Acknowledgements We thank IMPA of Rio de Janeiro (Brazil), and the organizing and scientific committees of the conference "New Trends on One-Dimensional Dynamics. Celebrating the 70th. Anniversary of Welington de Melo", held at IMPA in 2015. We also thank the financial support by the Program MATHAMSUD (France, Chile and Uruguay) and ANII (Uruguay) of the Project "PhySeCo", and by CSIC- Universidad de la República (Uruguay) of the Group Project N. 618, "Dynamical Systems".

References

1. J.F. Alves, V. Araújo, C.H. Vásquez, Stochastic stability of non-uniformly hyperbolic diffeomorphisms. Stoch. Dyn. **7**, 299–333 (2007)
2. V. Araújo, Attractors and time average for random maps. Ann. de l' Inst. Henry Poincaré. Non Linear Anal. **17**, 307–369 (2000)
3. V. Araujo, A. Tahzibi, Physical measures at the boundary of hyperbolic maps. Disc. Contin. Dyn. Syst. **20**, 849–876 (2008)
4. L. Arnold, *Random Dynamical Systems* (Springer, Berlin, 1978)
5. A. Ávila, J. Bochi, A generic C1 map has no absolutely continuous invariant probability measure. Nonlinearity **19**, 2717–2725 (2006)

6. H. Bahsoun, H. Hu, S. Vaienti, Pseudo-orbits, stationary measures and metastability. Dyn. Syst. Int. J. **29**, 322–336 (2014)

7. C. Bonatti, L.J. Díaz, M. Viana, in *Dynamics Beyond Uniform Hyperbolicity. A Global Geometric and Probabilistic Perspective*. Encyclopaedia of Mathematical Sciences, vol. 102 (Springer, Berlin, 2005)

8. M. Brin, Y. Kifer, Dynamics of Markov chains and stable manifolds for random diffeomorphisms. Ergod. Theory Dyn. Syst. **7**, 351–374 (1987)

9. J. Campbell, A. Quas, A generic C^1 expanding map has a singular SRB measure. Commun. Math. Phys. **221**, 335–349 (2001)

10. E. Catsigeras, On Ilyashenko's statistical attractors. Dyn. Syst. Int. J. **29**, 78–97 (2014)

11. E. Catsigeras, H. Enrich, SRB-like measures for C^0 dynamics. Bull. Polish Acad. Sci. Math. **59**, 151–164 (2011)

12. E. Catsigeras, H. Enrich, Equilibrium states and SRB-like measures of C^1-expanding maps of the circle. Portugal. Math. **69**, 193–212 (2012)

13. E. Catsigeras, S. Troubetzkoy, Pseudo-physical measures for typical continuous maps of the interval, pp. 1–36 (2017). arXiv:1705.10133 [math.DS]

14. J.P. Eckmann, D. Ruelle, Ergodic theory of chaos and strange attractors. Rev. Mod. Phys. **57**, 617 (1985)

15. D. Faranda, J.M. Freitas, P. Guiraud, S. Vaienti, Statistical properties of random dynamical systems with contracting direction. J. Phys. A Math. Theor. **49**, 204001 (2016)

16. A Friedman, Stochastic differential equations and applications, vol. 1, 2 (Academic Press, New York, 1975)

17. T. Golenishcheva-Kutuzova, V. Kleptsyn, Convergence of theKrylov-Bogolyubov procedure in Bowans example. Mat Zametki. **82**, 678–689 (2007) (Trans. Math. Notes **82**,608–618, 2007)

18. N. Ikeda, S. Watanabe, *Stochastic differential equations and diffusion processes* (North-Holland and Kodansha, Amsterdam, 1981)

19. Y. Kifer, On small random perturbations of some smooth dynamical systems. Math. USSR-Izv. **8**, 1083–1107 (1974)

20. Y. Kifer, Large deviations in dynamical systems and stochastic processes. Trans Am. Math. Soc. **321**, 505–524 (1990)

21. Y. Kifer, *Ergodic Theory of Random Transformations* (Birkhäuser, Boston, 1986)

22. Y. Kifer, *Random Perturbations of Dynamical Systems* (Birkhäuser, Boston, 1988)

23. F. Ledrappier, L.-S. Young, Entropy formula for random transformations. Prob. Theory Relat. Fields **80**, 217–240 (1988)

24. P.-D. Liu, M. Qian, Pesin's entropy formula and SBR measures of random diffeomorphisms. Sci. China (Ser. A) **36**, 940–956 (1993)

25. F.-D. Liu, M. Qian, *Smooth Ergodic Theory of Random Dynamical Systems*. Lecture Notes in Mathematics, vol. 1606 (Springer, Berlin, 1995)

26. Y.B. Pesin, Families of invariant manifolds corresponding to non-zero characteristic exponents. Math. USSR-Izv **10**, 1261–1305 (1976)

27. Y.B. Pesin, Lyapunov characteristic exponents and smooth ergodic theory. Russ. Math. Surv. **32**, 55–114 (1977)

28. S.M. Ulam, J. von Neumann, Random ergodic theorems (1945), cited in Ulam, S.M.: John von Neumann 1903–1957. Bull. Am. Math. Soc.**64**, 1–49 (1958)

29. M. Viana, Stochastic dynamics of deterministic systems. Lecture Notes 21 Brraz. Math Colloq., IMPA, Rio de Janeiro (1997)

30. L.-S. Young, Stochastic stability of hyperbolic attractors. Ergod. Theory Dyn. Syst. **6**, 311–319 (1986)

Blender-horseshoes in Center-unstable Hénon-like Families

Lorenzo J. Díaz and Sebastián A. Pérez

To Welington de Melo, in memoriam

Abstract A *blender-horseshoe* is a locally maximal transitive hyperbolic set that appears in dimension at least three carrying a distinctive geometrical property: its local stable manifold "behaves" as a manifold of topological dimension greater than the expected one (the dimension of the stable bundle). This property persists under perturbations turning this kind of dynamics an important piece in the global description of robust non-hyperbolic systems. In this paper, we consider a parameterized family of center-unstable Hénon-like of endomorphisms in dimension three and show how blender-horseshoes naturally occur in a specific parameter range.

Keywords Blender · Blender-horseshoe · Hénon-like families

2000 Mathematics Subject Classification. 37C45 · 37D30 · 37C29

This paper is part of the Ph.D. thesis of SP (PUC-Rio) supported by CNPq (Brazil). The authors thank the hospitality and support of Centro de Matemática of Univ. of Porto (Portugal). LJD is partially supported by CNE-Faperj, CNPq-grants (Brazil) and SP is partially supported by CMUP (UID/MAT/00144/2013) and PTDC/MAT-CAL/3884/2014, which are funded by FCT (Portugal) with national (MEC) and European structural funds through the programs COMPTE and FEDER, under the partnership agreement PT2020. The authors also thank the careful reading of an anonymous referee.

L. J. Díaz (✉)
Departamento de Matemática PUC-Rio, Marquês de São Vicente 225,
Gávea Rio de Janeiro 225453-900, Brazil
e-mail: lodiaz@mat.puc-rio.br

S. A. Pérez
Centro de Matemática da Universidade do Porto, Rua do Campo Alegre,
687, 4169-007 Porto, Portugal
e-mail: sebastian.opazo@fc.up.pt

© Springer Nature Switzerland AG 2019
M. J. Pacifico and P. Guarino (eds.), *New Trends in One-Dimensional Dynamics*, Springer Proceedings in Mathematics & Statistics 285,
https://doi.org/10.1007/978-3-030-16833-9_8

1 Introduction

Naively, a *blender* is a transitive hyperbolic set that appears in dimension at least three and whose special geometrical configuration implies that the "dimension" of its stable set is larger than the "expected" one. To be a bit more precise, recall that the *index* of a transitive hyperbolic set Λ, denoted by $\mathrm{ind}(\Lambda)$, is the dimension of its unstable bundle (by transitivity, the index is well defined). The leaves of the (local) stable sets of points in Λ have dimension $\mathrm{ind}(\Lambda)$, however the (local) stable set of the blender Λ behaves as a set of dimension $\mathrm{ind}(\Lambda) + 1$ (or greater). In practical terms and applications, blenders are dynamical "local plugs" which in some (semi-local or global) configurations carry further important properties of the dynamics (see the next paragraph). For an informal presentation of blenders and a discussion on their role in smooth dynamical systems we refer to [5] and [10, Chap. 6.2]. Blenders were introduced in [6] as a formalisation of the constructions in [11] in the context of bifurcations via heterodimensional cycles. In [6], blenders were used to construct new classes of robustly transitive diffeomorphisms. Later, blenders were used in several dynamical contexts: Generation of robust heterodimensional cycles [7] and homoclinic tangencies [8], stable ergodicity [1, 25], Arnold diffusion [20], and construction of nonhyperbolic measures [4], among others. Each of these applications involves a specific type of blender such as blender-horseshoes [8], symbolic blenders [2, 20], dynamical blenders [4], and super-blenders [1].

In the original definition in [6] the main emphasis is placed on the persistence of its geometrical configuration that was key to guarantee the robust transitivity of non-hyperbolic sets, see the discussion in [10, Chap. 6]. Although in many contexts the "original" blenders in [6] are shown to be very useful, a major con of them is that they fail to be locally maximal sets, this deficiency carries some constraints in their use and applications, for further discussions see Remark 2.5. This weakness was bypassed in [8] by introducing a special type of blenders, called *blender-horseshoes,* which are locally maximal and also conjugate to the standard Smale horseshoe, see Definition 2.3. These two additional useful properties can be explored to get additional relevant properties: blender-horsehoes are the key local plugs to get *robust heterodimensional cycles* and *robust homoclinic tangencies* in the C^1-topology, see [7] and [8]. In some cases, one can also get some extra "fractal-like" information about these blenders, see [12] and also [19]. Considering these aspects and also the use of blenders to get robust cycles in bifurcation theory, one can think of blender-horseshoes as a version of the so-called *thick horseshoes* introduced by Newhouse in the construction of robust homoclinic tangencies of surface diffeomorphisms, see [21].

In what follows, for simplicity and also considering the scope of this paper, our discussion is restricted to the three-dimensional case (adjustments to higher dimensions are straightforward). There are some settings where blender-horseshoes appear in a natural way. A first one is the bifurcation of *heterodimensional cycles* (i.e., there are a pair of saddles having indices one and two whose invariant manifolds meet cyclically). In this context, the occurrence of blender-horseshoes is related to the existence

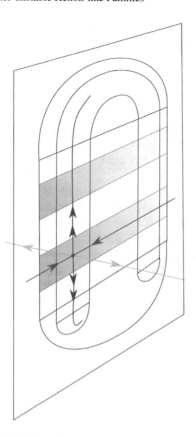

Fig. 1 Non-normally hyperbolic dynamics

of some non-normally hyperbolic dynamics that can be illustrated as follows. Think of a standard horseshoe defined on a "square" and "multiply" this dynamics by a "weak expansion" in the normal direction (to the square), see Fig. 1. In this way, one gets a hyperbolic set (of index two) contained in a non-normally hyperbolic (local) manifold.[1] Persistence of hyperbolicity implies that this horseshoe has continuations for small perturbations of the dynamics. However, since the horseshoe is contained in a non-normally hyperbolic square, the new horseshoes are in general not contained in a local surface. It turns out that appropriate perturbations of the initial dynamics provide blender-horseshoes. For a complete discussion of this construction (and also with explicit formulae) we refer to [9] (note that in [9] the term blenders is not used).

An interesting question is to provide explicit examples of maps (with an explicit analytic formula) exhibiting blender-horseshoes. This leads to the second ingredient of this paper, a family of endomorphisms so-called *center-unstable Hénon-like fami-*

---

[1]This means that in the "square" containing the horseshoe there is a direction whose expansion is greater than the expansion in the normal direction, the arrows in Fig. 1 describe this feature.

lies, see equation (1.1). We recall that in the two-dimensional case, Hénon-like maps are a fundamental ingredient in the study of homoclinic bifurcations which provide a "limit dynamics": there exists a sequence of bifurcation parameters providing a sequence of return maps at the homoclinic tangency converging to a Hénon-like map in suitable rescaled coordinates. This construction, known as *renormalisation scheme*, when performed at homoclinic tangencies allows to translate (robust) properties of the Hénon-like family to the dynamics of diffeomorphisms nearby the bifurcating one, for details see [22, Chap. 3]. Two remarkable examples of such portable properties are the persistence of homoclinic tangencies [22, Chap. 3] and the existence on strange attractors [17].

In view of the above discussion, it is natural to ask about renormalisation schemes and limit dynamics in heterodimensional settings. In this direction, in [13] it is considered a heterodimensional cycle involving a heteroclinic orbit corresponding to the tangential contact of the two-dimensional invariant manifolds of the saddles. This heteroclinic orbit is called a *heterodimensional tangency*, see [14]. In [13] it is provided a renormalisation scheme whose limit dynamics is a center-unstable Hénon-like family. This discussion justifies the following technical remark. On the one hand, the theory of homoclinic bifurcations and renormalisation schemes requires at least C^2-regularity of the diffeomorphisms.[2] On the other hand, the construction of robustly non-hyperbolic dynamics (robust cycles and tangencies) associated to heterodimensional cycles is mostly developed in the C^1-case.[3] Thus, an interesting problem is to develop these theories in higher regularity.

First, for direct approach dealing with perturbation of product dynamics (a hyperbolic part times the identity) we refer to [3]. On the other hand, bifurcations of heterodimensional tangencies seem to be an appropriate setting for obtaining robustly non-hyperbolic dynamics in high regularity, see for instance [16] where C^2-robust heterodimensional tangencies and C^2-robust heterodimensional cycles involving heterodimensional tangencies are obtained using blenders and the results of [23]. Our results are motivated by the ideas of [13], where blenders are generated at the bifurcation of heterodimensional cycles in high regularity topologies. More precisely, in [13] blenders are obtained for some (open) range of parameters of the center-unstable Hénon-like family and some applications (involving a renormalisation scheme) are given for the bifurcation of heterodimensional cycles in high regularity (in the spirit of [22]). In this paper, we prove that the blenders obtained in [13] are indeed blender-horseshoes. This step will allow (in further applications) to improve versions of [13, Theorem 1.4], getting robust cycles and robust tangencies in higher regularity (in the same spirit as in [7, 8]). In a forthcoming paper (see also [24]) we will introduce a renormalisation scheme for some non-transverse heterodimensional cycles (cycles

---

[2]Besides the regularity of the maps, necessary for the convergence of the renormalisation scheme, another key fractal-like ingredient is the thickness of a hyperbolic set, which has a radically different behaviour in the C^1 and C^2-topologies, see [26] and [18].

[3]The starting point of this progress is due to the development of a series of typically C^1-tools (started with Pugh's C^1 closing lemma and with Franks derivative perturbation lemma) that to the current date have no equivalents in C^r-topologies with $r > 1$. On the other hand, C^1-regularity is not sufficient to some results requiring control of the distortion.

with heterodimensional tangencies) converging to the center-unstable Hénon-like family (1.1) and state the persistence of cycles and tangencies (in higher regularity) after its bifurcation.

Finally, let us observe that [15] provides a quite complete numerical analysis of the center-unstable Hénon family in (1.1), showing strong numerical evidences of the occurrence of blenders in a parameter range wider than the one in [13] and illustrates the vanishing of these blenders beyond this range. We believe that the blenders detected in [15] are indeed blender-horseshoes.

It follows the main result of this paper.

Theorem 1 *Consider the center-unstable Hénon-like family of endomorphisms*

$$(1.1) \qquad G_{(\xi,\mu,\kappa,\eta)}(x, y, z) \overset{\text{def}}{=} (y, \mu + y^2 + \kappa\, y z + \eta\, z^2, \xi\, z + y), \quad \xi > 1.$$

Then there is $\varepsilon > 0$ such that for every

$$\bar{\nu} = (\xi, \mu, \kappa, \eta) \in \mathcal{O}_\varepsilon \overset{\text{def}}{=} (1.18, 1.19) \times (-10, -9) \times (-\varepsilon, \varepsilon)^2$$

the endomorphism $G_{\bar{\nu}}$ has a blender-horseshoe in the cube $\Delta \overset{\text{def}}{=} [-4, 4]^2 \times [-40, 22]$.

As a consequence, every diffeomorphism or endomorphism sufficiently C^1-close to $G_{\bar{\nu}}$ has a blender-horseshoe in Δ.

The consequence pointed out in the theorem arises from the C^1-persistence of blenders, see Remarks 2.4 and 2.9. Let us observe that this result is a version of [13, Theorem 1.1] where blenders are replaced by blender-horseshoes in a similar range of parameters.

This paper is organised as follows. In Sect. 2, we introduce the definitions of blender and blender-horseshoe and state the distinctive property of a blender-horseshoe (Lemmas 2.6 and 2.7). In Sect. 3, we prove Theorem 1.

2   Blenders and Blender-horseshoes

2.1   Blenders

The notion of a cu-blender (or simply blender) was introduced in [6], where were used to generate C^1-robust transitivity in the non-hyperbolic setting. The main virtue of a blender comes from its special internal geometry: a cu-blender is a transitive hyperbolic set whose (local) stable set robustly "behaves" as manifold of topological dimension larger than the dimension of its stable bundle. We now discuss the (axiomatic) definition of blenders in the three-dimensional case.

Definition 2.1 (cu-*Blender, Definition 3.1 in* [8]) Let $f : M \to M$ be a three-dimensional diffeomorphism. A transitive hyperbolic compact set Λ of index two of f is a cu-*blender* if there are a C^1-neighbourhood $\mathcal{U}$ of f and a C^1-open set $\mathcal{D}$ of embeddings of one-dimensional discs D into M such that for every $g \in \mathcal{U}$ and every disc $D \in \mathcal{D}$ the local stable manifold $W_{\text{loc}}^s(\Lambda_g)$ of the continuation Λ_g intersects D. The set $\mathcal{D}$ is called the *region of superposition* of the blender.

2.2 Blender-horseshoes

This kind of blenders was introduced in [8] as a mechanism for the generation of C^1-robust tangencies in dimension equal to or greater than three. Comparing with the standard blenders, blender-horseshoes satisfy the following additional property: they are locally maximal invariant sets conjugate to a complete shift of two symbols. These properties provide a complete description of its local stable manifold as well as a nice geometrical structure: the local stable manifold of a blender-horseshoe is the Cartesian product of a "fat Cantor set" by an "interval", see Remark 2.4. We now give the definition of a blender-horseshoe following [8, Sect. 3.2], for further details we refer to that paper. As the construction is local, we assume that the ambient space is $\mathbb{R}^3$. We start with some preliminary definitions.

For $a > 0$ consider the interval $I_a \overset{\text{def}}{=} [-a, +a]$ and for $x, y, z \in \mathbb{R}^+$ the cube

$$\Delta \overset{\text{def}}{=} I_x \times I_y \times I_z \subset \mathbb{R}^3.$$

We divide the boundary $\partial \Delta$ of Δ into three parts as follows:

$$\partial^s \Delta \overset{\text{def}}{=} \partial I_x \times I_y \times I_z, \quad \partial^{uu} \Delta \overset{\text{def}}{=} I_x \times \partial I_y \times I_z, \quad \partial^u \Delta \overset{\text{def}}{=} I_x \times \partial(I_y \times I_z).$$

Note that $\partial \Delta = \partial^s \Delta \cup \partial^u \Delta$ and $\partial^{uu} \Delta \subset \partial^u \Delta$.

Given $\theta > 0$ and $p \in \mathbb{R}^3$, define the s-, uu- and u-cone fields of size θ contained in $T_p \Delta$ as follows:

$$
\begin{aligned}
C_\theta^s(p) &\overset{\text{def}}{=} \left\{ (u, v, w) \in T_p\mathbb{R}^3 : \sqrt{v^2 + w^2} < \theta |u| \right\}, \\
(2.1) \qquad C_\theta^{uu}(p) &\overset{\text{def}}{=} \left\{ (u, v, w) \in T_p\mathbb{R}^3 : \sqrt{u^2 + w^2} < \theta |v| \right\}, \\
C_\theta^u(p) &\overset{\text{def}}{=} \left\{ (u, v, w) \in T_p\mathbb{R}^3 : |u| < \theta \sqrt{v^2 + w^2} \right\}.
\end{aligned}
$$

Note that $C_\theta^{uu}(p) \subset C_\theta^u(p)$.

Related to these cone fields, we define s_θ- and uu_θ-*discs* and u_θ-*strips* as follows:

- Let L be a regular curve. We say that L is an s_θ-*disc* if it is contained in Δ, $T_p L \subset C_\theta^s(p)$ for each $p \in L$, and its end-points are contained in different connected components of $\partial^s \Delta$. Similarly, we say that L is a uu_θ-*disc* if

$L \subset \mathbb{R} \times I_y \times \mathbb{R}$, $T_p L \subset C_\theta^{uu}(p)$ for each $p \in L$, and its end-points are contained in different connected components of $\mathbb{R} \times \partial I_y \times \mathbb{R}$.

- A surface $S \subset \Delta$ is a u_θ-*strip* if $T_p S \subset C_\theta^u(p)$ for every p in S and there exists a C^1-embedding $E : I_y \times J \to \Delta$ (where J is a subinterval of I_z) such that $E(I_y \times J) = S$ and $L(z) \overset{\text{def}}{=} E(I_y \times \{z\})$ is a uu$_\theta$-disc for every $z \in J$. The *width* of S, denoted by $w(S)$, is the infimum of the length of the curves in S which are transverse to C_θ^{uu} and join the two components of $E(I_y \times \partial J)$

Remark 2.2 (Right and left classes of uu-discs) In what follows, we fix $\theta, \vartheta > 0$. Note that every s_ϑ-disc W such that $(W \setminus \partial W)$ is contained in the interior of Δ defines two different (free) homotopy classes of uu$_\theta$-discs disjoint from W. This allows us to consider uu$_\theta$-discs *at the left* and *at the right* of W (corresponding to the two different homotopy classes), denoted by $\mathcal{U}_W^\ell$ and $\mathcal{U}_W^r$, respectively. The *right class* $\mathcal{U}_W^r$ (resp., *left class* $\mathcal{U}_W^\ell$) is the class containing the uu$_\theta$-disc $\{0\} \times I_y \times \{z^+\}$ (resp., containing the $\{0\} \times I_y \times \{z^-\}$). With a slight abuse of notation, we also denote by $\mathcal{U}_W^i$ the union of the uu-discs in $\mathcal{U}_W^i$, $i = r, \ell$.

Similarly, a u-strip S through Δ is *at the right* (resp. *at the left*) of W if it is foliated by uu-discs at the right (resp. *at the left*) of W.

We are now ready to recall the definition of a blender-horseshoe in [8].

Definition 2.3 (*Blender-horseshoe*) The maximal invariant $\Lambda_F \overset{\text{def}}{=} \cap_{i \in \mathbb{Z}} F^i(\Delta) \subset \text{int}(\Delta)$ of a (local) diffeomorphism $F : \Delta \to F(\Delta) \subset \mathbb{R}^3$ is a *blender-horseshoe* if conditions (**BH1**)–(**BH6**) below hold:

(**BH1**) s- *and* u-*legs* : There are a connected subsets $\mathcal{A}$ and $\mathcal{B}$ of Δ, called s-*legs of the blender*, with

$$\mathcal{A} \cap \mathcal{B} = \emptyset \quad \text{and} \quad (\mathcal{A} \cup \mathcal{B}) \cap \partial^{uu}\Delta = \emptyset$$

such that

$$F(\Delta) \cap (\mathbb{R} \times I_y \times \mathbb{R}) = F(\mathcal{A}) \cup F(\mathcal{B}) \subset (x^-, x^+) \times I_y \times \mathbb{R}.$$

Note that the sets $F(\mathcal{A})$ and $F(\mathcal{B})$ are the connected components of $F(\Delta) \cap (\mathbb{R} \times I_y \times \mathbb{R})$, they are called the u-*legs of the blender*. See Fig. 2.

(**BH2**) *Contracting and expanding invariant cone fields.* There exist $\theta, \vartheta > 0, \ell \in \mathbb{N}, c > 1$, and cone fields $C_\vartheta^s, C_\theta^u$, and C_θ^{uu} such that:

(i) *Strict invariance*: for every $p \in \mathcal{A} \cup \mathcal{B}$ we have that

$$DF_p^\ell(C_\vartheta^s(p)) \supset C_\vartheta^s(F^\ell(p)),$$
$$DF_p^\ell(C_\theta^u(p)) \subset C_\theta^u(F^\ell(p)), \quad \text{and} \quad DF_p^\ell(C_\theta^{uu}(p)) \subset C_\theta^{uu}(F^\ell(p)).$$

(ii) *Expansion/Contraction.* For every $v \in C_\vartheta^s(p)$ and every $w \in C_\theta^u(p)$ we have that

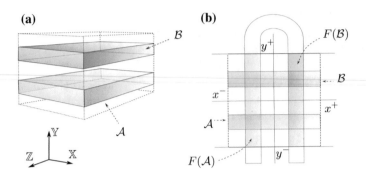

Fig. 2 **a** s-legs of the blender-horseshoes. **b** Projection of $F(\Delta) \cap (\mathbb{R} \times I_y \times \mathbb{R})$ in the plane $\mathbb{X}\mathbb{Y}$

$$|DF^{\ell}_p v| \leq c^{-1}|v| \quad \text{and} \quad |DF^{\ell}_p w| \geq c|w|.$$

Conditions (**BH1**) and (**BH2**) imply the existence of two fixed saddles $P \in \mathcal{A}$ and $Q \in \mathcal{B}$, called the *reference saddles* of Λ_F. We define the local stable manifolds of P and Q by

(2.2)　　　$W^s_{\mathrm{loc}}(R) \overset{\text{def}}{=}$ connected component of $W^s(R) \cap \Delta$ containing R,

where $R = P, Q$. These local stable manifolds are s-discs (in what follows we omit the dependence of θ and ϑ). Thus, either $\mathcal{U}^{\ell}_{W^s_{\mathrm{loc}}(P)} \cap \mathcal{U}^r_{W^s_{\mathrm{loc}}(Q)} \neq \emptyset$ or $\mathcal{U}^r_{W^s_{\mathrm{loc}}(P)} \cap \mathcal{U}^{\ell}_{W^s_{\mathrm{loc}}(Q)} \neq \emptyset$. We assume that the first case holds and denote by $\mathcal{U}^b \overset{\text{def}}{=} \mathcal{U}^{\ell}_{W^s_{\mathrm{loc}}(P)} \cap \mathcal{U}^r_{W^s_{\mathrm{loc}}(Q)}$. The family of discs $\mathcal{U}^b$ is called the *superposition region* of the blender-horseshoe. We say that a uu-disc is *in between* if it is contained $\mathcal{U}^b$. Similarly, a u-strip is *in between* if it is foliated by uu-discs in between.

(**BH3**) *Markov partition.* The connected components of $F^{-1}(\Delta) \cap \Delta$ are the sets

$$\mathbb{A} \overset{\text{def}}{=} F^{-1}(F(\mathcal{A}) \cap \Delta) \quad \text{and} \quad \mathbb{B} \overset{\text{def}}{=} F^{-1}(F(\mathcal{B}) \cap \Delta),$$

which satisfy

$$\mathbb{A} \cup \mathbb{B} \subset I_x \times (y^-, y^+) \times (z^-, z^+), \quad F(\mathbb{A}) \cup F(\mathbb{B}) \subset (x^-, x^+) \times I_y \times \mathbb{R}.$$

(**BH4**) uu-*discs through the local stable manifolds of P and Q:* Let L and L' be uu-discs such that $L \cap W^s_{\mathrm{loc}}(P) \neq \emptyset$ and $L' \cap W^s_{\mathrm{loc}}(Q) \neq \emptyset$. Then

$$L \cap \overline{\left(\partial^u \Delta \setminus \partial^{uu} \Delta\right)} = \emptyset, \quad L' \cap \overline{\left(\partial^u \Delta \setminus \partial^{uu} \Delta\right)} = \emptyset.$$

(**BH5**) *Positions of images of* uu-*discs:* Let L be a uu-disc in Δ and consider

$$L_\mathcal{C} \overset{\text{def}}{=} L \cap \mathcal{C}, \quad \mathcal{C} = \mathcal{A}, \mathcal{B}.$$

By (**BH1**) and (**BH2**), $F(L_\mathcal{C})$ is a uu-discs in $I_x \times I_y \times \mathbb{R}$. The relative position of $F(L_\mathcal{C})$ obeys the following rules:

(1) if $L \in \mathcal{U}^r_{W^s_{\text{loc}}(P)}$ then $F(L_\mathcal{A}) \in \mathcal{U}^r_{W^s_{\text{loc}}(P)}$,

(2) if $L \in \mathcal{U}^\ell_{W^s_{\text{loc}}(P)}$ then $F(L_\mathcal{A}) \in \mathcal{U}^\ell_{W^s_{\text{loc}}(P)}$,

(3) if $L \in \mathcal{U}^r_{W^s_{\text{loc}}(Q)}$ then $F(L_\mathcal{B}) \in \mathcal{U}^r_{W^s_{\text{loc}}(Q)}$,

(4) if $L \in \mathcal{U}^\ell_{W^s_{\text{loc}}(Q)}$ then $F(L_\mathcal{B}) \in \mathcal{U}^\ell_{W^s_{\text{loc}}(Q)}$,

(5) if $L \in \mathcal{U}^\ell_{W^s_{\text{loc}}(P)}$ or $L \cap W^s_{\text{loc}}(P) \neq \emptyset$ then $F(L_\mathcal{B}) \in \mathcal{U}^r_{W^s_{\text{loc}}(P)}$, and

(6) if $L \in \mathcal{U}^r_{W^s_{\text{loc}}(Q)}$ or $L \cap W^s_{\text{loc}}(Q) \neq \emptyset$ then $F(L_\mathcal{A}) \in \mathcal{U}^\ell_{W^s_{\text{loc}}(Q)}$.

(**BH6**) *Positions of images of* uu-*discs in* $\mathcal{U}^b$: Let L be a uu-disc in Δ such that $L \in \mathcal{U}^b$, then either $F(L_\mathcal{A})$ or $F(L_\mathcal{B})$ is contained in $\mathcal{U}^b$.

Figure 3 illustrates a prototypical blender-horseshoe.

We now pointed out some consequences of conditions (**BH1**)–(**BH6**), see [8, Sect. 3.2.4] for more details.

Remark 2.4

- The existence of the invariant (contracting or expanding) cone fields in (**BH2**) implies the hyperbolicity (and partial hyperbolicity) of the set Λ_F: the set Λ_F is hyperbolic and partially hyperbolic with a dominated splitting

$$T_{\Lambda_F}(\mathbb{R}^3) = E^s \oplus E^{cu} \oplus E^{uu},$$

where E^s and $E^u = E^{cu} \oplus E^{uu}$ are the stable and unstable bundles of Λ_F, respectively.

- From (**BH1**)–(**BH2**), one gets that $\{\mathbb{A}, \mathbb{B}\}$ is a Markov partition generating Λ_F. Therefore, the dynamics of F in Λ_F is hyperbolic and conjugate to the full shift of two symbols. In particular, the set Λ_F contains exactly two fixed points of F, $P \in \mathbb{A}$ and $Q \in \mathbb{B}$.

- Since Λ_F is locally maximal, we have that

$$W^s_{\text{loc}}(\Lambda_F) \overset{\text{def}}{=} \bigcap_{n \in \mathbb{N}} F^{-n}(\Delta) = \bigcup_{x \in \Lambda_F} W^s_{\text{loc}}(x) \subset W^s(\Lambda_F),$$

where $W^s_{\text{loc}}(x)$ is the connected component of $W^s(x) \cap \Delta$ containing x. We can write the local stable manifold $W^s_{\text{loc}}(\Lambda_F)$ as the Cartesian product of a Cantor set, say C, by an interval. This Cantor set is "fat" in the following sense: the projection of C in the center-unstable direction contains (open) intervals. See Fig. 3b.

- Conditions (**BH1**)–(**BH6**) are C^1-open. Hence if Λ_F is a blender-horseshoe of F then the continuation Λ_G of Λ_F is a blender-horseshoe for every G sufficiently C^1-close to F (with the same reference cube Δ).

Fig. 3 **a** Prototypical
blender-horseshoe. **b**
Projection in plane $\mathbb{YZ}$ of a
uu-disc L in the region of
superposition of the
blender-horseshoe $\mathcal{U}^b$

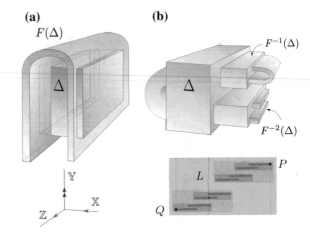

Remark 2.5 (Blenders vs. blender-horseshoes) In the introduction, we briefly compare blenders and blender-horseshoes. First, let us note that blender-horseshoes are dynamical blenders in the sense of [4]: there is an open family of uu-discs $\mathcal{D}$ such that the image of any disc of the family $\mathcal{D}$ contains a disc of $\mathcal{D}$. This invariance property and the fact that the family $\mathcal{D}$ is intimately related to the locally maximal hyperbolic set Λ_F plays a key role for obtaining robust cycles and tangencies (which are related to the hyperbolic set Λ_F). Moreover, as we saw in Remark 2.4, the hyperbolic set Λ_F is conjugate to the full shift of two symbols and some fractals properties of Λ_F can be obtained, as in [9, 12].

The next lemma states the *distinctive property* of a blender-horseshoe.

Lemma 2.6 (Lemma 3.13 in [4]) *For every $L \in \mathcal{U}^b$ it holds $L \cap W^s_{\text{loc}}(\Lambda_F) \neq \emptyset$.*

Proof. Consider $L = L'_0 \in \mathcal{U}^b$. By condition (BH6), $F(L)$ contains a disc $L'_1 \in \mathcal{U}^b$. We let $F^{-1}(L'_1) = L_1 \subset L$. We inductively define $L_n \subset L$ and $L'_n \in \mathcal{U}^b$ for $n > 1$ as follows. Assuming defined $L'_{n-1} \in \mathcal{U}^b$ and $L_{n-1} \subset L_0$ with $L'_{n-1} \subset F(L'_{n-2})$ and $F^{-n+1}(L'_{n-1}) = L_{n-1}$, we consider $L'_n \in \mathcal{U}^b$ contained in $F(L'_{n-1})$ and let $F^{-n}(L'_n) = L_n \subset L$. The sequence (L_n) is nested and hence $\emptyset \neq \bigcap_n L_n \subset L$. By construction, $\bigcap_n L_n \subset W^s_{\text{loc}}(\Lambda_F)$. $\square$

We also have the following refinement of the above lemma.

Lemma 2.7 *Every u-strip in between intersects transversely $W^s(P)$.*

Proof. Note that $F^{-1}(W^s_{\text{loc}}(P)) \cap \Delta$ consists of two connected components. We denote by W^s_0 the connected component that does not contain P. Note that this set is an s-disc. Observe that there is $\alpha > 0$ such that every u-strip S with $w(S) > \alpha$ intersets W^s_0 transversely. Conditions (**BH2**) and (**BH6**) imply that the width of a u-strip $S \subset \Delta$ in between grows exponentially after iterations by F (for simplicity let us assume that ℓ in (**BH2**) is $\ell = 1$): there is $c' > 1$ (independent of the strip) such that there are two possibilities, either $F(S)$ intersects (transversely) $W_{\text{loc}}^s(P)$ or $F(S)$ contains a u-strip S' in between such that $w(S') > c'w(S)$.

Take now a u-strip $S = S_0$ in between. If $S \cap W_0^s \neq \emptyset$ we are done. Otherwise we consider $F(S)$. If $F(S)$ intersects either W_0^s or $W_{\text{loc}}^s(P)$ we are also done. Otherwise we get a new u-strip S_1 in between contained in $F(S_0)$ with $w(S_1) > c'w(S_0)$. We now argue inductively, at some step we get a first n such that either $F(S_n)$ intersects W_0^s or $W_{\text{loc}}^s(P)$ or $w(S_n) > \alpha$ and hence S_n intersects W_0^s. In both cases, we are done. This proves the lemma. $\qquad\square$

2.2.1 Blender-horseshoes for Endomorphisms

For endomorphisms the blender horseshoe are defined as in the case of diffeomorphisms.

Definition 2.8 (*Blender-horseshoes for endomorphisms*) The maximal invariant set $\Lambda_G := \bigcap_{i\in\mathbb{Z}} G^i(\Delta) \subset \text{int}(\Delta)$ of an endomorphism $G : \Delta \to \mathbb{R}^3$ is a *blender-horseshoes* if G satisfies the conditions (**BH1**)–(**BH6**).

Remark 2.9 (Continuations of blender-horseshoes for endomorphisms) Assume that the endomorphism G has a blender-horseshoe in Δ. Then every diffeomorphism or endomorphism F such that $F|_\Delta$ is sufficiently close to $G|_\Delta$ has a blender-horseshoe in Δ.

3 Proof of Theorem 1

Theorem 1 is a consequence of following result and Remark 2.9.

Theorem 3.1 *For every* $(\xi, \mu) \in \mathcal{P} \overset{\text{def}}{=} (1.18, 1.19) \times (-10, -9)$, *the endomorphism*

$$G_{(\xi,\mu,0,0)}(x, y, z) = (y, \mu + y^2, \xi z + y)$$

has a blender-horseshoe in $\Delta = [-4, 4]^2 \times [-40, 22]$.

The proof of this theorem involves some preliminary steps. First, for the endomorphisms $G_{\xi,\mu} \overset{\text{def}}{=} G_{(\xi,\mu,0,0)}$, where $(\xi, \mu) \in \mathcal{P}$, we study their hyperbolic fixed points and their invariant manifolds. As we will see, these fixed points will be the reference saddles of the blender-horseshoe of $G_{\xi,\mu}$ in Δ.

3.1 Hyperbolic Fixed Points of $G_{\xi,\mu}$

We calculate the hyperbolic fixed points of $G_{\xi,\mu}$ and their invariant manifolds.

Lemma 3.2 *For every* $(\xi, \mu) \in \mathcal{P}$, *the endomorphism* $G_{\xi,\mu}$ *has two hyperbolic fixed saddles* $P_{\xi,\mu} = (p_{\xi,\mu}, p_{\xi,\mu}, \tilde{p}_{\xi,\mu})$ *and* $Q_{\xi,\mu} = (q_{\xi,\mu}, q_{\xi,\mu}, \tilde{q}_{\xi,\mu})$ *in* Δ, *where*

$$
\begin{aligned}
p_{\xi,\mu} &= \mu + (p_{\xi,\mu})^2 = (1 - \xi)\,\tilde{p}_{\xi,\mu}, \quad p_{\xi,\mu} = p_\mu = \frac{1 - (1 - 4\mu)^{1/2}}{2}, \\
q_{\xi,\mu} &= \mu + (q_{\xi,\mu})^2 = (1 - \xi)\,\tilde{q}_{\xi,\mu}, \quad q_{\xi,\mu} = q_\mu = \frac{1 + (1 - 4\mu)^{1/2}}{2}.
\end{aligned}
$$
(3.1)

Proof. A simple calculation shows that $P_{\xi,\mu} = (p_\mu, p_\mu, \tilde{p}_{\xi,\mu})$ and $Q_{\xi,\mu} = (q_\mu, q_\mu, \tilde{q}_{\xi,\mu})$ are the two solutions of $G_{\xi,\mu}(x, y, z) = (x, y, z)$. Using Eq. (3.1) and that $(\xi, \mu) \in \mathcal{P}$, we get the following estimates for the coordinates of $P_{\xi,\mu}$ and $Q_{\xi,\mu}$:

$$
\begin{aligned}
-2.71 &< p_\mu < -2.5, \quad 13 < \tilde{p}_{\xi,\mu} < 15, \\
3.5 &< q_\mu < 3.71, \quad -20.6 < \tilde{q}_{\xi,\mu} < -18.4.
\end{aligned}
$$
(3.2)

Thus, $P_{\xi,\mu}, Q_{\xi,\mu} \in \Delta$. We observe that the eigenvalues of $DG_{\xi,\mu}(P_{\xi,\mu})$, and $DG_{\xi,\mu}(Q_{\xi,\mu})$ are, respectively,

$$
\begin{aligned}
\lambda^s(P_{\xi,\mu}) = 0, \quad \lambda^{cu}(P_{\xi,\mu}) = \xi, \quad \lambda^{uu}(P_{\xi,\mu}) = 2\,p_\mu, \\
\lambda^s(Q_{\xi,\mu}) = 0, \quad \lambda^{cu}(Q_{\xi,\mu}) = \xi, \quad \lambda^{uu}(Q_{\xi,\mu}) = 2\,q_\mu,
\end{aligned}
$$

with respective eigenvectors

$$
\begin{aligned}
v^s(P_{\xi,\mu}) &= (1,0,0), \ v^{cu}(P_{\xi,\mu}) = (0,0,1), \ v^{uu}(P_{\xi,\mu}) = \big(2\,p_\mu - \xi, 2\,p_\mu(2p_\mu - \xi), 2\,p_\mu\big), \\
v^s(Q_{\xi,\mu}) &= (1,0,0), \ v^{cu}(Q_{\xi,\mu}) = (0,0,1), \ v^{uu}(Q_{\xi,\mu}) = \big(2\,q_\mu - \xi, 2\,q_\mu(2q_\mu - \xi), 2\,q_\mu\big).
\end{aligned}
$$

As $\xi > 1$ and $|\lambda^{uu}(P_{\xi,\mu})| = 2\,|p_\mu| > 5$ and $|\lambda^{uu}(Q_{\xi,\mu})| = 2\,|q_\mu| > 7$, we have that $P_{\xi,\mu}$ and $Q_{\xi,\mu}$ are hyperbolic fixed points of $G_{\xi,\mu}$ for every $(\xi, \mu) \in \mathcal{P}$, ending the proof of the lemma. $\square$

Remark 3.3 (Invariant directions and foliations) For $R = P, Q$ consider the eigenspaces

$$
E^s(R_{\xi,\mu}) \stackrel{\text{def}}{=} \mathbb{R} \times \{(0,0)\} \quad \text{and} \quad E^{cu}(R_{\xi,\mu}) \stackrel{\text{def}}{=} \{(0,0)\} \times \mathbb{R},
$$

associated to the eigenvalues $\lambda^s(R_{\xi,\mu}) = 0$ and $\lambda^{cu}(R_{\xi,\mu}) = \xi > 1$, and consider the straight lines through $R_{\xi,\mu}$:

$$
\{R_{\xi,\mu} + (t,0,0) : t \in \mathbb{R}\} \quad \text{and} \quad \{R_{\xi,\mu} + (0,0,t) : t \in \mathbb{R}\}.
$$

These lines are, respectively, tangent to the eigenspaces $E^s(R_{\xi,\mu})$ and $E^{cu}(R_{\xi,\mu})$ at $R_{\xi,\mu}$, and invariant by $G_{\xi,\mu}$:

$$
G_{\xi,\mu}\big(R_{\xi,\mu} + (t,0,0)\big) = R_{\xi,\mu}, \quad G_{\xi,\mu}\big(R_{\xi,\mu} + (0,0,t)\big) = R_{\xi,\mu} + (0,0,\xi t),
$$

for every $t \in \mathbb{R}$. Moreover,

$$(3.3) \qquad W^s(R_{\xi,\mu}) = \{R_{\xi,\mu} + (t, 0, 0) : t \in \mathbb{R}\}, \quad R = P, Q.$$

We define the *center unstable manifold* of $R_{\xi,\mu}$ by

$$(3.4) \qquad W^{cu}(R_{\xi,\mu}) \stackrel{\text{def}}{=} \{R_{\xi,\mu} + (0, 0, t) : t \in \mathbb{R}\}, \quad R = P, Q.$$

Consider the endomorphism of $\mathbb{R}^2$ obtained by projecting $G_{\xi,\mu}$ into the $\mathbb{Y}\mathbb{Z}$-plane,

$$(3.5) \qquad g_{\xi,\mu} : \mathbb{R}^2 \to \mathbb{R}^2, \quad g_{\xi,\mu}(y, z) \stackrel{\text{def}}{=} (\mu + y^2, \xi z + y).$$

This endomorphism preserves the foliation $\mathcal{F} = \{\{y\} \times \mathbb{R} : y \in \mathbb{R}\}$. In particular, for $r = p, q$, the leaves

$$W^{cu}_{\xi,\mu}(r_\mu, \tilde{r}_{\xi,\mu}) \stackrel{\text{def}}{=} \{(r_\mu, \tilde{r}_{\xi,\mu} + t) : t \in \mathbb{R}\},$$

are invariant by $g_{\xi,\mu}$.

3.2 The Legs of the Blender-horseshoe

In this section, we will concentrate on property **(BH1)** of blender-horseshoes. The definitions of s- and u-legs involve some preliminary constructions that we describe below.

For $\mu \in (-10, -9)$, consider the points

$$(3.6) \quad a_\mu \stackrel{\text{def}}{=} -\sqrt{4 - \mu}, \quad b_\mu \stackrel{\text{def}}{=} -\sqrt{-4 - \mu}, \quad c_\mu \stackrel{\text{def}}{=} \sqrt{-4 - \mu}, \quad d_\mu \stackrel{\text{def}}{=} \sqrt{4 - \mu}.$$

Note that if $\mu \in (-10, -9)$ it holds

$$(3.7) \qquad -\sqrt{14} < a_\mu = -d_\mu < -\sqrt{13}, \quad -\sqrt{6} < b_\mu = -c_\mu < -\sqrt{5}.$$

Consider the intervals $I_\mu \stackrel{\text{def}}{=} [a_\mu, b_\mu]$ and $J_\mu \stackrel{\text{def}}{=} [c_\mu, d_\mu]$. The choice of the parameter μ and the estimates in (3.7) imply that

$$(3.8) \qquad I_\mu = [a_\mu, b_\mu] \subset (-4, 0) \quad \text{and} \quad J_\mu = [c_\mu, d_\mu] \subset (0, 4).$$

Consider the sub-cubes of Δ defined by

$$(3.9) \quad \mathcal{A}_{\xi,\mu} \stackrel{\text{def}}{=} [-4, 4] \times I_\mu \times [-40, 22], \quad \mathcal{B}_{\xi,\mu} \stackrel{\text{def}}{=} [-4, 4] \times J_\mu \times [-40, 22].$$

From (3.8) it follows

$$\mathcal{A}_{\xi,\mu} \cap \mathcal{B}_{\xi,\mu} = \emptyset \quad \text{and} \quad (\mathcal{A}_{\xi,\mu} \cup \mathcal{B}_{\xi,\mu}) \cap \partial^{uu}\Delta = \emptyset.$$

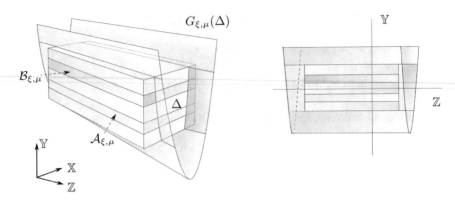

Fig. 4 The blender-horseshoe of $G_{\xi,\mu}$

Remark 3.4 If $\mu \in (-10, -9)$ then $p_\mu \in (a_\mu, b_\mu)$, $q_\mu \in (c_\mu, d_\mu)$, and thus $P_{\xi,\mu} \in$ interior$(\mathcal{A}_{\xi,\mu})$ and $Q_{\xi,\mu} \in$ interior$(\mathcal{B}_{\xi,\mu})$.

Hence the sets $\mathcal{A}_{\xi,\mu}$ and $\mathcal{B}_{\xi,\mu}$ satisfy the first part of condition (**BH1**). To prove that $G_{\xi,\mu}(\mathcal{A}_{\xi,\mu})$ and $G_{\xi,\mu}(\mathcal{B}_{\xi,\mu})$ satisfy the second part of (**BH1**), as in the case of the boundary of Δ, we split the boundary of $\mathcal{A}_{\xi,\mu}$ as follows. Let

$$\partial^{uu}\mathcal{A}_{\xi,\mu} \stackrel{\text{def}}{=} [-4, 4] \times \partial I_\mu \times [-40, 22],$$

$$\partial^{u}\mathcal{A}_{\xi,\mu} \stackrel{\text{def}}{=} [-4, 4] \times \partial\big(I_\mu \times [-40, 22]\big),$$

$$\partial^{s}\mathcal{A}_{\xi,\mu} \stackrel{\text{def}}{=} \partial([-4, 4]) \times I_\mu \times [-40, 22].$$

Note that $\partial\mathcal{A}_{\xi,\mu} = \partial^{u}\mathcal{A}_{\xi,\mu} \cup \partial^{s}\mathcal{A}_{\xi,\mu}$ and $\partial^{uu}\mathcal{A}_{\xi,\mu} \subset \partial^{u}\mathcal{A}_{\xi,\mu}$. Analogously, we split the boundary of $\mathcal{B}_{\xi,\mu}$.

Remark 3.5 We observe that for $\mathcal{C} = \mathcal{A}, \mathcal{B}$ it holds that

$$\overline{\partial\mathcal{C}_{\xi,\mu} \setminus (\partial^{uu}\mathcal{C}_{\xi,\mu} \cup \partial^{s}\mathcal{C}_{\xi,\mu})} \subset \partial^{u}\Delta \setminus \partial^{uu}\Delta, \quad (\xi, \mu) \in \mathcal{P}.$$

Roughly, these relations between the boundaries say that the "front" and "rear cover" of $\mathcal{A}_{\xi,\mu}$ and $\mathcal{B}_{\xi,\mu}$ are contained in the "front" and "rear cover" of Δ, respectively, (see Fig. 4).

Lemma 3.6 *For every $(\xi, \mu) \in \mathcal{P}$ it holds*

(a) $G_{\xi,\mu}(\Delta) \cap (\mathbb{R} \times [-4, 4] \times \mathbb{R}) = G_{\xi,\mu}(\mathcal{A}_{\xi,\mu}) \cup G_{\xi,\mu}(\mathcal{B}_{\xi,\mu})$,
(b) $G_{\xi,\mu}(\mathcal{A}_{\xi,\mu}) \cup G_{\xi,\mu}(\mathcal{B}_{\xi,\mu}) \subset (-4, 4) \times [-4, 4] \times \mathbb{R}$.

Proof. We begin showing the equality of the item a). Keeping in mind Remark 3.5, the inclusion "$\subset$" is obtained from the relations (see Fig. 4):

$$G_{\xi,\mu}(\mathcal{A}_{\xi,\mu}) \cap G_{\xi,\mu}(\mathcal{B}_{\xi,\mu}) = \emptyset, \quad G_{\xi,\mu}\big(\Delta \setminus (\mathcal{A}_{\xi,\mu} \cup \mathcal{B}_{\xi,\mu})\big) \cap \Delta = \emptyset,$$

(3.10)

$$(\xi, \mu) \in \mathcal{P}.$$

The reciprocal inclusion "⊃" follows from the relation:

(3.11) $$G_{\xi,\mu}\big(\partial^{uu}\mathcal{A}_{\xi,\mu} \cup \partial^{uu}\mathcal{B}_{\xi,\mu}\big) \subset \{|y| = 4\}, \quad (\xi, \mu) \in \mathcal{P}.$$

To get the first relation in (3.10), it is sufficient to study the projections of $G_{\xi,\mu}(\mathcal{A}_{\xi,\mu})$ and $G_{\xi,\mu}(\mathcal{B}_{\xi,\mu})$ in the plane $\mathbb{XY}$. We denote such projection by Π_3.

Claim 3.7 *For every* $(\xi, \mu) \in \mathcal{P}$ *it holds* $\Pi_3(G_{\xi,\mu}(\mathcal{A}_{\xi,\mu})) \cap \Pi_3(G_{\xi,\mu}(\mathcal{B}_{\xi,\mu})) = \emptyset$.

Proof. Let $(\xi, \mu) \in \mathcal{P}$, then we have that

$$\Pi_3(G_{\xi,\mu}(\mathcal{A}_{\xi,\mu})) = \{(y, \mu + y^2) : y \in I_\mu\}, \quad \Pi_3(G_{\xi,\mu}(\mathcal{B}_{\xi,\mu})) = \{(y, \mu + y^2) : y \in J_\mu\}.$$

From $I_\mu \cap J_\mu = \emptyset$ it follows that $\Pi_3(G_{\xi,\mu}(\mathcal{A}_{\xi,\mu})) \cap \Pi_3(G_{\xi,\mu}(\mathcal{B}_{\xi,\mu})) = \emptyset$, ending the proof of the claim. □

Remark 3.8 Equation (3.8) and the proof of the claim above also imply that

$$\Pi_3\big(G_{\xi,\mu}(\mathcal{A}_{\xi,\mu}) \cup G_{\xi,\mu}(\mathcal{B}_{\xi,\mu})\big) \subset (-4, 4) \times [-4, 4], \quad \text{for every } (\xi, \mu) \in \mathcal{P}.$$

We now prove (3.11) and the second part of (3.10). Since the endomorphisms $G_{\xi,\mu}$ collapse the $\mathbb{X}$-direction, it is sufficient to study the corresponding projections in the plane $\mathbb{YZ}$. For this, consider the sets

$$\Gamma_\mu \overset{\text{def}}{=} \big([-4, a_\mu) \cup (b_\mu, c_\mu) \cup (d_\mu, 4]\big) \times [-40, 22],$$

$$C_\mu^1 \overset{\text{def}}{=} \{a_\mu\} \times [-40, 22], \quad C_\mu^2 \overset{\text{def}}{=} \{b_\mu\} \times [-40, 22],$$

$$C_\mu^3 \overset{\text{def}}{=} \{c_\mu\} \times [-40, 22], \quad C_\mu^4 \overset{\text{def}}{=} \{d_\mu\} \times [-40, 22].$$

Note that Γ_μ, C_μ^1, C_μ^2, C_μ^3, and C_μ^4 are, respectively, the projections on the plane $\mathbb{YZ}$ of the sets

$$\Delta \setminus (\mathcal{A}_{\xi,\mu} \cup \mathcal{B}_{\xi,\mu}),$$
$$\partial^{uu}\mathcal{A}_{\xi,\mu} \cap \{y = a_\mu\}, \quad \partial^{uu}\mathcal{A}_{\xi,\mu} \cap \{y = b_\mu\},$$
$$\partial^{uu}\mathcal{B}_{\xi,\mu} \cap \{y = c_\mu\}, \quad \partial^{uu}\mathcal{B}_{\xi,\mu} \cap \{y = d_\mu\}.$$

Recall the definition of the endomorphism $g_{\xi,\mu}$ in (3.5).

Claim 3.9 *For every* $(\xi, \mu) \in \mathcal{P}$ *it holds that*

(a') $g_{\xi,\mu}(\Gamma_\mu) \cap \big([-4, 4] \times [-40, 22]\big) = \emptyset$,

(b') $g_{\xi,\mu}(C_\mu^1 \cup C_\mu^4) \subset \{y = 4\}$, *and*

(c') $g_{\xi,\mu}(C_\mu^2 \cup C_\mu^3) \subset \{y = -4\}$.

Proof. Consider the projection $\pi_2(y, z) \stackrel{\text{def}}{=} y$. It is easy to check the following equalities:

$$\pi_2\Big(g_{\xi,\mu}\big([-4, a_\mu) \times [-40, 22]\big)\Big) = (4, \mu + 16],$$

$$\pi_2\Big(g_{\xi,\mu}\big((b_\mu, c_\mu) \times [-40, 22]\big)\Big) = [\mu, -4),$$

$$\pi_2\Big(g_{\xi,\mu}\big((d_\mu, 4] \times [-40, 22]\big)\Big) = (4, \mu + 16].$$

Recalling that $\mu \in (-10, -9)$ we get item (a'). From Remark 3.3 and equation (3.6) it follows

- $g_{\xi,\mu}$ preserves the foliation $\mathcal{F} = \{\{y\} \times \mathbb{R} : y \in \mathbb{R}\}$, and
- $\mu + a_\mu^2 = \mu + d_\mu^2 = -(\mu + b_\mu^2) = -(\mu + c_\mu^2) = 4$.

These two facts imply items (b') and (c'). This ends the proof of the claim. $\square$

The proof of item (a) of the lemma is now complete. Finally, item (b) follows directly from Remark 3.8. The proof of the lemma is now complete. $\square$

3.3 Contracting/Expanding Invariant Cone Fields

In this section, we study the condition (**BH2**) of a blender-horseshoe involving invariance, contraction, and expansion of the cone fields in (2.1). This condition is a consequence of the following lemma.

Lemma 3.10 *Let* $0 < \vartheta < 1$ *and* $\theta = 1/2$. *Then, for every* $(\xi, \mu) \in \mathcal{P}$ *and every* $p \in \mathcal{A}_{\xi,\mu} \cup \mathcal{B}_{\xi,\mu}$ *the following holds:*

(i) $\mathcal{C}_\vartheta^s\big(G_{\xi,\mu}(p)\big) \subset D(G_{\xi,\mu})_p\big(\mathcal{C}_\vartheta^s(p)\big),$
(ii) $D(G_{\xi,\mu})_p\big(\mathcal{C}_\theta^u(p)\big) \subset \mathcal{C}_\theta^u\big(G_{\xi,\mu}(p)\big),$
(iii) $D(G_{\xi,\mu})_p\big(\mathcal{C}_\theta^{uu}(p)\big) \subset \mathcal{C}_\theta^{uu}\big(G_{\xi,\mu}(p)\big),$
(iv) $DG_{\xi,\mu}|_{\mathcal{C}_\theta^u}$ *is uniformly expanding and* $DG_{\xi,\mu}|_{\mathcal{C}_\vartheta^s}$ *is uniformly contracting for every* ϑ *sufficiently small.*

Proof. Consider $p = (x, y, z) \in \mathcal{A}_{\xi,\mu} \cup \mathcal{B}_{\xi,\mu}$ and $\mathbf{v} = (u, v, w) \in T_p \Delta$, write

$$(u_1, v_1, w_1) \stackrel{\text{def}}{=} D(G_{\xi,\mu})_p(\mathbf{v}) = (v, 2\, y\, v, v + \xi\, w).$$

Recalling (3.9) and (3.7), we have that if $(x, y, z) \in \mathcal{A}_{\xi,\mu} \cup \mathcal{B}_{\xi,\mu}$ then $y \in I_\mu \cup J_\mu$ and thus $|y| > \sqrt{5}$, for every $\mu \in (-10, -9)$.

The items of the lemma are proved in the following claims.

Claim 3.11 *(Item (i))* *Let* $0 < \vartheta < 1$. *For every* $\mathbf{v} \in \partial \mathcal{C}_\vartheta^s(p) \setminus \{\bar{0}\}$ *we have*
$$D(G_{\xi,\mu})_p(\mathbf{v}) \in \Big(\overline{\mathcal{C}_\vartheta^s(G_{\xi,\mu}(p))}\Big)^c.$$

Proof. If $\mathbf{v} \in \partial C_\vartheta^s(p) \setminus \{\bar{0}\}$ then $\vartheta\left(\sqrt{v^2 + w^2}\right) = |u|$. Since $|y| > \sqrt{5}$, we get that

$$\sqrt{v_1^2 + w_1^2} \geq |v_1| \geq 2\,|y|\,|v| > 2\sqrt{5}\,|v| = 2\sqrt{5}\,|u_1| > |u_1| > \vartheta\,|u_1|.$$

Therefore $D(G_{\xi,\mu})_p(\mathbf{v}) \notin \overline{C_\vartheta^s(G_{\xi,\mu}(p))}$, proving the claim. $\qquad\square$

Claim 3.12 *(Item (ii))* For every $\mathbf{v} \in C_{1/2}^u(p)$ it holds $D(G_{\xi,\mu})_p(\mathbf{v}) \in C_{1/2}^u\big(G_{\xi,\mu}(p)\big)$.

Proof. Since $|y| > \sqrt{5}$, we have that

$$\sqrt{v_1^2 + w_1^2} \geq |v_1| = 2\,|y|\,|v| > 2\sqrt{5}\,|v| > 2\,|u_1|,$$

proving the claim. $\qquad\square$

Claim 3.13 *(Item (iii))* For every $\mathbf{v} \in C_{1/2}^{uu}(p)$ it holds $D(G_{\xi,\mu})_p(\mathbf{v}) \in C_{1/2}^{uu}(G_{\xi,\mu}(p))$.

Proof. We need to check that

$$\sqrt{u^2 + w^2} < \frac{1}{2}\,|v| \quad \Rightarrow \quad \sqrt{u_1^2 + w_1^2} < \frac{1}{2}\,|v_1|.$$

Note that $\sqrt{u^2 + w^2} < \frac{1}{2}\,|v|$ implies that $|w| < \frac{1}{2}\,|v|$, and hence

$$u_1^2 + w_1^2 = v^2 + (v + \xi w)^2 \leq 2\,v^2 + 2\xi\,|v|\,|w| + \xi^2\,|w|^2 \leq \left(2 + \xi + \left(\frac{\xi}{2}\right)^2\right)v^2.$$

Now $\xi \in (1.18, 1.19)$ implies that

$$\left(2 + \frac{\xi}{2} + \left(\frac{\xi}{2}\right)^2\right) < 4$$

and hence

$$u_1^2 + w_1^2 < 4v^2.$$

Thus, since $p = (x, y, z) \in \mathcal{A}_{\xi,\mu} \cup \mathcal{B}_{\xi,\mu}$ implies that $|y| > \sqrt{5}$, it follows

$$2\sqrt{u_1^2 + w_1^2} < 4\,|v| < 2\,|y|\,|v| = |v_1|,$$

proving the claim. $\qquad\square$

Claim 3.14 *(Item (iv))* $DG_{\xi,\mu}|_{C_{1/2}^u}$ *is uniformly expanding and, if ϑ is small enough,* $DG_{\xi,\mu}|_{C_\vartheta^s}$ *is uniformly contracting.*

Proof. The uniform contraction of the cone field C_ϑ^s for small ϑ follows from the fact that $D(G_{\xi,\mu})_p$ is an endomorphism whose eigenspace associated the eigenvalue 0 is spanned by $(1, 0, 0)$.

To see that $DG_{\xi,\mu}$ uniformly expands the vectors in $C_{1/2}^u$ consider the norm

$$|(u, v, w)|_* \overset{\text{def}}{=} \max\left\{|u|, \sqrt{v^2 + w^2}\right\}.$$

Take $\mathbf{v} = (u, v, w) \in C_{1/2}^u(p)$ and write $D(G_{\xi,\mu})_p(\mathbf{v}) = (u_1, v_1, w_1) = (v, 2\, y\, v, v + \xi\, w)$. We claim that if $\mathbf{v} \in C_{1/2}^u(p)$ then $|D(G_{\xi,\mu})_p \mathbf{v}|_* > |\mathbf{v}|_*$. By compactness, this implies that $|D(G_{\xi,\mu})_p \mathbf{v}|_* > c_0 |\mathbf{v}|_*$, for some uniform $c_0 > 1$. Note that the Euclidean norm $||\cdot||$ and $|\cdot|_*$ are equivalent, hence there is $\kappa > 1$ such that $\kappa^{-1}||\mathbf{v}|| \leq |\mathbf{v}|_* \leq \kappa||\mathbf{v}||$. The number ℓ in **(BH2)** is the first ℓ_0 with $c_0^{\ell_0} > \kappa$.

We now prove that $|D(G_{\xi,\mu})_p \mathbf{v}|_* > |\mathbf{v}|_*$. Note that for $\mathbf{v} = (u, v, w) \in C_{1/2}^u(p)$ we have $|\mathbf{v}|_* = \sqrt{v^2 + w^2}$ and

$$(3.12) \qquad v_1^2 + w_1^2 = 4\, v^2\, y^2 + (v + \xi\, w)^2 \geq 4\, v^2\, y^2 + v^2 - 2\, \xi\, |v|\, |w| + \xi^2\, w^2.$$

We divide the proof into two cases: $(6.5)\, |v| \geq |w|$ and $(6.5)\, |v| \leq |w|$. If $(6.5)\, |v| \geq |w|$, using that $\xi \in (1.18, 1.19)$ and $|y| > \sqrt{5}$, we get that

$$(3.13) \qquad 4\, v^2\, y^2 - 2\, \xi\, |v|\, |w| \geq (20 - 13\, \xi)\, v^2 > 4\, v^2 \geq 0.$$

Equations (3.12) and (3.13) immediately imply that

$$v_1^2 + w_1^2 > 5\, v^2 + \xi^2\, w^2 > v^2 + w^2.$$

Hence, $|D(G_{\xi,\mu})_p \mathbf{v}|_* > |\mathbf{v}|_*$, proving the first case. Similarly, if $(6.5)\, |v| \leq |w|$ then

$$v_1^2 + w_1^2 \geq 4\, y^2\, v^2 + \xi^2\, w^2 - 2\, \xi\, |v|\, |w| + v^2 > 4y^2 v^2 + \xi^2\, w^2 - 2\, \xi\, (6.5)^{-1}\, w^2 + v^2.$$

Condition $\xi \in (1.18, 1.19)$ implies that

$$\xi^2 - 2\, \xi\, (6.5)^{-1} > 1.$$

Thus

$$v_1^2 + w_1^2 \geq v^2 + w^2.$$

Thus, $|D(G_{\xi,\mu})_p \mathbf{v}|_* > |\mathbf{v}|_*$. This ends the proof of the claim. $\square$

The proof of the lemma is now complete. $\square$

Remark 3.15 For each $p = (x, y, z) \in \mathbb{R}^3$ we identify $T_p\mathbb{R}^3$ with $\mathbb{R}^3$ and consider the canonical basis $\{\mathbf{i}, \mathbf{j}, \mathbf{k}\}$. Note that $D(G_{\xi,\mu})_p(\mathbf{i}) = \mathbf{0}$, $D(G_{\xi,\mu})_p(\mathbf{j}) = \mathbf{i} + 2\, y\, \mathbf{j} + \mathbf{k}$, and $D(G_{\xi,\mu})_p(\mathbf{k}) = \xi\, \mathbf{k}$. In particular, $\langle D(G_{\xi,\mu})_p(\mathbf{j}), \mathbf{j}\rangle < 0$ (resp. > 0) if $y < 0$ (resp. $y > 0$). As a consequence, for every $\theta > 0$ and every $p \in \mathcal{A}_{\xi,\mu}$, the derivative

$D(G_{\xi,\mu})_p$ sends the semi-positive cone $C_\theta^{uu}(p) \cap \{y > 0\}$ (resp. semi-negative cone) into the semi-space $\{y < 0\}$ (resp. $\{y > 0\}$). When $p \in \mathcal{B}_{\xi,\mu}$ the derivative $D(G_{\xi,\mu})_p$ maps the semi-positive cone $C_\theta^{uu}(p) \cap \{y > 0\}$ (resp. semi-negative cone) into $\{y > 0\}$ (resp. $\{y > 0\}$).

3.4 The Markov Partition

To define the Markov partition in Condition (**BH3**) we need some preliminary constructions.

For $(\xi, \mu) \in \mathcal{P}$ consider the auxiliary straight lines $R_{\xi,\mu}^1, R_{\xi,\mu}^2$ in the plane $\mathbb{YZ}$ defined by the equations and depicted in Fig. 5,

$$R_{\xi,\mu}^1 \stackrel{\text{def}}{=} \{(y, z_\xi^1(y)) : z_\xi^1(y) = \xi^{-1}(22 - y), \, y \in \mathbb{R}\},$$
$$R_{\xi,\mu}^2 \stackrel{\text{def}}{=} \{(y, z_\xi^2(y)) : z_\xi^2(y) = \xi^{-1}(-40 - y), \, y \in \mathbb{R}\}.$$

Recall the definition of the intervals $I_\mu = [a_\mu, b_\mu]$ and $J_\mu = [c_\mu, d_\mu]$ in (3.8). Consider the auxiliary parallelogram $\mathbb{A}_{\xi,\mu}$ in the plane $\mathbb{YZ}$ whose boundary consists of the following segments (see Fig. 5):

$$Ł_{\xi,\mu}^1 \stackrel{\text{def}}{=} \{(y, z_\xi^1(y)) : y \in I_\mu\}, \quad Ł_{\xi,\mu}^2 \stackrel{\text{def}}{=} \{a_\mu\} \times [z_\xi^2(a_\mu), z_\xi^1(a_\mu)],$$
$$Ł_{\xi,\mu}^3 \stackrel{\text{def}}{=} \{(y, z_\xi^2(y)) : y \in I_\mu\}, \quad Ł_{\xi,\mu}^4 \stackrel{\text{def}}{=} \{b_\mu\} \times [z_\xi^2(b_\mu), z_\xi^1(b_\mu)].$$

Analogously, consider the parallelogram $\mathbb{B}_{\xi,\mu}$ in the plane $\mathbb{YZ}$ bounded by

$$\tilde{Ł}_{\xi,\mu}^1 \stackrel{\text{def}}{=} \{(y, z_\xi^1(y)) : y \in J_\mu\}, \quad \tilde{Ł}_{\xi,\mu}^2 \stackrel{\text{def}}{=} \{c_\mu\} \times [z_\xi^2(c_\mu), z_\xi^1(c_\mu)],$$
$$\tilde{Ł}_{\xi,\mu}^3 \stackrel{\text{def}}{=} \{(y, z_\xi^2(y)) : y \in J_\mu\}, \quad \tilde{Ł}_{\xi,\mu}^4 \stackrel{\text{def}}{=} \{d_\mu\} \times [z_\xi^2(d_\mu), z_\xi^1(d_\mu)].$$

Remark 3.16 Since $(\xi, \mu) \in \mathcal{P}$, it follows that $\mathbb{A}_{\xi,\mu}$ and $\mathbb{B}_{\xi,\mu}$ are contained in $(-4, 0) \times (-40, 22)$ and $(0, 4) \times (-40, 22)$, respectively. By the definitions of $\mathbb{A}_{\xi,\mu}$ and $\mathbb{B}_{\xi,\mu}$, it holds that

$$g_{\xi,\mu}(\partial \mathbb{A}_{\xi,\mu}) = g_{\xi,\mu}\left(\cup_{i=1}^4 Ł_{\xi,\mu}^i\right) = \partial([-4, 4] \times [-40, 22]),$$

$$g_{\xi,\mu}(\partial \mathbb{B}_{\xi,\mu}) = g_{\xi,\mu}\left(\cup_{i=1}^4 \tilde{Ł}_{\xi,\mu}^i\right) = \partial([-4, 4] \times [-40, 22]),$$

and thus

$$g_{\xi,\mu}(\mathbb{A}_{\xi,\mu}) = g_{\xi,\mu}(\mathbb{B}_{\xi,\mu}) = [-4, 4] \times [-40, 22].$$

We now show that the sets $\mathbb{A}_{\xi,\mu}$ and $\mathbb{B}_{\xi,\mu}$ (see Fig. 5).

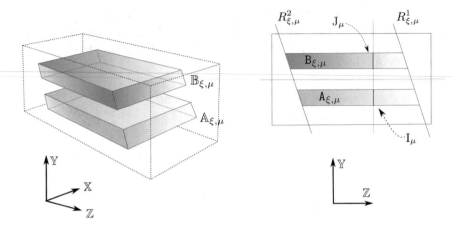

Fig. 5 The Markov partition of the blender-horseshoe

$$\mathbb{A}_{\xi,\mu} \stackrel{\text{def}}{=} [-4, 4] \times A_{\xi,\mu} \quad \text{and} \quad \mathbb{B}_{\xi,\mu} \stackrel{\text{def}}{=} [-4, 4] \times B_{\xi,\mu}.$$

form a Markov partition of the blender-horseshoe of $G_{\xi,\mu}$ in Δ. Observe first that

$$\mathbb{A}_{\xi,\mu} = (G_{\xi,\mu}|_\Delta)^{-1}\big(G_{\xi,\mu}(\mathcal{A}_{\xi,\mu}) \cap \Delta\big), \quad \mathbb{B}_{\xi,\mu} = (G_{\xi,\mu}|_\Delta)^{-1}\big(G_{\xi,\mu}(\mathcal{B}_{\xi,\mu}) \cap \Delta\big).$$

The next lemma completes the proof of condition (**BH3**).

Lemma 3.17 *For every* $(\xi, \mu) \in \mathcal{P}$ *the following holds*

(a) $\mathbb{A}_{\xi,\mu} \cup \mathbb{B}_{\xi,\mu} \subset [-4, 4] \times (-4, 4) \times (-40, 22)$,
(b) $G_{\xi,\mu}(\mathbb{A}_{\xi,\mu}) \cup G_{\xi,\mu}(\mathbb{B}_{\xi,\mu}) \subset (-4, 4) \times [-4, 4] \times \mathbb{R}$.

Proof. Item (a) follows from Remark 3.16. For item (b), note that Lemma 3.6 implies that

$$G_{\xi,\mu}(\mathbb{A}_{\xi,\mu}) \cup G_{\xi,\mu}(\mathbb{B}_{\xi,\mu}) \subset G_{\xi,\mu}(\mathcal{A}_{\xi,\mu}) \cup G_{\xi,\mu}(\mathcal{B}_{\xi,\mu}) \subset (-4, 4) \times [-4, 4] \times \mathbb{R},$$

completing of proof of lemma. □

3.5 uu-Discs Through the Local Stable Manifolds

We study Condition (**BH4**) of blender horseshoes about the relative position of the uu-discs through the local stable manifolds of $P_{\xi,\mu} = (p_\mu, p_\mu, \tilde{p}_{\xi,\mu})$ and $Q_\mu = (q_\mu, q_\mu, \tilde{q}_{\xi,\mu})$ with respect to the boundary of Δ. We reduce this analysis to the two dimensional case by projecting these discs on the plane $\mathbb{Y}\mathbb{Z}$. Consider the projection

Fig. 6 The lines $L^1_{\xi,\mu}$, $L^2_{\xi,\mu}$ and the projections in the plane $\mathbb{YZ}$ of the cube Δ and the uu-cones at $P_{\xi,\mu}$, $Q_{\xi,\mu}$

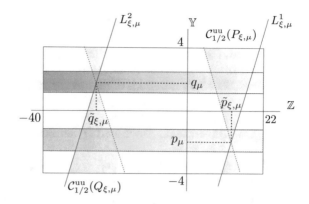

$$\Pi_1 : \mathbb{R}^3 \to \mathbb{R}^2, \quad \Pi_1(x, y, z) \stackrel{\text{def}}{=} (y, z).$$

Recalling the formulae for the stable manifolds $W^s(P_{\xi,\mu})$ and $W^s(Q_{\xi,\mu})$ in (3.3), we get $\Pi_1(W^s(P_{\xi,\mu})) = (p_\mu, \tilde{p}_{\xi,\mu})$ and $\Pi_1(W^s(Q_{\xi,\mu})) = (q_\mu, \tilde{q}_{\xi,\mu})$.

Consider the auxiliary straight lines in the plane $\mathbb{YZ}$ through $(p_\mu, \tilde{p}_{\xi,\mu})$ and $(q_\mu, \tilde{q}_{\xi,\mu})$:

$$L^1_{\xi,\mu} \stackrel{\text{def}}{=} \left\{ (y, z^1_{\xi,\mu}(y)) : z^1_{\xi,\mu}(y) = \frac{1}{2}(y - p_\mu) + \tilde{p}_{\xi,\mu}, \ y \in \mathbb{R} \right\},$$

$$L^2_{\xi,\mu} \stackrel{\text{def}}{=} \left\{ (y, z^2_{\xi,\mu}(y)) : z^2_{\xi,\mu}(y) = \frac{1}{2}(y - q_\mu) + \tilde{q}_{\xi,\mu}, \ y \in \mathbb{R} \right\}.$$

Note that $L^1_{\xi,\mu}$ and $L^2_{\xi,\mu}$ are contained in the boundary of $\Pi_1(\mathcal{C}^{uu}_{1/2}(P_{\xi,\mu}))$ and of $\Pi_1(\mathcal{C}^{uu}_{1/2}(Q_{\xi,\mu}))$, respectively. These conditions are depicted in Fig. 6. Thus **(BH4)** follows now from the next lemma.

Lemma 3.18 *For every* $(\xi, \mu) \in \mathcal{P}$ *it holds that*

$$L^1_{\xi,\mu} \cap (\Pi_1(\Delta) \cap \{z = 22\}) = \emptyset, \quad L^2_{\xi,\mu} \cap (\Pi_1(\Delta) \cap \{z = -40\}) = \emptyset.$$

Proof. To prove the lemma it is enough to check that

$$z^1_{\xi,\mu}(4) < 22 \quad \text{and} \quad z^2_{\xi,\mu}(-4) > -40, \quad \text{for every } (\xi, \mu) \in \mathcal{P}.$$

The choice of parameters (ξ, μ) and the estimates of $p_\mu, q_\mu, \tilde{p}_{\xi,\mu}, \tilde{q}_{\xi,\mu}$ in (3.2), lead directly to these inequalities. $\qquad\square$

3.6 Position of Images of uu-Discs

We now study the relative positions of the images of uu-discs contained in Δ in Condition (**BH5**). We see that this condition follows from the one-dimensional dynamics on the unstable center manifolds of the saddles $P_{\xi,\mu}$ and $Q_{\xi,\mu}$, recall (3.4).

3.6.1 One-Dimensional Associated Dynamics

Recall that $P_{\xi,\mu} = (p_\mu, p_\mu, \tilde{p}_{\xi,\mu})$ and $Q_\mu = (q_\mu, q_\mu, \tilde{q}_{\xi,\mu})$ and that the restriction of $G_{\xi,\mu}$ to the one-dimensional center unstable manifolds $W^{\mathrm{cu}}(P_{\xi,\mu})$, $W^{\mathrm{cu}}(Q_{\xi,\mu})$ in (3.4) is just an affine multiplication by $\xi > 1$, see Remark 3.3. Denote by $\phi^r_{\xi,\mu}$ the restriction map $G_{\xi,\mu}|_{W^{\mathrm{cu}}(R_{\xi,\mu}) \cap \Delta}$, $r = p, q$ and $R = P, Q$, that is given by

$$\phi^r_{\xi,\mu} : [-40, 22] \to \mathbb{R}, \quad \phi^r_{\xi,\mu}(z) \stackrel{\mathrm{def}}{=} \xi z + r_\mu = \xi z + (1 - \xi)\tilde{r}_{\xi,\mu}, \quad r = p, q,$$

where we use the relation $r_\mu = (1 - \xi)\tilde{r}_{\xi,\mu}$. For $r = p, q$, consider the interval $\mathrm{I}^r_{\xi,\mu} \stackrel{\mathrm{def}}{=} [\alpha^r_{\xi,\mu}, \beta^r_{\xi,\mu}]$, where

$$\alpha^r_{\xi,\mu} \stackrel{\mathrm{def}}{=} \xi^{-1}(-40 - (1 - \xi)\tilde{r}_{\mu,\xi}), \quad \beta^r_{\xi,\mu} \stackrel{\mathrm{def}}{=} \xi^{-1}(22 - (1 - \xi)\tilde{r}_{\xi,\mu}).$$

Note that $\phi^r_{\xi,\mu}(\mathrm{I}^r_{\xi,\mu}) = [-40, 22]$ and $\phi^r_{\xi,\mu}(\tilde{r}_{\xi,\mu}) = \tilde{r}_{\xi,\mu} \in \mathrm{I}^r_{\xi,\mu}$.

Lemma 3.19 *Given a uu-disc L contained in Δ let $L_{\mathcal{C}_{\xi,\mu}} \stackrel{\mathrm{def}}{=} L \cap \mathcal{C}_{\xi,\mu}$, with $\mathcal{C} = \mathcal{A}, \mathcal{B}$. Then $G_{\xi,\mu}(L_{\mathcal{C}_{\xi,\mu}})$ satisfies (**BH5**).*

Proof. We first show item (1) of (**BH5**). Items (2), (3), and (4) are obtained similarly and their proofs will be omitted.

From (2.2) and (3.3), the local stable manifolds of $P_{\xi,\mu}$ and $Q_{\xi,\mu}$ are given by

(3.14)
$$W^s_{\mathrm{loc}}(P_{\xi,\mu}) = \left\{ (t + p_{\xi,\mu}, p_{\xi,\mu}, \tilde{p}_{\xi,\mu}) : -4 - p_{\xi,\mu} \leq t \leq 4 - p_{\xi,\mu} \right\},$$
$$W^s_{\mathrm{loc}}(Q_{\xi,\mu}) = \left\{ (t + q_{\xi,\mu}, q_{\xi,\mu}, \tilde{q}_{\xi,\mu}) : -4 - q_{\xi,\mu} \leq t \leq 4 - q_{\xi,\mu} \right\}.$$

Given a uu-disc $L \subset \Delta$ consider the intersections

$$X^L_\mu \stackrel{\mathrm{def}}{=} L \cap \left(\Delta \cap \{y = p_\mu\} \right) = L_{\mathcal{A}_{\xi,\mu}} \cap \left(\Delta \cap \{y = p_\mu\} \right) = (x_\mu, p_\mu, z_\mu),$$
$$\bar{X}^L_\mu \stackrel{\mathrm{def}}{=} L \cap \left(\Delta \cap \{y = q_\mu\} \right) = L_{\mathcal{B}_{\xi,\mu}} \cap \left(\Delta \cap \{y = q_\mu\} \right) = (\bar{x}_\mu, q_\mu, \bar{z}_\mu).$$

Remark 3.20 Recall the definitions of the right and left classes of uu-discs $\mathcal{U}^r_W$ and $\mathcal{U}^\ell_W$, respectively, in Remark 2.2. Using (3.14) we have the following:

- $L \in \mathcal{U}^\ell_{W^s_{\mathrm{loc}}(P_{\xi,\mu})}$ iff $z_\mu < \tilde{p}_{\xi,\mu}$ and $L \in \mathcal{U}^r_{W^s_{\mathrm{loc}}(P_{\xi,\mu})}$ iff $z_\mu > \tilde{p}_{\xi,\mu}$,
- $L \in \mathcal{U}^\ell_{W^s_{\mathrm{loc}}(Q_{\xi,\mu})}$ iff $\bar{z}_\mu < \tilde{q}_{\xi,\mu}$ and $L \in \mathcal{U}^r_{W^s_{\mathrm{loc}}(Q_{\xi,\mu})}$ iff $\bar{z}_\mu > \tilde{q}_{\xi,\mu}$.

To prove (1) in (**BH5**), take any $L \in \mathcal{U}^r_{W^s_{\text{loc}}(P_{\xi,\mu})}$. We will see that $G_{\xi,\mu}(L_{\mathcal{A}_{\xi,\mu}}) \in \mathcal{U}^r_{W^s_{\text{loc}}(P_{\xi,\mu})}$. By Remark 3.20, the point $X^L_\mu = (x_\mu, p_\mu, z_\mu)$ satisfies $z_\mu > \tilde{p}_{\xi,\mu}$. Note that

$$G_{\xi,\mu}(X^L_\mu) = \big(p_\mu, p_\mu, \phi^p_{\xi,\mu}(z_\mu)\big) = \big(p_\mu, p_\mu, \xi z_\mu + (1-\xi)\tilde{p}_{\xi,\mu}\big).$$

Since $z_\mu > \tilde{p}_{\xi,\mu}$ it follows that $\phi^p_{\xi,\mu}(z_\mu) > \tilde{p}_{\xi,\mu}$. Remark 3.20 now implies that $G_{\xi,\mu}(L_{\mathcal{A}_{\xi,\mu}}) \in \mathcal{U}^r_{W^s_{\text{loc}}(P_{\xi,\mu})}$.

Since items (5) and (6) of (**BH5**) are analogous we just prove item (5). We just need to check that if $L \in \mathcal{U}^r_{W^s_{\text{loc}}(P_{\xi,\mu})}$ or $L \cap W^s_{\text{loc}}(P_{\xi,\mu}) \neq \emptyset$ then $G_{\xi,\mu}(L_{\mathcal{B}_{\xi,\mu}}) \in \mathcal{U}^r_{W^s_{\text{loc}}(P_{\xi,\mu})}$.

Remark 3.21 Consider the projection $\Pi_1(x, y, z) = (y, z)$ and note that

$$\Pi_1\big(L \cap \{y \geq p_\mu\}\big) \subset \Gamma_{\xi,\mu} \overset{\text{def}}{=} \big\{(y, z) : z \geq z^*_{\xi,\mu}(y)\big\},$$

see Fig. 7. Moreover, $\Pi_1(L_{\mathcal{B}_{\xi,\mu}}) \subset \Gamma_{\xi,\mu} \cap \Pi_1(\mathcal{B}_{\xi,\mu})$.

Note that the worst case to prove (5) in (**BH5**) occurs when L is contained in the plane $\mathbb{YZ}$ and equal to the straight line $L^*_{\xi,\mu}$ in the plane $\mathbb{YZ}$ through $(p_\mu, \tilde{p}_{\xi,\mu})$ given by

$$(3.15) \qquad L^*_{\xi,\mu} \overset{\text{def}}{=} \Big\{(y, z^*_{\xi,\mu}(y)) : z^*_{\xi,\mu}(y) = -\frac{1}{2}(y - p_\mu) + \tilde{p}_{\xi,\mu}, \ y \in \mathbb{R}\Big\}.$$

Consider the segment of $L^*_{\xi,\mu}$ given by (see Fig. 7)

$$\gamma_{\xi,\mu} \overset{\text{def}}{=} \big\{(y, z^*_{\xi,\mu}(y)) : y \in \mathbb{J}_\mu\big\} \subset L^*_{\xi,\mu} \cap \Pi_1(\mathcal{B}_{\xi,\mu})$$

and the point $\tilde{z}_{\xi,\mu}$ defined by

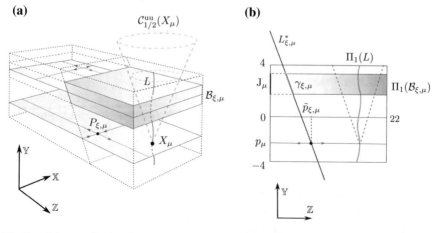

Fig. 7 **a** L is a uu-disc in $\mathcal{U}^r_{W^s_{\text{loc}}(P_{\xi,\mu})}$. **b** Projection of L in the plane $\mathbb{YZ}$

$$(3.16) \qquad\qquad g_{\xi,\mu}(\gamma_{\xi,\mu}) \cap \{y = p_\mu\} = \{(p_\mu, \tilde{z}_{\xi,\mu})\},$$

where the endomorphism $g_{\xi,\mu}$ obtained by projecting $G_{\xi,\mu}$ into the plane $\mathbb{YZ}$ defined in (3.5). By Remark 3.20 to get $G_{\xi,\mu}(L_{\mathcal{B}_{\xi,\mu}}) \in \mathcal{U}^r_{W^s_{loc}(P_{\xi,\mu})}$ it is sufficient to show that $\tilde{z}_{\xi,\mu} > \tilde{p}_{\xi,\mu}$.

Claim 3.22 *It holds* $\tilde{z}_{\xi,\mu} > \tilde{p}_{\xi,\mu}$ *for every* $(\xi, \mu) \in \mathcal{P}$.

Proof. The intersection (3.16) is defined by the conditions

$$(p_\mu, \tilde{z}_{\xi,\mu}) = (y^2 + \mu, \xi\, z^*_{\xi,\mu}(y) + y), \quad y > 0.$$

Recalling the definition of $z^*_{\xi,\mu}(y)$ in (3.15) we get

$$\tilde{z}_{\xi,\mu} = \xi\, \tilde{z}^*_{\xi,\mu}(\sqrt{p_\mu - \mu}) + \sqrt{p_\mu - \mu} = \frac{\xi}{2}\, p_\mu + \left(1 - \frac{\xi}{2}\right)\sqrt{p_\mu - \mu} + \xi\, \tilde{p}_{\xi,\mu}.$$

Hence

$$\tilde{z}_{\xi,\mu} - \tilde{p}_{\xi,\mu} = \frac{\xi}{2}\, p_\mu + \left(1 - \frac{\xi}{2}\right)\sqrt{p_\mu - \mu} + (\xi - 1)\, \tilde{p}_{\xi,\mu}.$$

The estimates in (3.2) and the choice of $(\xi, \mu) \in \mathcal{P}$ imply that

$$\frac{\xi}{2}\, p_\mu > -1.6065, \quad \left(1 - \frac{\xi}{2}\right)\sqrt{p_\mu - \mu} > 1.014, \quad (\xi - 1)\, \tilde{p}_{\xi,\mu} > 2.34.$$

These inequalities imply that $\tilde{z}_{\xi,\mu} - \tilde{p}_{\xi,\mu} > 0$, proving the claim. $\qquad\square$

The proof of the lemma is now complete. $\qquad\square$

3.7 Position of Images of uu-Discs in Between

Condition (**BH6**) is given by Lemma 3.23 below. First, recall the definition of the family of discs in between $\mathcal{U}^b \overset{\text{def}}{=} \mathcal{U}^\ell_{W^s_{loc}(P)} \cap \mathcal{U}^r_{W^s_{loc}(Q)}$.

Lemma 3.23 *Consider any* $L \in \mathcal{U}^b$. *Then either* $G_{\mu,\xi}(L_{\mathcal{A}_{\xi,\mu}})$ *or* $G_{\mu,\xi}(L_{\mathcal{B}_{\xi,\mu}})$ *contains a* uu-*disc in* $\mathcal{U}^b$.

Proof. Consider $L \in \mathcal{U}^b$. By item (2) in (**BH5**), if $G_{\xi,\mu}(L_{\mathcal{A}_{\xi,\mu}}) \in \mathcal{U}^r_{W^s_{loc}(Q_{\xi,\mu})}$ then $G_{\xi,\mu}(L_{\mathcal{A}_{\xi,\mu}}) \in \mathcal{U}^b$ and we are done. Similarly, by item (3) in (**BH5**), if $G_{\xi,\mu}(L_{\mathcal{B}_{\xi,\mu}}) \in \mathcal{U}^\ell_{W^s_{loc}(P_{\xi,\mu})}$ then $G_{\xi,\mu}(L_{\mathcal{B}_{\xi,\mu}}) \in \mathcal{U}^b$ and we are done. Thus in what follows we argue by contradiction assuming that:

(a) $G_{\xi,\mu}(L_{\mathcal{A}_{\xi,\mu}}) \in \mathcal{U}^\ell_{W^s_{loc}(Q_{\xi,\mu})}$ or intersects $W^s_{loc}(Q_{\xi,\mu})$ and
(b) $G_{\xi,\mu}(L_{\mathcal{B}_{\xi,\mu}}) \in \mathcal{U}^r_{W^s_{loc}(P_{\xi,\mu})}$ or intersects $W^s_{loc}(P_{\xi,\mu})$.

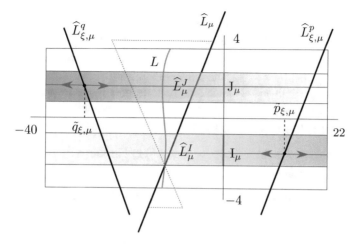

Fig. 8 The segments $\widehat{L}_\mu^a$ and $\widehat{L}_\mu^b$ and the lines $\widehat{L}_{\xi,\mu}^p$ and $\widehat{L}_{\xi,\mu}^q$

To prove the lemma we need some auxiliary constructions. Consider the point $Y_\mu^L = (x_\mu, a_\mu, z_\mu) \stackrel{\text{def}}{=} L \cap \{y = a_\mu\}$, where a_μ is defined in (3.6). In the plane $\mathbb{YZ}$, take the auxiliary straight line $\widehat{L}_\mu$ through (a_μ, z_μ) given by (see Fig. 8)

$$\widehat{L}_\mu \stackrel{\text{def}}{=} \left\{ (y, z_\mu^a(y)) \colon z_\mu^a(y) = \frac{1}{2}(y - a_\mu) + z_\mu, \ y \in \mathbb{R} \right\}.$$

Observe that $\widehat{L}_\mu \subset \partial \Pi_1 \big(\mathcal{C}_{1/2}^{uu}(Y_\mu^L) \big)$. Consider the sub segments of $\widehat{L}_\mu$ given by (see Fig. 8)

$$\widehat{L}_\mu^I \stackrel{\text{def}}{=} \left\{ (y, z_\mu^a(y)) : y \in \mathrm{I}_\mu \right\} \quad \text{and} \quad \widehat{L}_\mu^J \stackrel{\text{def}}{=} \left\{ (y, z_\mu^a(y)) : y \in \mathrm{J}_\mu \right\}.$$

Recall that $P_{\xi,\mu} = (p_\mu, p_\mu, \tilde{p}_{\xi,\mu})$ and $Q_\mu = (q_\mu, q_\mu, \tilde{q}_{\xi,\mu})$ and consider the straight lines $\widehat{L}_{\xi,\mu}^p$ and $\widehat{L}_{\xi,\mu}^q$ contained in $\partial \Pi_1 \big(\mathcal{C}_{1/2}^{uu}(P_{\xi,\mu}) \big)$ and $\partial \Pi_1 \big(\mathcal{C}_{1/2}^{uu}(Q_{\xi,\mu}) \big)$, respectively, given by

$$\widehat{L}_{\xi,\mu}^p \stackrel{\text{def}}{=} \left\{ (y, z^p(y)) \colon z^p(y) = \frac{1}{2}(y - p_\mu) + \tilde{p}_{\xi,\mu}, \ y \in \mathbb{R} \right\},$$

$$\widehat{L}_{\xi,\mu}^q \stackrel{\text{def}}{=} \left\{ (y, z^q(y)) \colon z^q(y) = -\frac{1}{2}(y - q_\mu) + \tilde{q}_{\xi,\mu}, \ y \in \mathbb{R} \right\}.$$

Finally, consider the following subsets of Δ

$$\Sigma_{\xi,\mu}^p \stackrel{\text{def}}{=} \left([-4, 4] \times \widehat{L}_{\xi,\mu}^p \right) \cap \Delta, \quad \Sigma_{\xi,\mu}^q \stackrel{\text{def}}{=} \left([-4, 4] \times \widehat{L}_{\xi,\mu}^q \right) \cap \Delta.$$

Observe that $\Delta \setminus \Sigma^r_{\xi,\mu}$, $r = p, q$, consists of two connected components. We let $\Delta^Q_{\xi,\mu,\mathrm{right}}$ the connected component of $\Delta \setminus \Sigma^q_{\xi,\mu}$ containing $P_{\xi,\mu}$ and by $\Delta^Q_{\xi,\mu,\mathrm{left}}$ the other component. Similarly, we let $\Delta^P_{\xi,\mu,\mathrm{left}}$ the connected component of $\Delta \setminus \Sigma^p_{\xi,\mu}$ containing $Q_{\xi,\mu}$ and by $\Delta^P_{\xi,\mu,\mathrm{right}}$ the other component.

After these preliminary constructions, we are now ready to prove the lemma. Note that by Remark 3.15 "$G_{\xi,\mu}\big([-4, 4] \times \widehat{L}^I_\mu\big)$ is at the left of $G_{\xi,\mu}(L_{\mathcal{A}_{\xi,\mu}})$" and "$G_{\xi,\mu}\big([-4, 4] \times \widehat{L}^J_\mu\big)$ is at the right of $G_{\xi,\mu}(L_{\mathcal{B}_{\xi,\mu}})$". Therefore

- condition (a) implies that $G_{\xi,\mu}\big([-4, 4] \times \widehat{L}^I_\mu\big) \subset \mathrm{closure}(\Delta^Q_{\xi,\mu,\mathrm{left}})$,
- condition (b) implies that $G_{\xi,\mu}\big([-4, 4] \times \widehat{L}^J_\mu\big) \subset \mathrm{closure}(\Delta^P_{\xi,\mu,\mathrm{right}})$.

We now see that these two conditions cannot hold simultaneously. Consider $\omega^I_{\xi,\mu}, \omega^J_{\xi,\mu} \in \mathbb{Z}$ given by

$$g_{\xi,\mu}\big(\widehat{L}^I_\mu\big) \cap \{y = q_\mu\} = (q_\mu, \omega^I_{\xi,\mu}) \quad \text{and} \quad g_{\xi,\mu}\big(\widehat{L}^J_\mu\big) \cap \{y = p_\mu\} = (p_\mu, \omega^J_{\xi,\mu}).$$

Arguing as in Claim 3.22, we get

$$\omega^I_{\xi,\mu} = \frac{\xi}{2}(q_\mu - a_\mu) + \xi z_\mu + q_\mu \quad \text{and} \quad \omega^J_{\xi,\mu} = \xi z_\mu + a_\mu.$$

On the other hand, our assumptions and Remark 3.20 imply that $\omega^I_{\xi,\mu} \leq \tilde{q}_{\xi,\mu}$ and $\omega^J_{\xi,\mu} \geq \tilde{p}_{\xi,\mu}$. Thus

$$|\tilde{q}_{\xi,\mu} - \tilde{p}_{\xi,\mu}| \leq |\omega^I_{\xi,\mu} - \omega^J_{\xi,\mu}| \leq \left(\frac{\xi}{2} + 1\right)|q_\mu - a_\mu| \leq 12.16,$$

where the last inequality follows from the estimates in (3.2) and (3.7). Since, also by (3.2), we have that $|\tilde{q}_{\xi,\mu} - \tilde{p}_{\xi,\mu}| \in [31.4, 35.6]$ we derive a contradiction, completing the proof of the lemma. $\qquad \Box$

References

1. A. Avila, S. Crovisier, A. Wilkinson, Diffeomorphisms with positive metric entropy. Publ. Math. Inst. Hautes Études Sci. **124**, 319–347 (2016)
2. P.G. Barrientos, Y. Ki, A. Raibekas, Symbolic blender-horseshoes and applications. Nonlinearity **27**, 2805–2839 (2014)
3. P.G. Barrientos, A. Raibekas, Robustly non-hyperbolic transitive symplectic dynamics. Amer. Inst. Math. Sci. **38**(12), 5993–6013 (2018)
4. J. Bochi, C. Bonatti, L.J. Díaz, Robust criterion for the existence of nonhyperbolic ergodic measures. Comm. Math. Phys. **344**, 751–795 (2016)
5. C. Bonatti, S. Crovisier, L.J. Díaz, A. Wilkinson, What is. . . a blender? Notices Amer. Math. Soc. **63**, 1175–1178 (2016)
6. C. Bonatti, L.J. Díaz, Persistent nonhyperbolic transitive diffeomorphisms. Ann. of Math. **143**(2), 357–396 (1996)

7. C. Bonatti, L.J. Díaz, Robust heterodimensional cycles and C^1-generic dynamics. J. Inst. Math. Jussieu **7**, 469–525 (2008)
8. C. Bonatti, L.J. Díaz, Abundance of C^1-robust homoclinic tangencies. Trans. Amer. Math. Soc. **364**, 5111–5148 (2012)
9. C. Bonatti, L.J. Díaz, M. Viana, Discontinuity of the Hausdorff dimension of hyperbolic sets, Comptes Rendus Acad. Sci. Paris Sér. I Math. 320, 713–718 (1995)
10. C. Bonatti, L.J. Díaz, M. Viana, *Dynamics beyond uniform hyperbolicity: A global geometric and probabilistic perspective*, vol. 102, (Springer Science & Business Media, 2006)
11. L.J. Díaz, Robust nonhyperbolic dynamics and heterodimensional cycles. Ergod. Theory Dyn. Syst. **15**, 291–315 (1995)
12. L.J. Díaz, K. Gelfert, M. Gröger, T. Jäger, Hyperbolic graphs: critical regularity and box dimension. Trans. Amer. Math. Soc. (2017). http://www.ams.org/journals/tran/0000-000-00/S0002-9947-2019-07454-8/
13. L.J. Díaz, S. Kiriki, K. Shinohara, Blenders in centre unstable Hénon-like families: with an application to heterodimensional bifurcations. Nonlinearity **27**, 353–378 (2014)
14. L.J. Díaz, A. Nogueira, E.R. Pujals, Heterodimensional tangencies. Nonlinearity **19**, 2543–2566 (2006)
15. S. Hittmeyer, B. Krauskopf, H. Osinga, K. Shinohara, Existence of blenders in a Hénon-like family: geometric insights from invariant manifold computations. Nonlinearity **31**(10), R239–R267 (2018)
16. S. Kiriki, T. Soma, C^2-robust heterodimensional tangencies. Nonlinearity **25**, 3277–3299 (2012)
17. L. Mora, M. Viana, Abundance of strange attractors. Acta Math. **171**, 1–71 (1993)
18. C.G. Moreira, There are no C^1-stable intersections of regular Cantor sets. Acta Math. **206**, 311–323 (2011)
19. C.G.T.D.A. Moreira, W.L.L.R. Silva, On the geometry of horseshoes in higher dimensions (2012). arXiv:1210.2623
20. M. Nassiri, E.R. Pujals, Robust transitivity in Hamiltonian dynamics. Ann. Sci. Éc. Norm. Supér. **4**(45), 191–239 (2012)
21. S.E. Newhouse, Nondensity of axiom A (a) on $\mathbb{S}^2$. Glob. Anal. **1**, 191–202 (1970)
22. J. Palis, F. Takens, *Hyperbolicity and Sensitive Chaotic Dynamics at Homoclinic Bifurcations: Fractal Dimensions and Infinitely Many Attractors in Dynamics* (Cambridge Studies in Advanced Mathematics, Cambridge University Press, 1995)
23. J. Palis, M. Viana, High dimension diffeomorphisms displaying infinitely many periodic attractors. Ann. Math. 207–250 (1994)
24. S.A. Pérez, C^r-stabilisation of non-transverse heterodimensional cycles, Ph.D. Thesis, Pontificia Universidade Catolica do Rio de Janeiro, Brazil (2016)
25. F. Rodriguez Hertz, M.A. Rodriguez Hertz, A. Tahzibi, R. Ures, New criteria for ergodicity and nonuniform hyperbolicity, Duke Math. J. **160**, 599–629 (2011)
26. R. Ures, Abundance of hyperbolicity in the C^1 topology. Ann. Sci. École Norm. Sup. **28**(4), 747–760 (1995)

On Slow Growth and Entropy-Type Invariants

Edson de Faria, Peter Hazard and Charles Tresser

Abstract We discuss a generalization of topological entropy in which the usual exponential growth-rate function is replaced by an arbitrary gauge function. This generalized topological entropy had previously been described by Galatolo in 2003—up to a choice of notation in the defining formulas—which in turn is essentially the same as that described by Zhao and Pesin in 2015 (that involves a re-parameterization of time). One of the main motivations for studying this new set of invariants comes from the need to distinguish maps with zero (standard) topological entropy. In such cases, if the dynamics is not equicontinuous, then there exists at least one gauge for which the corresponding generalized entropy is positive. After illustrating this simple qualitative criterion, we perform a more quantitative study of the growth of orbits in some low-dimensional examples of zero-entropy maps. Our examples include period-doubling maps in dimension one, and maps of the annulus built from circle homeomorphisms having an exceptional minimal set.

Keywords Growth · Topological entropy · Slow entropy · Equicontinuity · Topological complexity

2010 Mathematics Subject Classification Primary: 37B40 · Secondary: 26A12 · 37E05 · 37E20

This work has been partially supported by "Projeto Temático Dinâmica em Baixas Dimensões" FAPESP Grant no.2016/25053-8, FAPESP Grant no. 2015/17909-7, Projeto PVE CNPq 401020/2014-2, and CAPES Grant CSF-PVE-S - 88887.117899/2016-00.

E. de Faria (✉) · P. Hazard
Instituto de Matemática e Estatística, USP, São Paulo, SP, Brazil
e-mail: edson@ime.usp.br

P. Hazard
e-mail: pete@ime.usp.br

C. Tresser
301 West 118th Street #6B, New York, NY 10026, USA
e-mail: tresser.charles@gmail.com

M. J. Pacifico and P. Guarino (eds.), *New Trends in One-Dimensional Dynamics*, Springer Proceedings in Mathematics & Statistics 285,
https://doi.org/10.1007/978-3-030-16833-9_9

1   Introduction

The search for topological invariants that would help classify zero entropy maps goes back at least to the mid-1970's with the work of Goodman [15] who, inspired by the work of Kushnirenko [19], introduced the concept of *topological sequence entropy* (see also [20]). There are several such invariants, but in this paper we will only consider the *generalized entropy* proposed by Galatolo [14], and later by Zhao and Pesin [28] under the name of *scaled entropy* (in a slightly different formulation that involves a re-parameterization of time).

Recall that, for metric spaces, topological entropy can be computed in at least three different ways—either as the exponential growth rate in n of the maximal cardinality of (n, ϵ)-separated sets, or as the exponential growth rate in n of the minimum cardinality of (n, ϵ)-spanning sets (in the limit as $\epsilon \to 0$) or, in a yet more subtle way, via coverings. Galatolo observed in [14] that if one replaces the exponential gauge by other gauges, the resulting growths associated to these three quantities (still exist and) are the same. This equality allows him to conclude that, just like topological entropy, *these generalized entropies are topological invariants*, a conclusion similarly found in [28].

The generalized entropies studied in the present paper appear in various guises in the literature, under a variety of different names. Thus, Zhao and Pesin [28] use the expression 'scaled entropy' for the general case. We will often use the expression 'slow entropy' when the growth is a priori known (or at least expected) to be sub-exponential—in agreement with the terminology employed by Katok and Thouvenot in [18]. Likewise, the expression 'fast entropy' can be used when the growth is known (or at least expected) to be super-exponential. As it turns out, fast entropy happens generically in low smoothness [7, 8]. In this paper, however, we are only interested in slow entropy, since all examples we treat here are maps with zero topological entropy.

Measuring the precise growth rate of spanning or separated sets of a given dynamical system can be a formidable task. This is one of the main reasons why we focus on a few simple, computable examples. A much easier task is to determine whether there exists *some* gauge γ (a growth-rate function) for which the corresponding γ-entropy is positive. Here, a simple criterion is available: a topological dynamical system has *zero growth of all orders* (i.e., has zero γ-entropy for every gauge γ) if and only if it is *equicontinuous*. This was observed by Galatolo in [14] as an immediate consequence of a result on *topological complexity* proved by Blanchard, Host and Maass in [3]. Note that every isometry of a metric space is equicontinuous. Examples of dynamical systems that are *not* equicontinuous include, among many others: (i) smooth endomorphisms with an unstable periodic point (either a source or a saddle); in particular interval maps with a repelling periodic orbit, embeddings of the two-disk possessing a saddle such as Hénon maps, or their generalizations to dimensions greater than two; (ii) circle homeomorphisms with Bohl-Denjoy invariant Cantor sets [4, 11]; (iii) continuous twist-maps whose rotation set is an arc with positive length.

Going beyond these simple qualitative examples, our goal here is to be a bit more quantitative and provide actual estimates on the slow-entropy of the following examples:

- Period-doubling interval maps, along the cascade of period-doubling bifurcations, especially the accumulation point at the boundary of zero topological entropy;
- Certain bi-Hölder homeomorphisms of the 2-dimensional annulus (or 2-torus) with zero topological entropy and without periodic points.

The study of the *former* set of examples uses a generalization, for sub-exponential growth, of a famous result due to Misiurewicz and Szlenk [22], expressing the topological entropy of a piecewise-monotone interval map as the exponential growth rate of the lap numbers of the iterates of the map. For the statement of this more general version, see Sect. 4.1. We consider smooth enough unimodal maps at the boundary of zero topological entropy satisfying the functional fixed-point equation introduced in [6, 13, 25]. (We only need the weakest of proofs of existence of the fixed point).

Concerning the *latter* family of examples, our original motivation for studying such maps was an attempt at understanding what happens in situations lying 'between' Katok's Theorem [16], stating that, if a $C^{1+\alpha}$ *diffeomorphism* of a compact surface does not possess a horseshoe then it has zero topological entropy (so for maps with sufficient regularity, complicated dynamics requires the existence of a horseshoe), and Rees's example [23] of a *minimal homeomorphism* of the 2-torus with positive topological entropy (thus showing that for maps with little or no regularity, complicated dynamical behaviour may occur without a horseshoe).

For both sets of examples above, our results are merely sketched in the present paper, but full details will be given in [9].

2 Slow Entropy

We start with an informal description of slow entropy and then proceed to a formal definition. Suppose $f : X \to X$ is a continuous self-map of a compact metric space X. It is well-known since Bowen [5] and independently Dinaburg [12] that the *topological entropy* of f is a non-negative extended real number that measures the *exponential growth rate* in n of the maximal size of a set of ϵ-distinguishable[1] orbit segments $x, f(x), \ldots, f^{n-1}(x)$ of length n, in the limit as $\epsilon \to 0$. The original definition, due to Adler, Konheim and McAndrew [1], is given in terms of (open) coverings (see below) and is valid for continuous self-maps of arbitrary compact Hausdorff spaces. However, the Bowen-Dinaburg definition is oftentimes more practical, and in fact applicable also to uniformly continuous maps of non-compact metric spaces.

---

[1]We say that two orbit segments $x, f(x), \ldots, f^{n-1}(x)$ and $y, f(y), \ldots, f^{n-1}(y)$ are *ϵ-distinguishable* if there exists $0 \le j \le n-1$ such that $d(f^j(x), f^j(y)) > \epsilon$, where d is the metric of X.

The concept of *generalized entropy*, or γ-*entropy* is obtained simply by replacing the notion of 'exponential growth rate' in the above definition by 'γ-growth rate', where $\gamma \colon \mathbb{R}^+ \to \mathbb{R}^+$ is a non-decreasing monotone function such that $\gamma(t) \to +\infty$ as $t \to +\infty$. Thus, ordinary topological entropy corresponds to the case when $\gamma(t) = e^t$. We refer to γ as the (entropy) *gauge*, or (entropy) *growth function*. Since here we are only interested in gauges γ that grow less than the exponential, we sometimes use the expression *slow entropy* when referring generically to such γ-entropies.

Let us now present the formal definitions. Most of what follows is standard, and the details on what *is* standard can be found in [27]. What is *not* standard is kept here to a bare minimum—much more will be given in [9].

2.1 Definition Via Covers

Let X be a compact Hausdorff topological space. When we refer to a *cover* of X, we mean a collection $\mathcal{A} \subseteq \mathcal{P}(X)$ of (not necessarily open) subsets of X such that $X = \bigcup_{A \in \mathcal{A}} A$. Given two covers $\mathcal{A}$ and $\mathcal{B}$ of X we denote their *join* by $\mathcal{A} \vee \mathcal{B}$, i.e., $\mathcal{A} \vee \mathcal{B} = \{A \cap B \ : \ A \in \mathcal{A}, \ B \in \mathcal{B}\}$.

Let $f \colon X \to X$ be a continuous map, and let $\mathcal{A}$ be a cover. We define $f^{-1}(\mathcal{A})$, as usual, to be the cover of X consisting of sets $f^{-1}(A)$, $A \in \mathcal{A}$. Then $f^{-1}(\mathcal{A} \vee \mathcal{B}) = f^{-1}(\mathcal{A}) \vee f^{-1}(\mathcal{B})$.

Given a cover $\mathcal{A}$ of X, define $N(\mathcal{A})$ to be the minimal cardinality of all finite subcovers of $\mathcal{A}$. If there are no such subcovers we set $N(\mathcal{A}) = \infty$. This minimal cardinality satisfies several simple but important properties, such as: (i) $N(\mathcal{A} \vee \mathcal{B}) \le N(\mathcal{A})N(\mathcal{B})$ (sub-multiplicativity); (ii) $N(\mathcal{A}) \le N(\mathcal{B})$ whenever $\mathcal{B}$ is a refinement[2] of $\mathcal{A}$ (monotonicity); (iii) $N(f^{-1}\mathcal{A}) \le N(\mathcal{A})$, and equality holds whenever f is surjective (invariance under surjective maps).

For each $n \ge 1$, define

$$\mathcal{A}_f^n = \bigvee_{k=0}^{n-1} f^{-k}\mathcal{A} \ . \tag{2.1}$$

Given a cover $\mathcal{A}$ of X, define the γ-topological entropy of f with respect to $\mathcal{A}$ by

$$h^\gamma(f, \mathcal{A}) = \limsup_{n \to \infty} \frac{\log N(\mathcal{A}_f^n)}{\log \gamma(n)} \ . \tag{2.2}$$

Note that $h^\gamma(f, \mathcal{A}) \le h^\gamma(f, \mathcal{B})$ whenever $\mathcal{B}$ is a refinement of $\mathcal{A}$.

Remark 2.1 When $\gamma(t) = \exp(t)$, the above lim sup can be replaced by lim. This follows by observing that the sequence $a_n = \log N\left(\bigvee_{i=0}^{n-1} f^{-i}\mathcal{A}\right)$ is subadditive, and then invoking the standard subadditivity lemma.

---

[2]A cover $\mathcal{A}$ is said to be a *refinement* of the cover $\mathcal{B}$ if every member of $\mathcal{A}$ is a subset of a member of $\mathcal{B}$.

Definition 2.1 The γ-*topological entropy of* $f : X \to X$ *is given by*

$$h^\gamma(f) = \sup h^\gamma(f, \mathcal{A}) , \qquad (2.3)$$

where the supremum is taken over all *open* covers $\mathcal{A}$ of X.

Of course, the supremum always exists, though it may be infinite.

The following proposition is a slight extension of [27, Theorem 7.2].

Proposition 2.1 *If* γ *is any given gauge, then* γ-*entropy is a topological-conjugacy invariant. More precisely, let* $f : X \to X$ *and* $g : Y \to Y$ *be continuous, and let* $\phi : X \to Y$ *be a continuous surjection such that* $\phi \circ f = g \circ \phi$. *Then* $h^\gamma(f) \geq h^\gamma(g)$. *In particular, if* ϕ *is a homeomorphism then* $h^\gamma(f) = h^\gamma(g)$

The following is a straightforward generalization of [1, Theorem 4].

Proposition 2.2 *Let* $f : X \to X$ *be a continuous map of a compact Hausdorff space. Then for each gauge* γ *we have the following facts.*

(1) If Λ *is a closed* f-*invariant subset of* X, *then* $h^\gamma(f|_\Lambda) \leq h^\gamma(f)$.
(2) If $X = \Lambda_1 \cup \Lambda_2 \cup \cdots \cup \Lambda_N$, *where each* Λ_j *is closed and* f-*invariant, then* $h^\gamma(f) = \max_{1 \leq j \leq N} h^\gamma(f|_{\Lambda_j})$

Everything so far holds true for an arbitrary gauge. However, another desirable property for our generalized entropy is that it should grow as we iterate the map (or at the very least not decrease). In order to establish such a property, we need to impose some condition on the gauge.

Definition 2.2 We say that the gauge γ is *very good* if for all $n \in \mathbb{N}$,

$$\Gamma(n) = \lim_{m \to \infty} \frac{\log \gamma(mn)}{\log \gamma(m)} \qquad (2.4)$$

exists and is finite. We say that γ is *eventually very good* if the above limit exists for all $n \in \mathbb{N}$ sufficiently large.

Proposition 2.3 *If* γ *is a very good gauge, then for each positive integer* n, *we have* $h^\gamma(f^n) = \Gamma(n)h^\gamma(f)$.

For a proof of Proposition 2.3, and more, see [9].

Examples. Besides the standard exponential gauge, there are plenty of very good gauges. Indeed, every gauge of the form $\gamma_{r,s,t,C}(n) = n^r \exp\{Cn^s (\log n)^t\}$, where $C > 0$ and r, s, t are non-negative real numbers (not simultaneously zero), is a very good gauge. If $\Gamma_{r,s,t,C}(n)$ denotes the limit in (2.4) when $\gamma = \gamma_{r,s,t,C}$, then an easy calculation shows that $\Gamma_{r,s,t,C}(n) = n^s$ when $s > 0$, and $\Gamma_{r,s,t,C}(n) = 1$ when $s = 0$. This family of gauges includes in particular all exponential gauges ($r = t = 0, s = 1$) and power-law gauges ($s = t = 0, r > 0$). All gauges appearing in the present paper are of this form.

2.2 Definitions Via Separated and Spanning Sets

Just as with ordinary topological entropy, γ-entropy can be defined à la Bowen and Dinaburg, in terms of *separated* or *spanning* sets.

Given a metric space (X, d) let $K \subset X$ be compact, and let $f : X \to X$ be a continuous map. A set $E \subset K$ is (n, ϵ)-*separated for* K *with respect to* f if for all pairs of distinct points $x, y \in E$, there exists an integer $0 \le i < n$ such that $d(f^i(x), f^i(y)) \ge \epsilon$. We will denote by $S_f(n, \epsilon, K)$ the maximal cardinality of an (n, ϵ)-separated set for K with respect to f. The set E is (n, ϵ)-*spanning for* K *with respect to* f if for each $x \in K$, there exists $y \in E$ such that $d(f^i(x), f^i(y)) < \epsilon$ for all $0 \le i < n$. We denote by $R_f(n, \epsilon, K)$ the minimal cardinality of an (n, ϵ)-spanning set for K with respect to f. When X is compact and $K = X$, we write $S_f(n, \epsilon)$ for $S_f(n, \epsilon, X)$ and $R_f(n, \epsilon)$ for $R_f(n, \epsilon, X)$, respectively. The following is classical [27, Sect. 7.2].

Proposition 2.4 *For each $\epsilon' > \epsilon > 0$, $n \in \mathbb{N}$ and compact K, we have*

(i) $S_f(n, \epsilon', K) \le S_f(n, \epsilon, K)$,
(ii) $R_f(n, \epsilon', K) \le R_f(n, \epsilon, K)$,
(iii) $R_f(n, \epsilon, K) \le S_f(n, \epsilon, K) \le R_f(n, \epsilon/2, K)$.

This implies that the following limits exist (though they may be infinite) and are equal, for every given gauge γ:

$$h_d^\gamma(f, K) = \lim_{\epsilon \to 0} \limsup_{n \to \infty} \frac{\log S_f(n, \epsilon, K)}{\log \gamma(n)} = \lim_{\epsilon \to 0} \limsup_{n \to \infty} \frac{\log R_f(n, \epsilon, K)}{\log \gamma(n)} . \quad (2.5)$$

Accordingly, we formulate the following definition.

Definition 2.3 We define the γ-*entropy* of the map f with respect to the metric d to be the non-negative extended real number given by

$$h_d^\gamma(f) = \sup_K h_d^\gamma(f, K) , \quad (2.6)$$

where the supremum is taken over all compact subsets K of X.

The following can be found in [27, Corollary 7.7.1].

Proposition 2.5 *Let (X, d) be a compact metric space, and let $f : X \to X$ be a continuous map. For each positive δ, denote by $\mathcal{A}_\delta$ the cover of X by all balls of radius δ. Then for each $\epsilon > 0$ and each $n \in \mathbb{N}$, we have*

$$N \left(\bigvee_{i=0}^{n-1} f^{-i} \mathcal{A}_{2\epsilon} \right) \le R_f(n, \epsilon) \le S_f(n, \epsilon) \le N \left(\bigvee_{i=0}^{n-1} f^{-i} \mathcal{A}_{\epsilon/2} \right) . \quad (2.7)$$

This yields the following result.

Corollary 2.1 *If (X, d) is a compact metric space and $f : X \to X$ is a continuous map, then for each gauge γ we have $h_d^\gamma(f) = h^\gamma(f)$.*

This immediately implies that, for a continuous map $f : X \to X$ of a compact metrizable space X and any metric d generating its topology, the γ-entropy of f is independent of d. Here are some simple situations in which the computation of γ-entropy is trivial.

Proposition 2.6 (Eventually weak contractions) *Let (X, d) be a compact metric space and $f : X \to X$ a continuous mapping. Let K be a compact subset of X on which f is eventually weakly contracting, i.e., there exists some $k \in \mathbb{N}$ such that*

$$d(f^k(x), f^k(y)) \le d(x, y) \quad \forall x, y \in K . \tag{2.8}$$

Then $h_d^\gamma(f, K) = 0$ for every gauge function γ.

Remark 2.2 In particular, isometries have no growth of any order.

A similar argument also gives the following.

Proposition 2.7 *Let (X, d) be a compact metric space and $f : X \to X$ a continuous mapping. Let K be a compact subset of X for which $\lim_{k \to \infty} \operatorname{diam}(f^k(K)) = 0$. Then $h_d^\gamma(f, K) = 0$ for every gauge function γ.*

For these basic facts and more, see [9].

3 Equicontinuity Versus Slow Entropy

As we have seen in Sect. 2.2, Remark 2.2, every isometry of a compact metric space has zero growth of all orders. In this section, we give a simple characterization of maps which have zero growth of all orders. This characterization is due to Galatolo, who observed in [14] that such maps are precisely those that have zero topological complexity in the sense of Blanchard, Host and Maass [3] (see Definition 3.1).

Recall that a continuous dynamical system $f : X \to X$ acting on a metric space (X, d) is said to be *equicontinuous* if for any $\epsilon > 0$ there exists $\eta > 0$ such that if x and y belong to X then $d(x, y) < \eta$ implies that $d(f^n(x), f^n(y)) < \epsilon$ for all $n \ge 0$. One can also speak of *local equicontinuity* (as when studying, for instance, the dynamics of a rational map in its Fatou set—see [21]), but this need not concern us here.

Definition 3.1 The *topological complexity function* of the finite cover $\mathcal{A}$ of X with respect to the dynamical system $f : X \to X$ is the non-decreasing function

$$\operatorname{comp}(\mathcal{A}, n) = N(\mathcal{A}_f^n) . \tag{3.1}$$

The following result, given by Blanchard, Host and Maas in [3, Proposition 2.2] offers a characterization of equicontinuity in terms of the combinatorics of open covers.

Proposition 3.1 *The following two statements are equivalent:*

(i) The dynamical system $f : X \to X$ is equicontinuous;

(ii) For each finite open cover $\mathcal{A}$, there exists $k \geq 0$ such that $\text{comp}(\mathcal{A}, n) \leq k$.

As a straightforward consequence of this result, Galatolo deduced the following in [14, Proposition 20].

Proposition 3.2 *The dynamical system $f : X \to X$ is equicontinuous if and only if for each gauge γ the equality $h^\gamma(f) = 0$ holds true.*

Note that every eventually weak contraction (hence every isometry) is equicontinuous, so Proposition 3.2 implies Proposition 2.6. More generally, every map which is topologically conjugate to an eventually weak contraction is equicontinuous as well. This includes translations on compact abelian groups; in particular, every *adding machine*[3] is equicontinuous.

With the characterization of zero growth of all orders given by Proposition 3.2 at hand, Galatolo obtained in [14, Theorem 21] the following result.

Galatolo's Theorem. *The quadratic unimodal map q_∞ at the boundary of positive topological entropy is not equicontinuous, so that there exists some gauge γ such that $h^\gamma(q_\infty) > 0$.*

It is well-known that the map q_∞ is such that the closure of the forward orbit of its critical point is an invariant Cantor set K_∞, and that the action of q_∞ on K_∞ is conjugate to the *dyadic adding machine* (see for instance [10, Proposition 4.5, p. 242]). Hence the positive slow entropy guaranteed by Galatolo's Theorem is 'happening' away from the Cantor set K_∞. Indeed, the proof of the above theorem uses, besides Proposition 3.2, the fact that q_∞ has expanding periodic orbits (of periods given by powers of two) lying in the gaps of K_∞.

It should be clear that one can use the criterion given by Proposition 3.2 to produce many other examples of dynamical systems having positive slow entropy. It is often quite easy to decide whether a system is or is not equicontinuous. Here is a list of well-known examples.

Theorem 3.1 *The following dynamical systems are not equicontinuous.*

(i) Smooth self-maps $f : M \to M$ of a compact manifold having an unstable periodic orbit.

---

[3]By an *adding machine* we mean a translation in the compact abelian group arising as the inverse limit of a sequence of homomorphisms $\ldots \to \mathbb{Z}_{m_{k-1}} \to \mathbb{Z}_{m_k} \to \cdots \mathbb{Z}_{m_2} \to \mathbb{Z}_{m_1}$, where each cyclic group $\mathbb{Z}_{m_k}$ is given the discrete topology, and for each $k \geq 1$, m_{k-1} divides m_k, and the homomorphism $\mathbb{Z}_{m_k} \to \mathbb{Z}_{m_{k-1}}$ is reduction modulo m_{k-1}. The translation map is induced by the add-one map $x \mapsto x + 1$ in each $\mathbb{Z}_{m_k}$. When $m_k = 2^k$ for all k, the adding machine is called *dyadic*. See [2] for a thorough discussion of general adding machines and references therein.

(ii) Circle homeomorphisms $f : \mathbb{T}^1 \to \mathbb{T}^1$ with Bohl-Denjoy invariant Cantor sets[4].

(iii) Skew products (or twist maps) of the 2-torus $\mathbb{T}^2 = \mathbb{R}^2/\mathbb{Z}^2$, say $f : \mathbb{T}^2 \to \mathbb{T}^2$, of the form $f(x, y) = (x + \alpha, y + \varphi(x)) \pmod{\mathbb{Z}^2}$, where $\alpha \in \mathbb{R}$ and $\varphi : [0, 1] \to \mathbb{R}$ is continuous, $\varphi(0) = 0$ and $\varphi(1) = m \in \mathbb{Z}$, $m \neq 0$.

In particular, each of these systems has positive γ-entropy for some gauge γ.

Proof The arguments establishing failure of equicontinuity in cases (i), (ii) and (iii) are straightforward. We give the argument only for case (ii), and leave the other two cases as exercises for the reader. If $f : \mathbb{T}^1 \to \mathbb{T}^1$ is a map in (ii), we know that f has a wandering interval $J \subset \mathbb{T}^1$. Let $J_n = f^n(J)$ for each $n \in \mathbb{Z}$, and let $0 < \epsilon < |J|$. Since the intervals J_n are pairwise disjoint, we have $|J_n| \to 0$ as $|n| \to \infty$. Given $\delta > 0$, choose $n \in \mathbb{N}$ so large that J_{-n} has length less than δ. Then the endpoints of J_{-n} are less than δ apart, and yet their images under f^n are the endpoints of J, which are more than ϵ apart. This shows that f is not equicontinuous. $\square$

Note that $\mathrm{Per}(f) = \emptyset$ for all maps in (ii), and the same is true for all maps in (iii) for which α is irrational. Also, it is clear that the quadratic unimodal map q_∞ introduced above is of type (i). Thus, Galatolo's Theorem is a special case of Theorem 3.1. Regarding the latter, we could of course have added several other systems to our list, such as continuous twist maps on the two-dimensional annulus (or the two-dimensional torus) with a rotation set of positive length, skew-products over translations on other compact abelian groups, etc. In each case, failure of equicontinuity is fairly easy to prove.

Regarding topological complexity, we wish to point out that, in [20], Li and Shen proved the following interesting result. Let f be a C^3 unimodal map with a non-flat critical point c, and suppose that f has a Cantor attractor. Then for each open cover $\mathcal{U}$ of $\omega(c)$, the topological complexity function satisfies the inequality $\mathrm{comp}(\mathcal{U}, n) \leq Cn \log n$ for some constant $C > 0$ that may depend on $\mathcal{U}$. From this it easily follows that, if γ is any super-polynomial gauge, then $h^\gamma(f|_{\omega(c)}) = 0$. Thus, if such a unimodal map has super-polynomial growth, then such growth is coming from *outside* the ω-limit set of its critical point. As we shall see in Sect. 4.2, the period-doubling map (or the quadratic unimodal map q_∞ in Galatolo's Theorem) does have super-polynomial growth, and therefore it perfectly illustrates this situation.

4 Slow Growth for One-Dimensional Maps

The qualitative results of Sect. 3, obtained via a simple criterion for non-equicontinuity, are obviously not very satisfactory. We need quantitative methods and/or results that allow us to compute exactly, or at least estimate, the γ-growth rate of a system for a given gauge γ. It seems quite hard to find methods that work in ample generality,

---

[4]In other words, circle homeomorphisms having an *exceptional minimal set*. These go back to Poincaré in the C^0 category. Bohl [4] was the first to construct C^1 diffeomorphisms with this property, and later Denjoy [11] constructed $C^{1+\alpha}$ examples of this type.

but for one-dimensional systems—more precisely, for *piecewise-monotone interval maps*—a very useful method is available, provided γ is a *very good gauge* in the sense of Definition 2.2. In such cases, it suffices to find (through whatever means) the γ-growth rate in n of the *total number of turning points* of the n-th iterate of the map. This fact is a generalization of a well-known theorem due to Misiurewicz and Szlenk [22], and its precise statement is given in Sect. 4.1.

In Sect. 4.2, we examine an interesting non-trivial example: an interval unimodal map at the boundary of chaos—the *period-doubling fixed point*, or any map topologically conjugate to it. We obtain some good estimates on the growth of maximal separated sets, or equivalently of the number of critical points, as we iterate the map. The precise growth-rate is super-polynomial but sub-exponential.

4.1 Slow Growth and Lap Numbers

Let us denote by I the interval $[-1, 1]$. Given a piecewise (strictly) monotone map $f : I \to I$, let

$$\mathrm{crit}(f) = \{c \in I : f \text{ is not locally monotone at } c\} \ . \qquad (4.1)$$

Observe that this set is finite. Its elements are called *turning points* or *critical points*. Since the composition of locally monotone maps is locally monotone, for each $n \in \mathbb{N}$ we have

$$\mathrm{crit}(f^n) = \bigcup_{0 \le k < n} f^{-k}(\mathrm{crit}(f)) \ . \qquad (4.2)$$

For each $k \in \mathbb{N}$, let

$$C_f(k) = f^{-k}(\mathrm{crit}(f)) \setminus f^{-k+1}(\mathrm{crit}(f)). \qquad (4.3)$$

Then $\mathrm{crit}(f^n) = \bigcup_{0 \le k < n} C_f(k)$. We call $c_f(k) = \mathrm{Card}\, C_f(k)$ the *k-th cutting number* of f. Define a *lap* of f to be a maximal closed subinterval of I on which f is monotone. The collection of all laps of f^n is denoted $L_f(n)$. We call $\ell_f(n) = \mathrm{Card}\, L_f(n)$ the *n-th lap number* of f. Observe that since $\mathrm{crit}(f^n) \cap \partial I = \emptyset$ it follows that $\ell_f(n) = 1 + \sum_{0 \le k < n} c_f(k)$.

Theorem 4.1 *Assume that γ is eventually very good, the γ-topological conditional entropy of f^n with respect to $L_f(n)$ satisfies $h^\gamma(f^n | L_f(n)) = o(\Gamma(n))$, and either*

(i) $\lim_{n \to \infty} \Gamma(n) = +\infty$; *or*
(ii) *if U_ϵ denotes the ϵ-neighbourhood of the critical set $\mathrm{crit}(f)$ in I then*

$$\lim_{\epsilon \to 0} \limsup_{n \to \infty} \frac{\log \mathrm{card}(U_\epsilon \cap \mathrm{crit}(f^n))}{\log \mathrm{card}(\mathrm{crit}(f^n))} = 0 \ .$$

Then

$$h^\gamma(f) = \limsup_{n \to \infty} \frac{\log \ell_f(n)}{\log \gamma(n)} . \tag{4.4}$$

This is the promised generalization, for very good gauges, of Misiurewicz-Szlenk's theorem [22]. The proof is a straightforward adaptation of their original argument, and it will be given in [9].

4.2   Maps at the Accumulation of Period-Doubling

In this section we examine the period-doubling map $f : I \to I$ (where, as before, $I = [-1, 1]$) from the point of view of slow entropy. The map f is the unique quadratic unimodal map (meaning a unimodal map with a quadratic turning point) arising as the solution to the functional equation

$$f(x) = -\frac{1}{\alpha} f \circ f(-\alpha x) , \quad \text{for all } x \in I , \tag{4.5}$$

where $0 < \alpha < 1$ is the unique fixed point of f in the interior of I. Cascades of period-doubling bifurcations in smooth one-parameter families of unimodal maps typically terminate in a map topologically conjugate to f. Thus, unimodal maps topologically conjugate to f are said to *lie at the accumulation of period-doubling*. (See [26] for more details.)

Although f is a map with zero topological entropy, we will show that f exhibits *super-polynomial growth of maximal separated sets* (we will shorten this expression to *super-polynomial growth*). Since slow entropy is a topological invariant, the same result is also true for the quadratic polynomial q_∞, as well as any other map which is topologically conjugate to f. Thus we will obtain the following.

Theorem 4.2 *Every quadratic unimodal map which lies at the accumulation of period-doubling has super-polynomial growth.*

The super-polynomial growth of f, and thus the above theorem, is a straightforward consequence of the following lemma.

Lemma 4.1 *Let $f : I \to I$ be the period-doubling map. Then the maximal cardinality of an (n, ϵ)-separated set for f, namely $S_f(n, \epsilon)$, has super-polynomial growth in n. More precisely, there exists a constant $C > 0$ such that, for each sufficiently small $\epsilon > 0$, we have*

$$\limsup_{n \to \infty} \frac{\log S_f(n, \epsilon)}{\log n} \geq C \log \frac{1}{\epsilon} . \tag{4.6}$$

Proof Consider the central interval $J_0 = [-\alpha, \alpha] \subset I$, which is invariant under f^2. Denote by J_1 the unique component of $f^{-1}(J_0)$ which lies to left of the critical

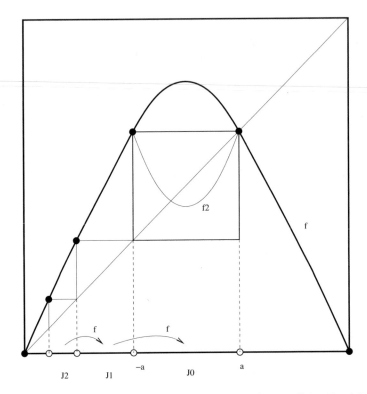

Fig. 1 The graph of the period-doubling map f. Here J_0 denotes the central interval and $J_1, J_2, \ldots,$ denotes the sequence of left-preimages of J_0

point 0; note that J_1 is in fact adjacent to J_0 (their common endpoint being $-\alpha$). Inductively, define J_n as the unique component of $f^{-1}(J_{n-1})$ which lies to left of 0. It follows inductively that the intervals J_n and J_{n-1} are adjacent, J_n being to the left of J_{n-1}, and $f(J_n) = J_{n-1}$ (see Fig. 1).

Now let us fix $0 < \epsilon < |J_1|$ and $n \geq 1$, and let $E_0 \subset J_0$ be an (n, ϵ)-separated set for $f^2|_{J_0} : J_0 \to J_0$ with maximal cardinality $S_{f^2|_{J_0}}(n, \epsilon) = S_{f^2}(n, \epsilon, J_0)$. Since $f^2|_{J_0}$ is linearly conjugate to f with linear scaling given by $\alpha = |J_0|/|I|$ (as we can see from (4.5)), it follows that

$$\text{Card}(E_0) = S_{f^2}(n, \epsilon, J_0) = S_f\left(n, \frac{\epsilon}{\alpha}\right). \tag{4.7}$$

Define $E_k = f^{-k}(E_0) \cap J_k$ for all k, and let $G = \bigcup_{k=0}^n E_{2k}$.

Claim *G is a $(2n, \epsilon)$-separated set for $f^2 : I \to I$.*

To prove this claim, let $x, y \in G$ be any two distinct elements. If, on the one hand, there exists $0 \leq k \leq n$ such that $x, y \in E_{2k}$, then the points $x^* = f^{2k}(x)$ and $y^* = f^{2k}(y)$ both belong to E_0, so x^* and y^* get ϵ-separated after at most n iterates under f^2. This means that x and y get ϵ-separated after at most $k + n \leq 2n$ iterates under f^2 in this case. If, on the other hand, $x \in E_{2j}$ and $y \in E_{2k}$ with, say, $j > k$, then $y^* = f^{2k}(y) \in E_0$, whereas $x^* = f^{2k}(x) \in E_{2(j-k)}$. But then

$$|x^* - y^*| \geq \text{dist}(J_{2(j-k)}, J_0) \geq |J_1| > \epsilon .$$

Hence x and y get ϵ-separated after $k < n$ iterates under f^2. This proves the claim. From the claim it follows that

$$\begin{aligned} S_{f^2}(2n, \epsilon) &\geq \text{Card}(G) > n\text{Card}(E_0) \\ &= nS_{f^2}(n, \epsilon, J_0) . \end{aligned} \tag{4.8}$$

Since every $(2n, \epsilon)$-separated set for $f^2 : I \rightarrow I$ is clearly a $(4n, \epsilon)$-separated set for $f : I \rightarrow I$, we also have

$$S_{f^2}(2n, \epsilon) \leq S_f(4n, \epsilon) . \tag{4.9}$$

Combining (4.7), (4.8) and (4.9), we get the inequality

$$S_f(4n, \epsilon) \geq nS_f(n, \frac{\epsilon}{\alpha}) . \tag{4.10}$$

Now, as long as ϵ/α is still smaller than $|J_1|$, we can repeat the argument with ϵ replaced by ϵ/α, and so on, by induction. More precisely, let $m \in \mathbb{N}$ be such that

$$\frac{\epsilon}{\alpha^m} < |J_1| \leq \frac{\epsilon}{\alpha^{m+1}} . \tag{4.11}$$

Given $k > m$, we apply (4.10) with $n = 4^{k-j}$ for each $j = 0, 1, \ldots, m - 1$ and get

$$S_f(4^{k-j}, \frac{\epsilon}{\alpha^j}) \geq 4^{k-j-1}S_f(4^{k-j-1}, \frac{\epsilon}{\alpha^{j+1}}) \quad \text{for } 0 \leq j \leq m - 1 .$$

From this we deduce that

$$S_f(4^k, \epsilon) \geq S_f(4^{k-m}, \frac{\epsilon}{\alpha^m}) \prod_{j=0}^{m-1} 4^{k-j-1} \geq 4^{-\frac{m(m+1)}{2}} \cdot 4^{mk} .$$

This last inequality can be re-written as

$$\frac{\log S_f(4^k, \epsilon)}{\log 4^k} \geq m - \frac{m(m+1)}{2k} . \tag{4.12}$$

Note from (4.11) that

$$m = \left\lceil \frac{\log \frac{|J_1|}{\epsilon}}{\log \frac{1}{\alpha}} \right\rceil \geq C \log \frac{1}{\epsilon}, \tag{4.13}$$

where $C = (2 \log \frac{1}{\alpha})^{-1}$, provided $0 < \epsilon < |J_1|^{2/\log \frac{1}{\alpha}}$. Hence, taking the lim sup as $k \to \infty$ in (4.12) and using (4.13), we finally get

$$\limsup_{n \to \infty} \frac{\log S_f(n, \epsilon)}{\log n} \geq \limsup_{k \to \infty} \frac{\log S_f(4^k, \epsilon)}{\log 4^k} \geq C \log \frac{1}{\epsilon}.$$

This establishes (4.6) and the lemma is proved.   □

The above result has the following immediate consequence. For each $s > 0$, consider the gauge function $\gamma_s(n) = n^s$.

Corollary 4.1 *Let* $f: I \to I$ *denote the period-doubling map. Then* $h^{\gamma_s}(f) = \infty$, *for all* $s > 0$.

Once again, it is clear that the same result is valid for any map which is topologically conjugate to f.

One can go much further than Theorem 4.6, getting not only a better lower bound for growth, but also an upper-bound. This is achieved by carefully analysing the growth of the *pre-images* of the critical point, or equivalently, the growth in n of the number of critical points of f^n, and then invoking Theorem 4.1.

Theorem 4.3 *Let* $f: I \to I$ *denote the period-doubling map. Then the following assertions hold.*

(i) *For all* $n \geq 1$,
$$\text{Card}\,(\text{crit}(f^n)) \leq n^{\frac{3}{2}} \exp\left\{ B_0 \,(\log n)^2 \right\},$$

where $B_0 = (\log 2)^{-1} - \frac{1}{2}\log 2 \approx 1.096$.
(ii) *For each* $\delta \in (0, 1)$ *there exists* $C_\delta > 0$ *such that for all* $n \geq 1$,
$$\text{Card}\,(\text{crit}(f^n)) \geq C_\delta n^{1-\delta} \exp\left\{ B_1 \,(\log n)^2 \right\},$$

where $B_1 = (4 \log 2)^{-1} \approx 0.3607$.

The proof of (a sharper version of) Theorem 4.3 will be given in [9].

Summarizing, as an immediate consequence of these results, we can state that the number of critical points of f^n grows super-polynomially but sub-exponentially with n, and in fact:

- We have $h^\gamma(f) < \infty$ for the gauge $\gamma(n) = n^{\frac{3}{2}} e^{B_0 (\log n)^2} = n^{\frac{3}{2} + B_0 \log n}$.
- For each $\delta \in (0, 1)$, we have $h^\gamma(f) > 0$ when we take γ to be the gauge $\gamma(n) = n^{1-\delta} \exp\left\{ B_1 (\log n)^2 \right\} = n^{1-\delta + B_1 \log n}$.

The full analysis of growth for the period-doubling map f, to be carried out in [9], will be based on a careful study of the *generating function* of the sequence $c_n =$ Card $(\text{crit}(f^n))$ which, perhaps not surprisingly, is intimately related via Milnor-Thurston theory to the *dynamical zeta-function* of f, which is well-known (see for instance [24, p. 888]).

5 Slow Growth for Two-Dimensional Maps

What else, besides the qualitative results of Sect. 3, can be said about the growth of orbits of two-dimensional zero-entropy homeomorphisms? Computing slow entropy in such broad generality is a difficult problem. The best we can do at this point is to investigate what happens in certain specific families of examples. In this spirit, we offer the following result, whose proof will be given in [9]. The theorem below is stated for self-maps of the annulus only, but such examples can be easily grafted onto any other surface. For each $\alpha \in (0, 1)$, let us denote by γ^α the subexponential gauge given by $\gamma^\alpha(n) = \exp(n^\alpha)$.

Theorem 5.1 *For each $\alpha \in (0, 1)$ and each $\epsilon > 0$ sufficiently small, there exists a bi-Hölder homeomorphism $F \colon \mathbb{T}^1 \times I \to \mathbb{T}^1 \times I$ of class $C^{1-\alpha-\epsilon}$ without periodic points such that $h^{\gamma^\alpha}(F) > 0$.*

The details will be given in [9]. A very rough sketch of the proof goes as follows.

(i) We start with a diffeomorphism $F_0 \colon \mathbb{T}^1 \times I \to \mathbb{T}^1 \times I$ of the form $f \times \text{id}$, where $f \colon \mathbb{T}^1 \to \mathbb{T}^1$ is a $C^{1+\beta}$ Denjoy example (for some $\beta > 0$), i.e., a circle diffeomorphism without periodic points having a Cantor minimal set K.

(ii) We write $\mathbb{T}^1 \setminus K = \bigcup_{n \in \mathbb{Z}} J_n$, where each J_n is an interval (gap). We may assume that $f(J_n) = J_{n+1}$ for all n (so that f is transitive on gaps). We also refer to the rectangles $R_n = J_n \times I$ as gaps.

(iii) We also consider a smooth, orientation-preserving horseshoe diffeomorphism $\phi \colon Q \to Q$ with support inside the square $Q = I \times I$ (so that $\phi|_{\partial Q} \equiv \text{id}$). The measurable Riemann mapping theorem yields us an isotopy between ϕ and the identity map.

(iv) For each $N \in \mathbb{N}$, we *slice* the above isotopy in such a way as to be able to write $\phi = \phi_{N,N} \circ \phi_{N-1,N} \circ \cdots \circ \phi_{1,N}$, where each $\phi_{j,N} \colon Q \to Q$ is a diffeomorphism which is Hölder-close to the identity (successively closer as N increases).

(v) We choose a sequence of natural numbers $N_1 < N_2 < \cdots < N_k < \cdots$, and for each N_k we perform the above slicing. In this fashion, for each $m \in \mathbb{N}$ we see that ϕ^m has been sliced as a composition of $N_1 + N_2 + \cdots + N_m$ diffeomorphisms. The choice of the sequence (N_k) will depend on what kind of growth (or gauge) we wish to achieve (i.e., on the value of α).

(vi) Next we build, in succession, an affine copy of each slice ϕ_{j,N_k} as a map $\varphi_n \colon R_n \to R_n$ of the appropriate gap R_n, where $n = j + \sum_{i=1}^{k-1} N_i$. Each φ_n is Hölder close to the identity (for a Hölder exponent smaller than $1 - \alpha$, as it turns out),and it extends to a global homeomorphism of the annulus, being the identity outside R_n.

(vii) Finally, we define F as a limit, $F = \lim_{n \to \infty} \varphi_n \circ \cdots \circ \varphi_2 \circ \varphi_1 \circ F_0$. Note that the maps φ_n have pairwise disjoint supports (in particular they commute with each other). We are able to show (using a quantitative form of the Arzelá-Ascoli theorem) that the limit exists as a bi-Hölder homeomorphism with Hölder exponent slightly smaller than $1 - \alpha$.

Acknowledgements We would like to thank IME-USP for their hospitality during the period in which this note was conceived and written, and the anonymous referee for their comments and suggestions. We also thank the editors, Maria José Pacífico and Pablo Guarino, for their encouragement while we were writing this paper, and for their patience in awaiting the final version.

References

1. R.L. Adler, A.G. Konheim, M.H. MacAndrew, Topological entropy. Trans. Am. Math. Soc. **114**(2), 309–319 (1965)
2. H. Bass, M.V. Otero-Espinar, D. Rockmore, C. Tresser, in *Cyclic Renormalization and Automorphism Groups of Rooted Trees*. Lecture Notes in Mathematics, vol. 1621. Springer (1995)
3. F. Blanchard, B. Host, A. Maass, Topological complexity. Ergod. Theory Dyn. Syst. **20**, 641–662 (2000)
4. P. Bohl, Über die Hinsichtlich der Unabhängigen und Abhängigen Variabeln Periodische Differentialgleichung Erster Ordnung. Acta Math. **40**, 321–336 (1916)
5. R. Bowen, Entropy for group endomorphisms and homogeneous spaces. Trans. Am. Math. Soc. **153**, 401–414 (1971)
6. P. Coullet, C. Tresser, Itérations d'endomorphismes et groupe de renormalisation. J. Phys. Colloq. **C5**, C5-25–C5-28 (1978)
7. E. de Faria, P. Hazard, C. Tresser, Infinite entropy is generic in Hölder and Sobolev spaces. C. R. Acad. Sci. Paris Sér. **I**(355), 1185–1189 (2017)
8. E. de Faria, P. Hazard, C. Tresser, in *Genericity of Infinite Entropy for Maps with Low Regularity* (2017). arXiv:1709.02431
9. E. de Faria, P. Hazard, C. Tresser, in *Growth, Entropy-Type Invariants and Regularity, (Parts I and II)* (In preparation)
10. W. de Melo, S. van Strien, in *One-Dimensional Dynamics*. Ergebnisse der Mathematik und ihrer Grenzgebiete (3), vol. 25 (Springer, Berlin, 1993)
11. A. Denjoy, Sur les courbes définies par les équations différentielles à la surface du tore. J. de Math. Pures et Appl. **11**, 333–376 (1932)
12. E.I. Dinaburg, The relation between topological entropy and metric entropy. Sov. Math. **11**, 13–16 (1970)
13. M.J. Feigenbaum, Quantitative universality for a class of nonlinear transformations. J. Stat. Phys. **19**, 133–141 (1978)
14. S. Galatolo, Global and local complexity in weakly chaotic dynamical systems. Discret. Cont. Dyn. Syst. **9**, 1607–1624 (2003)
15. T.N.T. Goodman, Topological sequence entropy. Proc. Lond. Math. Soc. **29**, 331–350 (1974)
16. A. Katok, Lyapunov exponents, entropy and periodic points for diffeomorphisms. Publ. Math I.H.É.S. **51**(1), 137–173 (1980)
17. A. Katok, B. Hasselblatt, in *Introduction to the Modern Theory of Dynamical Systems*. Encyclopedia of Mathematics and its Applications, vol. 54 (Cambridge University Press, 1995)

18. A. Katok, J.P. Thouvenot, Slow entropy type invariants and smooth realization of commuting measure-preserving transformations. Ann. Inst. H. Poincaré Probab. Statist. **33**(3), 323–338 (1997)
19. A.G. Kushnirenko, On metric invariants of entropy type. Russ. Math. Surv. **22**, 53–61 (1967)
20. S. Li, W. Shen, The topological complexity of Cantor attractors for unimodal interval maps. Trans. Am. Math. Soc. **368**, 659–688 (2016)
21. J. Milnor, in *Dynamics in One Complex Variable*. Annals of Mathematics Studies, vol. 160 (Princeton University Press, 2006)
22. M. Misiurewicz, W. Szlenk, in *Entropy of Piecewise Monotone Mappings*. Studia Math. Tome LXVII.1-4, pp. 45–63 (1980)
23. M. Rees, A minimal positive entropy homeomorphism of the 2-torus. J. Lond. Math. Soc. (2) **23**(3), 537–550 (1981)
24. D. Ruelle, Dynamical zeta functions and transfer operators. Not. Am. Math. Soc. 887–895 (2002)
25. C. Tresser, P. Coullet, Itérations d'endomorphismes et groupe de renormalisation. C. R. Acad. Sci. Paris **287A**, 577–580 (1978)
26. C. Tresser, P. Coullet, E. de Faria, Period doubling. Scholarpedia **9**(6), 3958 (2014). http://www.scholarpedia.org/article/Period_doubling
27. P. Walters, in *An Introduction to Ergodic Theory*. Graduate Texts in Mathematics, vol. 79 (Springer, 1982)
28. Y. Zhao, Y. Pesin, Scaled entropy for dynamical systems. J. Stat. Phys. **158**, 447–475 (2015)

Mixing Properties in Coded Systems

Jeremias Epperlein, Dominik Kwietniak and Piotr Oprocha

Abstract We show that topological mixing, weak mixing, the strong property P, and total transitivity are equivalent for coded systems (shift spaces presented by labeling the edges of a countable irreducible graphs by symbols from a finite alphabet). We provide an example of a topologically mixing coded system which cannot be approximated by any increasing sequence of topologically mixing shifts of finite type, has only periodic points of even period and each set of its generators consists of blocks of even length. We prove that such an example cannot be a synchronized system.

Keywords Symbolic dynamics · Shift spaces · Coded shifts · Synchronized shifts · Shifts of finite type · Sofic shift · Topological mixing · Strong property P · Specification property · Fischer cover

2010 Mathematics Subject Classification 37B10 · 37B20

J. Epperlein
Institute for Analysis and Center for Dynamics, Technische Universität Dresden, Zellescher Weg 12-14, 01069 Dresden, Germany
e-mail: jeremias.epperlein@tu-dresden.de

D. Kwietniak (✉)
Faculty of Mathematics and Computer Science, Jagiellonian University in Kraków, ul. Łojasiewicza 6, 30-348 Kraków, Poland
e-mail: dominik.kwietniak@uj.edu.pl
URL: http://www.im.uj.edu.pl/DominikKwietniak/

Institute of Mathematics, Federal University of Rio de Janeiro, Cidade Universitaria - Ilha do Fundão, Rio de Janeiro 21945-909, Brazil

P. Oprocha
Faculty of Applied Mathematics, AGH University of Science and Technology, al. A. Mickiewicza 30, 30-059 Kraków, Poland
e-mail: oprocha@agh.edu.pl

National Supercomputing Centre IT4Innovations, Division of the University of Ostrava, Institute for Research and Applications of Fuzzy Modeling, 30. dubna 22, 70103 Ostrava, Czech Republic

1 Introduction

We consider coded systems and their recurrence properties that are stronger than topological transitivity. We are interested in topological (weak) mixing, and properties like the strong property P, which is a variant of the specification property.

Recall that a shift space is a *coded system* if it can be presented by an irreducible directed graph whose edges are labeled by symbols from a finite alphabet $\mathcal{A}$. Here, "presented" means that the shift space is the closure in $\mathcal{A}^{\mathbb{Z}}$ of all bi-infinite sequences of symbols which are labels for bi-infinite paths in the graph. Equivalently, X is a coded system if it is a closure in $\mathcal{A}^{\mathbb{Z}}$ of the set of all bi-infinite sequences obtained by freely concatenating the words in a (possibly infinite) list of words over $\mathcal{A}$. Such a list is the set of *generators* of X.

Transitive sofic shifts are coded systems which can be presented by a finite irreducible directed graph. Coded systems were introduced by Blanchard and Hansel [5], who showed that any factor of a coded system is coded. This is a generalization of a well-known property of sofic shifts [19, Corollary 3.2.2].

It is natural to ask which properties of irreducible sofic shifts extend to coded systems. Here we prove that several properties known to be equivalent to topological mixing for sofic shifts remain equivalent for coded systems. We give an example showing that, quite surprisingly, another condition equivalent to topological mixing for sofic shifts does not need to hold for topologically mixing coded systems. The counter-intuitive nature of this example is in our opinion the most interesting feature of this paper, but we hope that our other results will fill a gap in the literature. Furthermore, there has been a resurgence in interest in coded systems in general, and their notable subclasses in particular (for example S-gap shifts [2, 9, 12], β-shifts [9, 23], Dyck shifts [20, 21]). Coded systems often provide a testing ground for further extensions (see [10], where the line of investigation initiated in [9] is developed and extended to non-symbolic systems). Therefore understanding the situation for coded systems may lead to solutions of more general problems.

In order to describe our results, note first that any coded system is topologically transitive (irreducible). Recall also that for a non trivial transitive sofic shift X the following stronger variants of transitivity (for their definitions, see the next section) are equivalent:

(a) X is topologically mixing;
(b) X has the strong property P;
(c) X is topologically weakly mixing;
(d) X is totally transitive;
(e) X has two periodic points with relatively prime primary periods;
(f) X has the periodic specification property.

It seems that this is a folklore theorem, but we could not find it in this form in the literature. In brief outline, here is the main idea of the proof using the terminology presented in [19]. First recall that for every shift space (a) implies all other properties. Furthermore, it is not hard to see that (a) or (b) implies (c) and the latter implies (d). For the proof of remaining implications we assume that X is a nontrivial sofic shift. Observe that every transitive sofic shift has a minimal, right-resolving, and follower separated presentation (e.g. see Theorem 3.3.2, Proposition 3.3.11, and Corollaries 3.3.19–20 in [19]). Using synchronizing words and any condition on the list (a)–(e) it is not hard to show that this presentation has two cycles with relatively prime lengths (as in the proof of Lemma 7.2 below). But then, using standard techniques (e.g. see [16]) we obtain that there is $N > 0$ such that for every $n > N$ there is a path of length n between any two vertices in this presentation. This immediately implies (f) and hence all other properties on the list hold (see also the proof of Lemma 7.1 below).

It is well known, that every shift space with the periodic specification property is synchronized (again, see the next section for details) and hence it is coded, but the converse is not true: there are synchronized systems without specification property, and coded systems which are not synchronized. It follows that (f) is no longer equivalent to (a)–(e) for shift spaces which are not sofic. For example all β-shifts are coded and topologically mixing but some are not synchronized and some are synchronized but do not have the specification property (see [8, 22]).

Here we examine the remaining possible connections between transitivity variants (a)–(e) for not necessarily sofic coded systems. Note that the proof of equivalence of properties (a)–(e) for shifts of finite type can be adapted for irreducible Markov shifts over countable alphabets. Furthermore, every coded system contains a dense subset which is a factor of an irreducible countable Markov shift. This suggests that properties (a)–(e) should remain equivalent for coded systems. And this is *almost* the case. For coded systems (a)–(d) are equivalent, but the condition (e) no longer follows, nor implies (a). To show that (e) is not a consequence of (a) we construct a topologically mixing coded system without a periodic point of odd period and hence without a generator of odd length. We note that such a system cannot contain a topologically mixing shift of finite type. Krieger [17] characterized coded systems as those shift spaces which contain an increasing sequence of irreducible shifts of finite type with dense union. Krieger's characterization is the best possible, in the sense that there exists an increasing sequence of sofic shifts whose closure is a shift space which is not a coded system [6]. It follows from our result that there are topologically mixing coded systems, which cannot be approximated from the inside by topologically mixing shifts of finite type as they do not contain any topologically mixing sofic shift[1]. We also present an example (suggested by a remark of one of the reviewers) showing that the mere existence of two periodic points with relatively prime primary periods is not enough to imply topological mixing even for synchronised systems ((e) does not imply (a)). The reason is that every cycle in an irreducible labelled graph presenting a coded system X leads to a periodic point in X, but the converse is not true if the graph is infinite. That is, to form a coded system we take the closure of the set of all bi-infinite sequences of symbols which are labels for bi-infinite paths

---

[1] Another example of this kind was obtained independently in [11]. Preprint version of [11] was posted on arXiv on July 29, 2015 as arXiv:1507.08048. Preprint of our paper was posted on March 10, 2015 as arXiv:1503.02838.

in the graph, thus some periodic points may appear as limit points. These periodic points may not correspond to a label of any closed path in the graph. In addition, we show that if X is a transitive but not totally transitive coded system, then for some prime p we can write $X = X_1 \cup \ldots \cup X_p$, where the X_j's are closed subsets of X cyclically permuted by the shift map σ, each X_j is σ^p invariant, $X_i \cap X_j$ is nowhere dense for $i \neq j$, and (X_j, σ^p) is topologically mixing.

Finally, we note that replacing (e) by (e*) the conditions (a)–(e*) are still equivalent for synchronized shift spaces. The new condition (e*) says that there are two cycles with relatively prime lengths *in the canonical presentation* of a synchronized shift space.

This paper is organized as follows: In the next section we set up the notation and terminology. In Sect. 3 we prove a structure theorem for topologically mixing coded systems and that total transitivity, topological weak mixing and topological mixing are equivalent for coded systems. Section 4 contains an example of a shift space which has the strong property P, but is not topologically mixing. In Sect. 5 we describe a topologically mixing coded system without a periodic point of odd period. In Sect. 6 we establish an equivalence of the strong property P and topological mixing for coded systems. In Sect. 7 we present some complementary results on synchronized systems. They imply that (a)–(e*) are equivalent for synchronized systems, thus our example does not have any synchronizing words. In the last section of this paper we present the example suggested by the reviewer.

The following theorem summarizes our results on the connections between variants of transitivity for coded systems. We give the proof in Section 8.

Theorem 1.1 *Let X be a non trivial coded system. Then the following conditions are equivalent:*

(a) X is topologically mixing;
(b) X has the strong property P;
(c) X is topologically weakly mixing;
(d) X is totally transitive.

Additionally, if X is synchronized, then any of the above conditions is equivalent to

(e) X can be canonically presented by an irreducible directed graph (Fischer cover) with two cycles of relatively prime lengths.*

Moreover, there exists a coded system $\mathbb{X}$ fulfilling (a)–(d), but not (e).*

2　Notation and Definitions

We assume the reader is familiar with elementary symbolic dynamics as in [19]. We fix a finite set $\mathcal{A}$ with at least two elements and call it the *alphabet*. Let $\mathcal{A}^{\mathbb{Z}}$ denote the set of bi-infinite (two-sided) sequences

$$x = (x_i)_{i \in \mathbb{Z}} = \ldots x_{-3} x_{-2} x_{-1} x_0 x_1 x_2 \ldots,$$

such that $x_i \in \mathcal{A}$ for all i. We equip $\mathcal{A}$ with the discrete topology and we consider $\mathcal{A}^{\mathbb{Z}}$ as a compact metric space in the product topology. The shift operator on $\mathcal{A}^{\mathbb{Z}}$ is denoted by σ. A *shift space* over the alphabet $\mathcal{A}$ is a shift-invariant subset of $\mathcal{A}^{\mathbb{Z}}$ which is closed in that topology. The set $\mathcal{A}^{\mathbb{Z}}$ itself is a shift space called the *full shift*. In this paper all shift spaces will be two-sided and transitive.

A *block* of length k is an element $w = w_1 w_2 \ldots w_k$ of $\mathcal{A}^k$. Throughout this paper "*a word*" is a synonym for "*a block*". The length of a block is denoted $|w|$. The set of all words over $\mathcal{A}$ is denoted by $\mathcal{A}^*$. Given $x \in \mathcal{A}^{\mathbb{Z}}$ and $i, j \in \mathbb{Z}$ with $i \leq j$ we write $x_{[i,j]}$ to denote the block $x_i x_{i+1} \ldots x_j$. We say that a block w occurs in x if $w = x_{[i,j]}$ for some $i, j \in \mathbb{Z}$. A language of a shift space X is the set $\mathcal{B}(X)$ of all blocks that occur in X. The set of blocks of length n in the language of X is denoted $\mathcal{B}_n(X)$. Similarly, the set of all words that occur in a point $x \in \mathcal{A}^{\mathbb{Z}}$ is denoted $\mathcal{B}(x)$. The *empty word* $\perp$ is the unique word of length 0. We write $\mathcal{A}^+$ for the set of nonempty words over $\mathcal{A}$.

A *central cylinder set* of a word $u \in \mathcal{B}_{2r+1}(X)$, where $r \in \mathbb{N}$, is the set $[u] \subset X$ of points from X in which the block u occurs starting at position $-r$, that is, $\{y \in X : y_{[-r,r]} = u\}$. Central cylinders (or *cylinders* for short) are open and closed subsets of X. The family

$$\{[x_{[-r,r]}] : r \in \mathbb{N}\}$$

of cylinder sets determined by a central subblock of x is a neighbourhood basis for a point $x \in X$. We use a multiplicative notation for *concatenation* of words, so that $w^n = w \ldots w$ (n-times) and $w^{\infty} = www \ldots \in \mathcal{A}^{\mathbb{N}}$.

Given a set of words $Q \subset \mathcal{A}^+$, we define $Q^0 = \{\perp\}$, and $Q^n = Q^{n-1} Q = \{uw : u \in Q^{n-1}, w \in Q\}$. We also let Q^+ denote the set of all possible finite concatenations of words from Q, that is, $Q^+ = \bigcup_{n=1}^{\infty} Q^n$. In particular, $Q \subset Q^+$.

By $Q^{\mathbb{Z}}$ we denote the set containing all possible bi-infinite concatenations of elements of Q, that is, $x \in Q^{\mathbb{Z}}$ if x can be partitioned into elements of Q.

Dynamical properties like those mentioned in (a)–(d) and (f) above are usually defined for a continuous map acting on a metric space. Here we define them in the language of symbolic dynamics.

A shift space X is:

(1) *transitive* if for any $u, v \in \mathcal{B}(X)$ there is $w \in \mathcal{B}(X)$ such that $uwv \in \mathcal{B}(X)$;
(2) *totally transitive* if for any $u, v \in \mathcal{B}(X)$ and any $n > 0$ there is $w \in \mathcal{B}(X)$ such that $uwv \in \mathcal{B}(X)$ and n divides $|uw|$;
(3) *topologically weakly mixing* if for any $u_1, v_1, u_2, v_2 \in \mathcal{B}(X)$ there are $w_1, w_2 \in \mathcal{B}(X)$ such that $u_1 w_1 v_1, u_2 w_2 v_2 \in \mathcal{B}(X)$ and $|u_1 w_1| = |u_2 w_2|$;
(4) *topologically mixing* if for every $u, v \in \mathcal{B}(X)$ there is $N > 0$ such that for every $n > N$ there is $w \in \mathcal{B}_n(X)$ such that $uwv \in \mathcal{B}(X)$.

We say that a shift space X has:

(1) the *strong property P* if for any $k \geq 2$ and any words $u_1, \ldots, u_k \in \mathcal{B}(X)$ with $|u_1| = \ldots = |u_k|$ there is an $n \in \mathbb{N}$ such that for any $N \in \mathbb{N}$ and function $\varphi : \{1, \ldots, N\} \to \{1, \ldots, k\}$ there are words $w_1, \ldots, w_{N-1} \in \mathcal{B}_n(X)$ such that $u_{\varphi(1)} w_1 u_{\varphi(2)} \ldots u_{\varphi(N-1)} w_{N-1} u_{\varphi(N)} \in \mathcal{B}(X)$;

(2) the *specification property* if there is an integer $N \geq 0$ such that for any $u, v \in \mathcal{B}(X)$ there is $w \in \mathcal{B}_N(X)$ such that $uwv \in \mathcal{B}(X)$.

Blanchard [7] introduced the strong property P in order to provide an easy to verify criterion for another property he was interested in: *uniform positive entropy*. The strong property P is easily seen to follow from the specification property. For the details we refer to [7]. Blanchard proved there that the strong property P implies topological weak mixing, and does not imply topological mixing. Therefore the strong property P is strictly weaker then the specification property.

It is convenient to rephrase the above definitions using the sets

$$N_\sigma([u], [v]) = \{\ell \in \mathbb{N} : uwv \in \mathcal{B}(X) \text{ for some } w \in \mathcal{A}^* \text{ such that } |uw| = \ell\},$$

where $u, v \in \mathcal{B}(X)$. For example a shift space X is topologically weakly mixing if for every $u, v \in \mathcal{B}(X)$ the set $N_\sigma([u], [v])$ contains arbitrarily long intervals of consecutive integers (see [14, Theorem 1.11]). A shift space X is transitive for σ^k where $k \in \mathbb{N}$ if and only if the set

$$N_{\sigma^k}([u], [v]) = \{\ell \in \mathbb{N} : uwv \in \mathcal{B}(X) \text{ for some } w \in \mathcal{A}^* \text{ such that } |uw| = k\ell\},$$

is non-empty for every $u, v \in \mathcal{B}(X)$.

If a dynamical system on a compact metric space is transitive, then there is a dense G_δ-set of points with dense orbit. In particular, in a transitive shift space X in every cylinder set there is a point x such that every block in $\mathcal{B}(X)$ occurs infinitely many times in x.

By a *countable graph* we mean a directed graph with at most countably many vertices and edges. A countable graph is *irreducible* if given any pair of its vertices, say (v_i, v_j), there is a path from v_i to v_j. A countable graph G is *labeled* if there is a *labeling* Θ which is simply a function from the set of edges of G to the alphabet $\mathcal{A}$. A labeling of edges extends, in an obvious way, to a labeling of all finite (respectively, infinite, bi-infinite) paths on G by blocks (respectively, infinite or bi-infinite sequences) over $\mathcal{A}$. The set Y_G of bi-infinite sequences constructed by reading off labels along a bi-infinite path on a labeled graph (G, Θ) is shift invariant, but usually it is not closed and therefore not a shift space. Nevertheless, its closure $X = \overline{Y_G}$ in $\mathcal{A}^\mathbb{Z}$ is a shift space, and we say that X is *presented* by (G, Θ). Any shift space admitting an irreducible presentation is a *coded system*. Irreducibility of the graph presentation implies that all coded systems are transitive. If there is a finite irreducible graph presenting a shift space X, then X is a transitive sofic shift.

A word w is *magic* for a coded system presented by (G, Θ) if $w \in \mathcal{B}(X)$ and all paths in G labeled by w end at the same vertex of G. A set of *generators* for a shift space X is a family of words $\mathcal{Q} \subset \mathcal{A}^*$ such that the language of X coincides with the set of all subblocks occurring in elements of $\mathcal{Q}^+$. Equivalently, X is the closure of $\mathcal{Q}^\mathbb{Z}$ in $\mathcal{A}^\mathbb{Z}$. Since the generators can be freely concatenated, every shift space possessing them must be transitive. Every coded system has a set of generators, and the converse also holds: if a shift space has a set of generators then it is a coded system. The class of transitive sofic shift spaces coincides with the class of shift spaces having a finite set of generators.

A *synchronizing word* for a shift space X is an element v of $\mathcal{B}(X)$ such that $uv, vw \in \mathcal{B}(X)$ for any blocks u, w over $\mathcal{A}$ imply $uvw \in \mathcal{B}(X)$. A *synchronized system* is a transitive shift space with a synchronizing word. Synchronized systems were introduced in [5]. Every synchronized system is coded, because if v is the synchronizing word for X, then $\{wv : vwv \in \mathcal{B}(X)\}$ is a set of generators for X. The uniqueness of the minimal right-resolving presentation known for sofic shifts extends to synchronized systems as outlined in [19, p. 451] (see also [24, p. 1241] and references therein). This special presentation is called the *Fischer cover*. Synchronized systems and their generalizations were extensively studied in [13].

Let $x_1, x_2, \ldots, x_n$ be positive integers. It is well-known that every sufficiently large integer can be represented as a non-negative integer linear combination of the x_i if and only if $\gcd(x_1, x_2, \ldots, x_n) = 1$. For a later reference we formulate an important consequence of this result as a remark.

Remark 2.0.1 Let $x_1, x_2, x_3, \ldots$ be positive integers. If $\gcd(x_1, x_2, x_3, \ldots) = k$, then every sufficiently large multiple of k can be represented as a non-negative integer linear combination of the x_i.

3 Total Transitivity Implies Topological Mixing for Coded Systems

We prove that total transitivity, topological weak mixing and topological mixing are equivalent for coded systems. This leads to a structure theorem for coded systems which are not totally transitive.

Theorem 3.1 *Suppose that X is a coded system and let $D \subset X$ be a closed set with nonempty interior such that $\sigma^k(D) = D$ for some $k > 0$. If the shift space (D, σ^k) is totally transitive, then it is topologically mixing and the set of periodic points of σ^k is dense in D.*

Proof Let G be an irreducible countable labeled graph presenting X. Since D has nonempty interior, for each $r \in \mathbb{N}$ large enough there is $w \in \mathcal{B}(X)$ of length $2r + 1$ such that the cylinder $[w] = \{x \in X : x_{[-r,r]} = w\}$ is contained in the interior of D.

We claim that the periodic points of (D, σ^k) are dense in D. Let V be a nonempty open subset of D. Then there is an open set $U \subset X$ with $V = U \cap D$. Without loss of generality we may assume that U is a cylinder set of some block $u \in \mathcal{B}(X)$. By total transitivity there is $s \in \mathcal{B}(X)$ and a path on G labeled by wsu such that the length of ws is jk for some integer j. We can join the last vertex on the path labeled by wsu with the first vertex of the same path by a path labeled by a word t. Therefore there is a bi-infinite periodic sequence $y \in X$ such that $y_{[-r,\infty)} = (wsut)^\infty$ for some words $s, t \in \mathcal{B}(X)$. But then $y \in [w] \subset D$ and $\sigma^{jk}(y) \in [u]$ because $|ws| = jk$. On the other hand $y \in D$, hence $\sigma^{jk}(y) \in D$ and then $\sigma^{jk}(y) \in U \cap D = V$. This shows that periodic points are dense in D under σ^k. Since every totally transitive dynamical system with dense set of periodic points is topologically weakly mixing (e.g. see Corollary 1.1 in [3]), it follows that (D, σ^k) is topologically weakly mixing.

We proceed to a proof of topological mixing of (D, σ^k). We recall that by [15, Lemma 3.1] a shift system (D, σ^k) is topologically mixing if for each cylinder v in some neighborhood basis of a point with dense orbit in D the set $N_{\sigma^k}([v], [v])$ is cofinite. Let $\bar{x} \in [w]$ be a point whose orbit is dense in D under σ^k. Given $a \geq r$ set $v_a = \bar{x}_{[-a,a]}$ and

$$[v_a] = \{x \in D : x_{[-a,a]} = v_a\}.$$

Note that $\bar{x} \in [v_a] \subset [w] \subset D$, hence the cylinder of v_a in X and in D coincide. Furthermore, $([v_a])_{a \geq r}$ is a neighborhood basis of $\bar{x}$.

It is now enough to show that the set $N_{\sigma^k}([v_a], [v_a])$ is cofinite. Let u be a word such that $v_a u$ is a labeling of a loop in G and let m be the length of $v_a u$. Without loss of generality we may assume that k divides m (we replace $v_a u$ by $(v_a u)^k$ if necessary). Denote the loop presenting $v_a u$ on G by ξ. By topological weak mixing of (D, σ^k) the set $N_{\sigma^k}([v_a], [v_a])$ contains a set of m consecutive integers, hence there is an integer $q > 0$ such that for each $i = 1, \ldots, m$ the graph G contains a path η_i labeled $v_a v_i v_a$ where $|v_i| = (q + i)k - |v_a|$. Since G is irreducible, for each $i = 0, 1, \ldots, m$ there exists a path γ_i in G such that the following path is a loop on G:

$$\pi = \xi \gamma_0 \eta_1 \gamma_1 \eta_2 \ldots \gamma_{m-1} \eta_m \gamma_m.$$

Let $p = |\pi|$. We claim that for every $j \geq 1$ and $i = 1, \ldots, m$ we have

$$p + (mj) + (q + i)k \in N_{\sigma}([v_a], [v_a]).$$

In order to show this, consider the labeling of the following path:

$$\eta_i \gamma_i \ldots \eta_m \gamma_m (\xi)^j \gamma_0 \eta_1 \gamma_1 \ldots \gamma_{i-1} \eta_i.$$

It starts and ends with v_a (it is a path on G because π and ξ are loops). This proves that $N_{\sigma^k}([v_a], [v_a])$ is cofinite. □

Corollary 3.2 *If a coded system X is totally transitive, then it is topologically mixing.*

Proof Take $D = X$ and apply Theorem 3.1. □

We will now describe the structure of coded systems which are not totally transitive. Banks proved in [3] that if a dynamical system (X, T) is transitive, but (X, T^k) is not transitive for some $k > 1$ then there is a *regular periodic decomposition of X of length k*, that is, one can find a finite cover $\{D_0, \ldots, D_{k-1}\}$ of X by non-empty regular closed sets with pairwise disjoint interiors such that $T(D_{i-1}) \subseteq D_{i \pmod k}$ for each $1 \leq i \leq k$. Recall that a set is *regular closed* if it is the closure of its interior. We say that a dynamical system (X, T) is *relatively mixing* with respect to regular periodic decomposition $\mathcal{D} = \{D_0, \ldots, D_{k-1}\}$ of X if (D_i, T^k) is topologically mixing for $0 \le i < k$. Observe that (X, T) is relatively mixing with respect to $\mathcal{D} = \{D_0, \ldots, D_{k-1}\}$ if and only if (D_0, T^k) is topologically mixing. It was proved in [3] that if there is an upper bound on the possible lengths of periodic decompositions of a transitive dynamical system then there exists a regular periodic decomposition $D_0, D_1, \ldots, D_{n-1}$ such that (D_i, T^n) is totally transitive for every $0 \le i < n$ (this decomposition is called *terminal*). In addition, in that case we also have $T^n(D_i) = D_i$ for every $0 \le i < n$, beacuse totally transitive maps are onto.

Theorem 3.3 *Every coded system is relatively mixing with respect to some regular periodic decomposition.*

Proof Let X be a coded system. By Theorem 3.1 and the result of Banks mentioned above, it suffices to show that there is an upper bound on the possible lengths of regular periodic decompositions of X.

Let G be an irreducible countable labeled graph presenting X and let k be the length of a cycle η in G. We claim that the length of a periodic decomposition of X can not be greater than k. On the contrary, assume that $D_0, \ldots, D_{n-1}$ is a regular periodic decomposition and $n > k$. Since D_0 is regularly closed, there is $r \in \mathbb{N}$ and a word w of length $2r + 1$ such that the cylinder $[w] = \{x : x_{[-r,r]} = w\} \subset D_0$. Since for each $i > 0$ the interior of a regular closed set D_i is disjoint with the interior of D_0, we have $D_i \cap [w] = \emptyset$ for each $i > 0$.

Since G is irreducible, there are paths π, γ with $|\pi| \ge |w|$ such that $\pi \eta \gamma$ is a cycle on G labeled $wu \in \mathcal{B}(X)$ for some word u. Repeating the path $\pi \eta \gamma$ if necessary we may assume that n divides $|\pi \eta \gamma| = |wu|$. Let a word $wu' \in \mathcal{B}(X)$ be the label of the path $\pi \eta \eta \gamma$ on G. We have $|wu'| = nj + k$ for some $j \ge 1$. Note that $wu'w \in \mathcal{B}(X)$ because $\pi \eta \eta \gamma \pi$ is a path on G and hence there is $x \in X$ with $x_{[-r,t]} = wu'w$ for some $t > r$. But then x and $\sigma^{nj+k}(x)$ both belong to $[w] \subset D_0$. On the other hand

$$\sigma^{nj+k}(x) \in \sigma^{nj+k}(D_0) = \sigma^k(D_0) = D_k.$$

Since $k < n$, we have $[w] \cap D_k = \emptyset$ which leads to a contradiction. $\qquad\square$

4 The Strong Property P Does Not Imply Topological Mixing

We construct a topologically weakly mixing but not topologically mixing shift Y with the strong property P. This shows that the strong property P and topological mixing are not equivalent in general. Note that Y can not be a coded system by Corollary 3.2. A similar example was first given by Blanchard [7], but our construction is much simpler.

Given $R \subset \mathbb{N}$ we follow [18] and define a *spacing shift* Ω_R as the set of all $x \in \{0, 1\}^{\mathbb{Z}}$ such that the condition $x_i = x_j = 1$ for some $i, j \in \mathbb{Z}$ with $i \ne j$ implies

$|i - j| \in R$. Elements of $\mathcal{B}(\Omega_R)$ are called R-*allowed blocks* (see [4, 18] for more properties of spacing shifts).

Theorem 4.1 *There is a shift space Y with the strong property P, which is not topologically mixing.*

Proof We construct a *spacing shift* with the desired properties. Below we write $\langle n \rangle_2$ for the binary representation of a positive integer n, that is,

$$\langle 1 \rangle_2 = 1, \ \langle 2 \rangle_2 = 10, \ \langle 3 \rangle_2 = 11, \ldots.$$

Also for $u = u_1 \ldots u_n \in \{0, 1\}^n$ we let $\Delta(u) = \{|i - j| : 1 \leq j < i \leq n, \ u_i = u_j = 1\}$.

Let $R = \mathbb{N} \setminus \{2^k : k \in \mathbb{N}\}$. Define $Y = \Omega_R$, and note that R is thick, thus Ω_R is nontrivial and topologically weakly mixing (see [18]). We claim that for every $k \in \mathbb{N}$, $L = 2^k$, $w = 0^{2L}$ and any family of R-allowed blocks $v_0, v_1, \ldots, v_t$ of length L, the block $u = v_0 w v_1 w v_2 \ldots v_{t-1} w v_t$ is also R-allowed. This clearly implies that the spacing shift Ω_R has the strong property P. A simple calculation yields that

$$\Delta(u) \subset \left(\{1, \ldots, L - 1\} \cap R \right) \cup \left(\bigcup_{m=0}^{\infty} \{(3m + 2)L + 1, \ldots, (3m + 4)L - 1\} \right).$$

It is enough to show that no power of 2 is in $\Delta(u)$.

Note that any

$$q \in \bigcup_{m=0}^{\infty} \{(3m + 2)L + 1, \ldots, (3m + 4)L - 1\}$$

can be written as $q = a + b$ where $a \in \{2^{k+1} + 1, \ldots, 2^{k+2} - 1\}$ and $b = 3m \cdot 2^k$. If $b = 0$ then clearly $q = a$ is not a power of 2, hence we may assume that $m > 0$. Then $\langle b \rangle_2 = \langle 3m \rangle_2 0^k$ and $\langle a \rangle_2 = 1 x_k \ldots x_0$, where not all x_i's are 0. Denote

$$\langle a + b \rangle_2 = y_l y_{l-1} \ldots y_k \ldots y_0.$$

Note that $l > k + 1$. If $x_i \neq 0$ for some $i = 0, 1, \ldots, k - 1$, then $y_i \neq 0$ and $a + b$ is not a power of 2. If $x_0 = x_1 = \ldots = x_{k-1} = 0$ and $x_k = 1$, then $a = 3 \cdot 2^k$. In that case, $a + b$ is also divisible by 3 and hence it is not a power of 2. Therefore $\Delta(u) \subset R$ and Ω_R has the strong property P.

On the other hand Ω_R is not topologically mixing because it is easy to see that $N([1]_R, [1]_R) = R$. Since R does not contain powers of 2, the set $N([1]_R, [1]_R)$ is not cofinite which is a necessary condition for topological mixing (see [4], cf. [18]). Here $[1]_R = \{x \in \Omega_R : x_0 = 1\}$ is a nonempty open subset of Ω_R. $\qquad \square$

5 A Topologically Mixing Coded System Without Periodic Points of Odd Period

We construct a topologically mixing coded system without periodic points of odd period and such that every set of generators for this system contains only words of even length.

Let $t = t_0 t_1 t_2 \ldots = 10010110\ldots$ be the Prouhet-Thue-Morse sequence (see [1]). Recall that it obeys $t_{2n} = t_n$ and $t_{2n+1} = 1 - t_n$. It is well-known that t is a cube-free sequence, in particular neither 000 nor 111 occur in t. Let $\mathcal{B}(t)$ be the set of all words occurring in t. It is well-known that $\mathcal{B}(t)$ is a language of a minimal and non-periodic shift space X_{TM}.

We first define auxiliary sets $L_n \subset \{0, 1\}^*$ for $n = 1, 2, \ldots$ and a sequence of words $\{a_k\}_{k=0}^{\infty}$. We begin by setting $a_0 = 01$ and $L_1 := \{a_0\}$. Assume that we have performed $n - 1$ steps of our construction ($n \in \mathbb{N}$). We are given the set L_{n-1} and $\{a_k\}_{k=0}^{\infty}$ is defined for indices $0, 1, \ldots, s_n - 1$, that is, s_n denotes the number of words in the sequence $\{a_k\}$ constructed up to the step n. In particular, we have $s_1 = 0$ and $s_2 = 1$. At each step $n \geq 2$ we enumerate the blocks in L_{n-1} starting from s_n, that is, we write

$$L_{n-1} = \{w_{s_n}, \ldots, w_{s_n + |L_{n-1}| - 1}\}.$$

We extend the sequence $\{a_k\}$ by adding words

$$a_j = 01110 t_{[0,4j-3]} 011110 w_j 011110 t_{[0,4j-1]} 01110.$$

for $j = s_n, \ldots, s_n + |L_{n-1}| - 1$. Then we set

$$L_n := \left\{ a_{s_n}, a_{s_n+1}, \ldots, a_{s_{n+1}-1} \right\} \cup \bigcup_{k=1}^{n} L_{n-1}^k,$$

where $L_{n-1}^k = \{w_1 w_2 \ldots w_k : w_j \in L_{n-1} \text{ for } j = 1, \ldots, k\}$. This completes the step n and our induction. Let $\mathcal{Q} := \{a_i ; i \in \mathbb{N}_0\}$.

We will call the words 01110 and 011110 *markers*. Note that a_0 is the only element of $\mathcal{Q}$ without markers, and since 111 is not in $\mathcal{B}(t)$ the positions of markers in a_k are unique and therefore we can identify positions of blocks $t_{[0,4j-3]}$ and $t_{[0,4j-1]}$. Hence knowing that $w \in \mathcal{Q}$ and the length of the longest subblock from $\mathcal{B}(t)$ in w between two markers (when $w \neq a_0$) we can uniquely determine k such that $w = a_k$.

Notice that

$$L_n \subset L_{n+1} \quad \text{and} \quad \mathcal{Q}^+ = \bigcup_{n=1}^{\infty} L_n. \tag{1}$$

Thus $\mathcal{Q}$ and $\bigcup_{n=1}^{\infty} L_n$ generate the same coded system denoted by $\mathbb{X}$.

Lemma 5.1 *The coded system $\mathbb{X}$ is topologically mixing.*

Proof Every block $u \in \mathcal{B}(\mathbb{X})$ is a subword of some concatenation of generators. Therefore it is enough to show that for any $k, \ell \in \mathbb{N}$ and $u_1, \ldots, u_k, v_1 \ldots v_\ell \in Q$ there is $M \in \mathbb{N}$ such that for all $m > M$ there is a block $w \in \{0, 1\}^m$ with $u_1 \ldots u_\ell w v_1 \ldots v_k \in \mathcal{B}(\mathbb{X})$.

Observe that by (1) there is $n \in \mathbb{N}$ such that $u_1, \ldots, u_k \in L_n$. Clearly, we may also assume that $n > k$. Then it follows directly from the construction that $u = u_1 \ldots u_k \in L_{n+1}$ and there is $q \in \mathbb{N}$ such that

$$a_q = 01110 t_{[0,4q-3]} 011110 u 011110 t_{[0,4q-1]} 01110 \in L_{n+2}.$$

Define $s_{2q-1} = t_{[0,4q-3]}$ and $s_{2q} = t_{[0,4q-1]}$. Set $M = 4q + 11$. If $m > M$ is odd, then we have

$$a_q a_0^{\frac{m-4q-11}{2}} v_1 \ldots v_k =$$

$$01110 s_{2q-1} 011110 \underbrace{u_1 \ldots u_\ell}_{u} \underbrace{011110 s_{2q} 01110 (01)^{\frac{m-4q-11}{2}}}_{w \in \{0,1\}^m} v_1 \ldots v_k \in Q^+.$$

For even $m > M$ we have

$$u a_0^{\frac{m}{2}} v_1 \ldots v_k = u (01)^{\frac{m}{2}} v_1 \ldots v_k \in Q^+.$$

This completes the proof, since $Q^+ \subset \mathcal{B}(\mathbb{X})$. □

Lemma 5.2 *If X is a non trivial coded system generated by a set Q and there is a word $w \in Q$ with odd length, then X contains a periodic point of odd primary period greater than one.*

Proof Since X is a nontrivial coded system generated by Q, there must be a non constant word u in Q^+ as otherwise X would consist of a single fixed point. Thus $uu \in Q^*$ is a non constant word of even length. Then the word uuw is a non constant word of odd length k in Q^+ whose infinite concatenation is a non fixed periodic point with an odd primary period dividing k. □

We are going to prove that $\mathbb{X}$ has no periodic points with odd period. For $n \in \mathbb{N}$ let Y_n be the coded system obtained by taking L_n as a set of generators. Since L_n is finite, Y_n is sofic. We show that an odd periodic point cannot occur in Y_n.

Lemma 5.3 *For every $n \in \mathbb{N}$ the sofic shift Y_n generated by the set L_n does not contain a periodic point with odd prime period.*

Proof Fix any $n \in \mathbb{N}$ and let $x \in Y_n$ be a periodic point with prime period q. Clearly, $q > 1$ because the lengths of runs of 0's and 1's in $\mathcal{B}(\mathbb{X})$ are bounded. If x does not contain any marker then $x = (01)^\infty$ and q is even. Thus we may assume that there are (infinitely many) markers in x. Let ℓ be the length of the longest block w from $\mathcal{B}(t)$ appearing in x between two markers. Then $\ell = 4k$ for some k, and there must be $j \in \mathbb{Z}$ such that $x_{[j, j+|a_k|-1]} = a_k$ (no word a_r with $r > k$ can appear in x, since then we would have $\ell \geq 4k + 4$). Because x has period q, we also have $x_{[j+q,\,j+q+|a_k|-1]} = a_k$. Let $w \in \mathcal{Q}^+$ be a word which contains $x_{[j,\,j+q+|a_k|-1]}$ as a subblock and which does not contain a block from $\mathcal{B}(t)$ of length greater than $4k$. Write $w = v_1 \ldots v_t$ where $v_j \in \mathcal{Q}$. By the above observation, we have that $x_{[j,\,j+|a_k|-1]} = x_{[j+q,\,j+q+|a_k|-1]} = a_k \in \mathcal{Q}$ must be among v_i. This immediately implies that $x_{[j,\,j+q-1]} \in \mathcal{Q}^+$ and since all words in $\mathcal{Q}$ have even length, we find that $q = |x_{[j,\,j+q-1]}|$ is even.   □

Finally we show that taking the closure of $\bigcup_{n=1}^{\infty} Y_n$ does not introduce periodic points. We will use the fact that the Prouhet-Thue-Morse sequence t is cube-free, that is, for every word w over $\{0, 1\}$ we have $www \notin \mathcal{B}(t)$.

Lemma 5.4 *Any element x of $\mathbb{X} \setminus \bigcup_{n=1}^{\infty} Y_n$ contains arbitrary long blocks from $\mathcal{B}(t)$. In particular, x cannot be periodic.*

Proof First, note that if $x \in \mathbb{X} \setminus \bigcup_{n=1}^{\infty} Y_n$ is periodic and contains arbitrary long blocks from t, then for some word w we would have $www \in \mathcal{B}(t)$ contradicting that t is cube-free. Now, assume that there is $x \in \mathbb{X} \setminus \bigcup_{n=1}^{\infty} Y_n$ such that the longest block from $\mathcal{B}(t)$ appearing in x has length at most $4k$ for some integer $k > 0$. Then x must contain infinitely many markers as subwords, as otherwise $x = \ldots a_0 a_0 w a_0 a_0 \ldots$ for some $w \in L_n$ and some n, thus $x \in Y_n$.

Therefore there exists an infinite set $J \subset \mathbb{Z}$ and a strictly increasing infinite sequence of integers $(n_i)_{i \in J}$ such that $x_{[j,\,j+4]} = 01110$ if and only if there is $i \in J$ such that $j = n_i$. The set J is the set of positions at which markers occur in x. Let m be the least integer such that $a_k \in L_m$.

We claim that every word $w_{s,i} := x_{[n_s,\ldots,n_i+4]}$ for any $s, i \in J$, $s < i$ is contained in $\mathcal{B}(Y_m)$. Note that by the definition each $w_{s,i}$ starts and ends with the marker 01110. Let j be the smallest positive integer, such that $w_{s,i} \in \mathcal{B}(Y_j)$. There must be some word $g \in L_j^+$ such that $g = bw_{s,i}c$ with $b, c \in \mathcal{B}(\mathbb{X})$. Either $j \leq m$, in which case we are done, or $j > m$ and g must contain a_ℓ for some $\ell > k$. Since $w_{s,i}$ starts and ends with 01110, the two longest blocks of symbols from $\mathcal{B}(t)$ occurring in a_ℓ must be contained in b and c and thus $w_{s,i}$ is already contained in the middle word $\xi = 01110u01110$ of a_ℓ, where $u \in L_{j-1}$. Note that ξ is an element of L_{j-1}^+. This contradicts the minimality of j. Therefore our claim holds. If J is bi-infinite, then $x \in Y_m$ which is a contradiction. Otherwise, either

$$a = \inf\{n_i : i \in J\} = \min\{n_i : i \in J\} > -\infty, \quad \text{or}$$
$$b = \sup\{n_i : i \in J\} = \max\{n_i : i \in J\} < \infty.$$

If $a > -\infty$, then $x_{(-\infty,\,a-1]}$ does not contain markers and $x_{[a,\,a+4]} = 011110$ which implies that $x_{(-\infty,\,a-1]} = \ldots a_0 a_0 a_0$. In the second case $b < \infty$ and we obtain that $x_{[b+1,\,\infty)} = a_0 a_0 a_0 \ldots$, hence both cases imply again that $x \in Y_m$ and this contradiction completes the proof.   □

Theorem 5.5 *There exists a coded system X which is topologically mixing, but does not have periodic points with odd periods. In particular, every set of generators for X contains only blocks of even length.*

Proof The shift $\mathbb{X}$ is topologically mixing by Lemma 5.1. Using Lemmas 5.3 and 5.4 we see that $\mathbb{X}$ does not contain a periodic point with odd prime period. Then it follows from Lemma 5.2 that every set of generators for $\mathbb{X}$ contains only blocks of even length.   □

6 The Strong Property P and Topologically Mixing Coded Systems

As we have seen before, the strong property P does not imply topological mixing. Clearly, the converse implication is not true either, since Blanchard [7] proved that the property P implies positive topological entropy and there are examples of topologically mixing shifts with zero topological entropy.

The purpose of this section is to show that every topologically mixing coded system has the strong property P.

Lemma 6.1 *Let X be a topologically mixing coded system, Q be its generator, and $k = \gcd\{|u| : u \in Q\}$. Then for each $u \in \mathcal{B}(X)$ with $|u| = 0 \bmod k$ there is a word $v \in Q^+$ such that $v = aub$ with $|a| = 0 \bmod k$ and $|b| = 0 \bmod k$.*

Proof Since X is topologically mixing, there is a word $v \in Q^+$ such that for some $\tilde{a}, \tilde{b}, z_1, \ldots, z_k \in \mathcal{B}(X)$ we have

$$v = \tilde{a} u z_1 u z_2 u z_3 \ldots u z_k \tilde{b}$$

with $|z_i| = 1 \bmod k$ for each $i = 1, \ldots, k$. Replacing v by v^k if necessary, we may assume that $|v| = 0 \bmod k$. There is $\ell \in \{0, \ldots, k - 1\}$ such that $|\tilde{a}| = -\ell \bmod k$. Let $a = \tilde{a} u z_1 u z_2 \ldots u z_\ell$ and $b = z_{\ell+1} u z_{\ell+2} \ldots u z_k \tilde{b}$ and observe that $|a| = 0 \bmod k$, $|b| = |v| - |a| - |u| = 0 \bmod k$ and $v = aub$.   □

Theorem 6.2 *If X is a topologically mixing coded system, then X has the strong property P.*

Proof Blanchard [7, Proposition 4] proved that a shift space X over $\mathcal{A}$ has the strong property P if for any integer p belonging to some infinite strictly increasing sequence of integers there exists an integer $q = q(p)$ such that for any $s \geq 2$ and any words $u_1, \ldots, u_s \in \mathcal{B}_p(X)$ there are words $w_1, \ldots, w_{s-1} \in \mathcal{B}_q(X)$ such that $u_1 w_1 u_2 \ldots u_{s-1} w_{s-1} u_s \in \mathcal{B}(X)$. Let Q be a set of generators of X and $k = \gcd\{|u| : u \in Q\}$.

We will show that Blanchard's criterion [7, Proposition 4] applies to any $p \in \{\ell \in \mathbb{N} : \ell = 0 \bmod k\}$. To this end, fix $p = 0 \bmod k$ and enumerate all blocks of length p by $v_1, \ldots, v_n$. We use Lemma 6.1 to obtain $a_1, \ldots, a_n, a'_1, \ldots, a'_n \in \mathcal{B}(X)$ such that

$$0 = |a_1| = \cdots = |a_n| = |b_1| = \cdots = |b_n| \bmod k$$

and $a_i v_i a_i' \in Q^+$ for $i \in \{1, \ldots, n\}$. By Remark 2.0.1 there exists N such that for any $j \geq N$ there is a word $\bar{e}_j \in Q^+$ with $|\bar{e}_j| = kj$. Let $|a_i| = \ell_i k$ and $|a_i'| = \ell_i' k$ where $\ell_i, \ell_i' \in \mathbb{N}$. Set $L = N + \max\{\ell_i, \ell_i' : i = 1, \ldots, n\}$. Define $j(i) = L - \ell_i \geq N$ and $j'(i) = L - \ell_i' \geq N$. Set $c_i = \bar{e}_{j(i)}$ and $d_i = \bar{e}_{j'(i)}$ for $i = 1, \ldots, n$. Thus we have found $c_1, \ldots, c_n \in Q^+$ and $d_1, \ldots, d_n \in Q^+$ such that $Lk = |c_i a_i| = |a_i' d_i|$ for $i = 1 \ldots, n$. Now let $u_1, \ldots, u_s$ be any words in $\mathcal{B}_p(X)$. Then there is a function $\phi \colon \{1, \ldots, s\} \to \{1, \ldots, n\}$ with $u_j = v_{\phi(j)}$ for $j = 1, \ldots, s$. Observe that $c_{\phi(j)}, d_{\phi(j)}, a_{\phi(j)} u_j a_{\phi(j)}' \in Q^+$, hence

$$a_{\phi(1)} u_1 \underbrace{(a_{\phi(1)}' d_{\phi(1)} c_{\phi(2)} a_{\phi(2)})}_{w_1} u_2 \underbrace{(a_{\phi(2)}' d_{\phi(2)} c_{\phi(3)} a_{\phi(3)})}_{w_2} u_3 \ldots c_{\phi(s)} a_{\phi(s)} u_s a_{\phi(s)}' \in Q^+.$$

Therefore $u_1 w_1 u_2 \ldots u_{s-1} w_{s-1} u_s \in \mathcal{B}(X)$. We set $q(p) = 2Lk$ and we obtain that X has the strong property P by [7, Proposition 4]. $\qquad \square$

7   Two Folklore Results

We finish the paper with two results which are probably folklore, but we were unable to find them in the literature so we attach them for completeness. Combining them with Corollary 3.2 we obtain that the stronger forms of transitivity mentioned in the Introduction are equivalent for synchronized systems.

Lemma 7.1 *Let X be a coded system presented by a labeled graph G. If there are two cycles on G with relatively prime lengths, then X is topologically mixing.*

Proof Denote by α_1 and α_2 two cycles on G with relatively prime lengths, $k_j = |\alpha_j|$, $j = 1, 2$. Let e_j be a vertex of some edge belonging to α_j for $j = 1, 2$. Let u_j be the label of α_j for $j = 1, 2$ read off traversing α_j from e_j. Take any words $w_1, w_2 \in \mathcal{B}(X)$. Since X is coded, we can find paths γ_1, γ_2 on G labeled, respectively, by w_1, w_2. Let a_2 be the initial vertex of γ_2, and b_1 be the terminal vertex of γ_1. Let ℓ_1 (ℓ_2) be the length of the shortest path π_1 (π_2) on G from b_1 to e_1 (from e_2 to a_2) and m be the length of the the shortest path ρ on G from e_1 to e_2. Let $v_1, v_2, z \in \mathcal{B}(X)$ be labels of π_1, π_2, ρ, respectively. It follows that for each $p, q \in \mathbb{N}$ the path

$$\pi(p, q) = \gamma_1 \pi_1 (\alpha_1)^p \rho (\alpha_2)^q \pi_2 \gamma_2$$

is labeled by $w_1 v_1 (u_1)^p z (u_2)^q v_2 w_2$. Since k_1 and k_2 are relatively prime, the set $\{pk_1 + qk_2 : p, q \in \mathbb{N}\}$ is cofinite. Therefore there is $N > 0$ such that if we fix any $n \geq N$ then we can find $p_n, q_n \in \mathbb{N}$ so that $n = \ell_1 + p_n k_1 + m + q_n k_2 + \ell_2$ and the path $\pi(p_n, q_n)$ on G has length $n + |w_1| + |w_2|$. Therefore for each $n \geq N$ the word $w^{(n)} = v_1 (u_1)^{p_n} z (u_2)^{q_n} v_2$ has length n and $w_1 w^{(n)} w_2 \in \mathcal{B}(X)$. It follows that X is topologically mixing. $\qquad \square$

Fiebig and Fiebig proved in [13] that every synchronized system X can be presented by a special labeled graph called the *Fischer cover*. Furthermore, they showed that the Fischer cover of X is unique up to an isomorphism of labeled graphs. The Fischer cover of a synchronized system may be characterized as the countable labeled graph (G, Θ) presenting X which is right-resolving, follower separated, and every synchronized word w for X is a magic word for (G, Θ). This generalizes the notion of a minimal, right-resolving, and follower separated presentation of sofic shifts. For details we refer the reader to [13] or [24].

Lemma 7.2 *A synchronized shift X is topologically mixing if and only if there are two cycles with relatively prime lengths in its Fischer cover.*

Proof The "if" part follows from Lemma 7.1. For the "only if" part assume that X is topologically mixing. Let w be a synchronizing word for X. Then w is a magic word for the Fischer cover (G, Θ) of X, that is, there is a vertex e of G such that every path labeled by w ends at e. Since X is topologically mixing there is $N \in \mathbb{N}$ and there are words u_1, u_2 with $|u_1| = N, |u_2| = N + 1$ such that $wu_1w, wu_2w \in \mathcal{B}(X)$. Because each path labeled by w ends at e, there are cycles in G labeled by u_1w and u_2w with relatively prime lengths.   □

8   Proof of Theorem 1.1

Finally, we can prove our main result.

Proof of Theorem 1.1 To prove equivalence of (a)–(d), it is enough to combine Corollary 3.2, Theorem 6.2 and [7] (the strong property P implies topological weak mixing). The equivalence of these conditions with (e*) follows from Lemma 7.2. Finally, the "moreover" part is a consequence of Theorem 5.5.   □

9   Additional Example

The following example (suggested by a referee) shows that in order to obtain a condition equivalent to (a)–(d) for synchronized systems it is necessary to strengthen the condition (e) to (e*). Consider the coded system over $\{0, 1, 2, 3, 4\}$ generated by the code words

$$220101(010101)^n444444 \quad \text{and} \quad 333001(001001)^n444444 \quad \text{for } n = 1, 2, \ldots.$$

Note that 444444 is a synchronizing word for this system, and it is clearly transitive, hence it is a synchronized system. It is also easy to see, that it contains two periodic points with relatively prime primary periods generated by 01 and 001. The shift space is not totally transitive: the set of return times to the cylinder of 333 contains only integers of the form $6n + 9$ for $n = 1, 2, \ldots$, so σ^2 is not transitive on our shift space. Thus no two cycles in any presentation of this shift have co-prime lengths, as this implies total transitivity by Theorem 1.1.

Acknowledgements The authors would like to thank the anonymous referee for careful reading and many remarks, which helped us to improve the paper. The research of P. Oprocha was supported by the Polish Ministry of Science and Higher Education from sources for science in the years 2013-2014, Grant No. IP2012 004272. The research of D. Kwietniak was supported by the National Science Centre (NCN) under grant no. DEC-2012/07/E/ST1/00185. The research of J. Epperlein was partly supported by the German Research Foundation (DFG) through the Cluster of Excellence (EXC 1056), Center for Advancing Electronics Dresden (cfaed).

References

1. J.-P. Allouche, J. Shallit, The ubiquitous Prouhet-Thue-Morse sequence. Sequences and their applications (Singapore, **1–16** (Springer Ser. Discrete Math. Theor. Comput. Sci, Springer, London, 1998), p. 1999
2. S. Baker, A. Ghenciu, Dynamical properties of S-gap shifts and other shift spaces. J. Math. Anal. Appl. **430**(2), 633–647 (2015)
3. J. Banks, Regular periodic decompositions for topologically transitive maps. Ergod. Theor. Dynam. Syst. **17**(3), 505–529 (1997)
4. J. Banks, T.D. Nguyen, P. Oprocha, B. Stanley, B. Trotta, Dynamics of spacing shifts. Discret. Contin. Dyn. Syst. **33**(9), 4207–4232 (2013)
5. F. Blanchard, G. Hansel, G. *Systèmes codés* (French) [Coded systems], Theoret. Comput. Sci. **44**(1), 17–49 (1986)
6. F. Blanchard, G. Hansel, *Systèmes codéss et limites de systèmes sofiques.* (French) [Coded systems and limits of sofic systems] C. R. Acad. Sci. Paris Sèr. I Math. **303**, 475–477 (1986)
7. F. Blanchard, *Fully positive topological entropy and topological mixing in Symbolic dynamics and its applications* (New Haven, CT, 1991), 95–105, Contemp. Math., **135**, Amer. Math. Soc., Providence, RI (1992)
8. J. Buzzi, Specification on the interval. Trans. Amer. Math. Soc. **349**(7), 2737–2754 (1997)
9. V. Climenhaga, D. Thompson, Intrinsic ergodicity beyond specification: β-shifts, S-gap shifts, and their factors. Israel J. Math. **192**(2), 785–817 (2012)
10. V. Climenhaga, D. Thompson, Intrinsic ergodicity via obstruction entropies. Ergod. Theor. Dynam. Syst. **34**(6), 1816–1831 (2014)
11. D. Dastjerdi, M. Dabbaghian, Mixing coded systems. Georgian Math. J., to appear. https://doi.org/10.1515/gmj-2017-0058
12. D. Dastjerdi, S. Jangjoo, Dynamics and topology of S-gap shifts. Topol. Appl. **159**(10), 2654–2661 (2012)
13. D. Fiebig, U.-R. Fiebig, Covers for coded systems. Symbolic dynamics and its applications (New Haven, CT, 1991), 139–179, Contemp. Math., 135, Amer. Math. Soc., Providence, RI (1992)
14. E. Glasner, Ergodic Theory via Joinings, Mathematical Surveys and Monographs, 101. American Mathematical Society, Providence, RI, 2003. xii+384 pp
15. G. Harańczyk, D. Kwietniak, P. Oprocha, A note on transitivity, sensitivity and chaos for graph maps. J. Difference Equ. Appl. **17**, 1549–1553 (2011)
16. D. Jabłoński, M. Kulczycki, Topological transitivity, mixing and nonwandering set of subshifts of finite type-a numerical approach. Int. J. Comput. Math. **80**, 671–677 (2003)
17. W. Krieger. On subshifts and topological Markov chains. Numbers, Information and Complexity. Eds. I. Althöfer et al. Kluwer, pp. 453-472 (2000)

18. K. Lau, A. Zame, On weak mixing of cascades. Math. Systems Theory. **6**, 307–311 (1972/73)
19. D. Lind, B. Marcus, *An introduction to symbolic dynamics* (Cambridge University Press, New York, 1995)
20. T. Meyerovitch, Tail invariant measures of the Dyck shift. Isr. J. Math. **163**(1), 61–83 (2008)
21. T. Meyerovitch, Gibbs and equilibrium measures for some families of subshifts. Ergod. Theor. Dynam. Syst. **33**(3), 934–953 (2013)
22. J. Schmeling, Symbolic dynamics for β-shifts and self-normal numbers. Ergod. Theor. Dynam. Syst. **17**(3), 675–694 (1997)
23. D.J. Thompson, Irregular sets, the β-transformation and the almost specification property. Trans. Amer. Math. Soc. **364**, 5395–5414 (2012)
24. K. Thomsen, On the ergodic theory of synchronized systems. Ergod. Theor. Dynam. Syst. **26**(4), 1235–1256 (2006)

Transversality for Critical Relations of Families of Rational Maps: An Elementary Proof

Genadi Levin, Weixiao Shen and Sebastian van Strien

In memory of our dear friend and colleague Welington de Melo.

Abstract In this paper we will give a short and elementary proof that critical relations unfold transversally in the space of rational maps.

Keywords Holomorphic dynamics · Rational maps · Transversality

1 Introduction

In this short paper we will give an elementary proof of some transversality properties for families of rational maps. We will consider the space $\mathbf{Rat}_d^\mu$ of rational maps of degree d with precisely ν critical points of multiplicities $(\mu_1, \mu_2, \ldots, \mu_\nu)$. In Theorem 2.1 we will show that this space of maps can be locally parametrised by critical values. Given $f \in \mathbf{Rat}_d^\mu$, let $\zeta = \zeta(f) \geq 0$ be the maximal number of critical points with pairwise disjoint infinite orbits and define $N = \nu - \zeta(f)$. In Theorem 3.2 we will show that if f is not a flexible Lattès map then one can organise the set of critical relations of f in the form

G. Levin
Hebrew University, Jerusalem, Israel
e-mail: genady.levin@mail.huji.ac.il

W. Shen
Shanghai Center for Mathematical Sciences, Fudan University, Shanghai, China
e-mail: wxshen@fudan.edu.cn

S. van Strien (✉)
Imperial College, London, UK
e-mail: s.van-strien@imperial.ac.uk

© The Author(s) 2019
M. J. Pacifico and P. Guarino (eds.), *New Trends in One-Dimensional Dynamics*, Springer Proceedings in Mathematics & Statistics 285,
https://doi.org/10.1007/978-3-030-16833-9_11

201

$$\{f^{m_k}(c_{i_k}) = f^{n_k}(c_{j_k}), k = 1, \ldots, N\}$$

so that the map

$$\mathbf{Rat}_d^{\mu} \ni g \mapsto \{\sigma(g^{m_k}(c_{i_k}(g))) - \sigma(g^{n_k}(c_{j_k}(g)))\}_{k=1}^N \qquad (1.1)$$

has maximal rank for g near f, where σ is any Möbius transformation with $\sigma(f^{m_k}(c_{i_k})) \neq \infty$. Property (1.1) obviously is a transversality condition.

In fact, the choice of critical relations is in general not unique, but as long as the selected collection is *full*, as made explicit in Definition 3.6 below, the maximal rank property holds.

Indeed, we should emphasise that some care is required in the choice of critical relations. For example, in the case of $f_t(z) = z^2 + t$ with $t = 0$, the derivative of $t \mapsto f_t^2(0) - f_t(0)$ vanishes at $t = 0$. The correct way of expressing transversal unfolding of the critical relation $f_t(0) = 0$ in (1.1) is by taking $m_1 = 1$ and $n_1 = 0$ in this equation, i.e. by asserting that the derivative $t \mapsto f_t(0) - 0$ is non-zero at $t = 0$.

In the unicritical case, transversal unfolding of critical relations in the pre-periodic case goes back to Douady and Hubbard [5] and Tsujii [32], see also [16, Remark 5.10]. Milnor and Thurston [28] and Sullivan, see [25, Theorem VI.4.2], proved a 'topological' version of transversality.

An abstract approach to transversality for finite type maps was developed by A. Epstein, see [8, 9], obtaining in Part 1 of [8] transversality within the Teichmüller deformation space $\mathrm{Def}_A^B(f)$, and in Sect. 5.4 in [8] the loci defined by critical relations within $\mathrm{Def}_A^B(f)$ is discussed. Part 2, and in particular Sect. 10, of [8] goes into a strategy for transferring the transversality results obtained in $\mathrm{Def}_A^B(f)$ to the space of rational functions. However, we were not able to find an explicit statement covering Theorem 3.2 or Theorem 3.3. Nevertheless, it is likely that the strategy in [8] can be executed to obtain statements similar to the ones in this paper.

Our results also hold in the setting of degenerate critical points and gives unfoldings of critical relations even when critical points share the same critical value. For this we use that $\mathbf{Rat}_d^{\mu}$ is a manifold and that $\mathbf{Rat}_d^{\mu} \ni f \mapsto (f(c_1), \ldots, f(c_v))$ has rank v, see Theorem 2.1.

In this short and self-contained paper we prove transversality following the approach developed by Levin in [17–20], see also [15]. The starting point of this paper are calculations from [18, 20] which show that if the transversality property (1.1) fails at $g = f$, then one can construct a non-zero integrable meromorphic quadratic differential that is invariant under push-forward by f, which in turn implies that f is a flexible Lattès example. Indeed, the main Theorem 3.3 can be proved as in [20], see Remark 5.1, although we shall provide a more direct and shorter proof in this paper.

The idea of using quadratic differentials appeared first in Thurston's characterization of post-critically finite branched covering of the 2-sphere [6]. It has been used in for example [7, 15] and this was also used in [33] to obtain a similar statement to ours for the quadratic case.

Theorem 3.2 was proved previously for the case that critical points are non-degenerate and eventually mapped into repelling periodic orbits, but never into a critical point, see [3, 30] and also [13, Theorem 4.8].

Transversality also holds in other settings. For example, if each critical point is mapped into a hyperbolic set, see [30], when a summability condition holds along the orbit of critical values, see [2, 20], for the unfolding of multipliers of periodic orbits, see [8, 18] and for a large class of interval maps, see [21].

As mentioned, the aim of this paper is to present a proof of transversality for rational maps with critical relations in a complete and readily accessible form.

In Sect. 6 we discuss corresponding results for polynomials.

2 Parametrising Rational Maps by Their Critical Values

Let $\mathbf{Rat}_d$ denote the collection of all rational maps of degree $d \geq 2$. This space is naturally parameterized by an open set in $P\mathbb{C}^{2d+1}$.

Given a non-ordered list $\boldsymbol{\mu} = (\mu_1, \mu_2, \ldots, \mu_\nu)$ with $\sum_{i=1}^{\nu} \mu_i = 2d - 2$, we say a rational map $f \in \mathbf{Rat}_d$ is in the class $\mathbf{Rat}_d^{\mu}$ if f has precisely ν distinct critical points $c_1, c_2, \ldots, c_\nu$ with multiplicities $\mu_1, \mu_2, \ldots, \mu_\nu$ respectively. Taking $\mathbf{1} = (1, \ldots, 1)$, $\mathbf{Rat}_d^{\mathbf{1}}$ corresponds to the space of rational maps with $2d - 2$ non-degenerate critical points.

Rational maps are not fully determined by their critical values (not even on small open subsets $W \subset \mathbf{Rat}_d^{\mu}$), because one can precompose a rational map by a Möbius transformation without changing its critical values. However one can find a neighbourhood W of f and a normalisation (based on precompositions with Möbius transformations) so that critical values parametrise all maps in W satisfying this normalisation:

Theorem 2.1 *For each $\boldsymbol{\mu}$, $\mathbf{Rat}_d^{\mu}$ is an embedded submanifold of dimension $\nu + 3$ of $\mathbf{Rat}_d$ and the functions defined by the critical values form a* **partial** *holomorphic local coordinate system, i.e. the map $\mathbf{Rat}_d^{\mu} \ni f \mapsto (f(c_1), \ldots, f(c_\nu))$ has rank ν and can be completed by 3 other coordinates to be a holomorphic coordinate system.*

Remark 2.1 Theorem 2.1 is not new. Similar statements are proved e.g. in [11, 19] (see also [12]) using the Measurable Riemann Mapping Theorem with dependence on parameters; the idea of those proofs goes back probably to [31]. Our proof borrows an idea of Douady and Sentenac [29, Appendix A], and is short and elementary. The case $\mu_\nu = d - 1$ corresponds to the polynomial case, which in some real cases was dealt with in [25, p. 120] and [29], see also [10].

Theorem 2.1 follows from Proposition 2.2 below. Assume without loss of generality (by post and pre composing f by Möbius transformations if necessary) that the critical points and the critical values avoid the point at ∞. Then for each $i = 1, 2, \ldots, \nu$,

$$f'(c_i) = f''(c_i) = \cdots = f^{(\mu_i)}(c_i) = 0, \ f^{(\mu_i+1)}(c_i) \neq 0.$$

Applying the Implicit Function Theorem to the maps $(g, \zeta_i) \mapsto g^{(\mu_i)}(\zeta_i)$ for (g, ζ_i) near (f, c_i), gives that there exists a neighborhood W of f in $\mathbf{Rat}_d$ and uniquely defined functions $\zeta_i \colon W \to \mathbb{C}$ which are holomorphic such that $\zeta_i(f) = c_i$ and $g^{(\mu_i)}(\zeta_i(g)) = 0, \ g^{(\mu_i+1)}(\zeta_i(g)) \neq 0$ for each $g \in W$. Replacing W by a smaller neighborhood, for each $g \in W$ the equation $g'(\zeta) = 0$ has μ_i solutions ζ (counting multiplicity) near c_i. It follows that for any $g \in W \cap \mathbf{Rat}_d^{\mu}$, $g'(\zeta) = 0$ has a unique solution near c_i (with multiplicity μ_i); hence $\zeta_i(g)$ is the only critical point of $g \in W \cap \mathbf{Rat}_d^{\mu}$ near c_i and it has multiplicity μ_i.

For $g \in W$, write

$$\zeta_i^0(g) = g(\zeta_i(g)), \ \zeta_i^1(g) = g'(\zeta_i(g)), \ \zeta_i^2(g) = g''(\zeta_i(g)), \ldots.$$

Thus $\zeta_i(g)$ is a critical point of g with multiplicity μ_i if and only if $\zeta_i^j(g) = 0$ for all $1 \leq j \leq \mu_i - 1$ (note that $g^{(\mu_i)}(\zeta_i(g)) = 0, \ g^{(\mu_i+1)}(\zeta_i(g)) \neq 0$ holds automatically for all $g \in W$). Define $G \colon W \to \mathbb{C}^{2d-2}$ by

$$g \to (\zeta_1^0(g), \zeta_1^1(g), \ldots, \zeta_1^{(\mu_1-1)}(g), \ldots, \zeta_v^0(g), \zeta_v^1(g), \ldots, \zeta_v^{(\mu_v-1)}(g)).$$

Since W has dimension $2d + 1$, Theorem 2.1 follows immediately from:

Proposition 2.2 *For each rational map f as above, the Jacobian of G has rank $2d - 2$ at $g = f$.*

This proposition also immediately implies:

Corollary 2.3 *Assume that all critical points of f are non-degenerate. Then there exists a neighbourhood W of f in $\mathbf{Rat}_d$ so that the critical points $c_1(g), \ldots, c_{2d-2}(g)$ of g depend holomorphically on $g \in W$ and the Jacobian of the map*

$$g \mapsto (g(c_1(g)), g(c_2(g)), \ldots, g(c_{2d-2}(g)))$$

has maximal rank at every $g \in W$.

2.1 Proof of Proposition 2.2

Proof of Proposition 2.2. Arguing by contradiction, assume that the assertion of the proposition is false. Then there exist complex numbers A_i^j, $1 \leq i \leq v, 0 \leq j < \mu_i$, not all equal to zero, such that all partial derivatives of the map

$$\mathbf{G}(g) = \sum_{i=0}^{v} \sum_{j=0}^{\mu_i-1} A_i^j \zeta_i^{(j)}(g)$$

are equal to zero at $g = f$. This means that for any holomorphic curve f_t in $\mathbf{Rat}_d$, passing through f at $t = 0$, the map $G(t) = \mathbf{G}(f_t)$ satisfies $G'(0) = 0$. Let us write

$$f_t(z) = \frac{\sum_{k=0}^d a_k(t)z^k}{\sum_{k=0}^d b_k(t)z^k} =: \frac{P_t(z)}{Q_t(z)},$$

where $a_k(t)$, $b_k(t)$ are holomorphic in a neighborhood of 0 and P_0 and Q_0 are coprime polynomials. For $1 \le i \le \nu$, $j = 0, \ldots, \mu_i - 1$ define $v_{i,j}(t) = \zeta_i^{(j)}(f_t)$. Then

$$v_{i,j}'(0) = \left(\frac{\sum_{k=0}^d a_k'(0)z^k Q_0(z) - \sum_{l=0}^d b_l'(0)z^l P_0(z)}{Q_0(z)^2} \right)^{(j)} \Bigg|_{z=c_i},$$

where we use $f^{(j+1)}(\zeta_i(f)) = f^{(j+1)}(c_i) = 0$. So

$$0 = G'(0) = \sum_{i,j} A_i^j v_{i,j}'(0)$$

$$= \sum_{i,j} A_i^j \left(\frac{\sum_{k=0}^d a_k'(0)z^k Q_0(z) - \sum_{l=0}^d b_l'(0)z^l P_0(z)}{Q_0(z)^2} \right)^{(j)} \Bigg|_{z=c_i}. \quad (2.1)$$

We claim that for any polynomial T, we have

$$\sum_{i,j} A_i^j \left(\frac{T(z)}{Q_0(z)^2} \right)^{(j)} \Bigg|_{z=c_i} = 0. \quad (2.2)$$

To see this, first notice that since $T_0(z) = \prod_{i=1}^{\nu}(z - c_i)^{\mu_i}$ has a zero at $z = c_i$ of multiplicity μ_i, the Eq. (2.2) holds for $T = T_0$ and $T = T_0 U$, where U is an arbitrary polynomial. Since $\deg(T_0) = 2d - 2$ and any polynomial can be written as $T_0 U + T$ where $\deg(T) < 2d - 2$ it therefore suffices to prove (2.2) in the case that $\deg(T) < 2d - 2$. For such a polynomial T, we can find polynomials R, S of degree at most $d - 1$ such that $T = RQ_0 - SP_0$, since P_0 and Q_0 are coprime and one of them has degree d. Choosing a_k, b_l suitably such that $R(z) = \sum_k a_k'(0)z^k$ and $S = \sum_l b_l'(0)z^l$ and applying (2.1), we obtain (2.2).

We shall now deduce from this equation that $A_i^j = 0$ for all i, j and thus obtain a contradiction. Indeed, (2.2) implies that for any polynomial V, we have

$$\sum_{i,j} A_i^j V^{(j)}(c_i) = 0.$$

Fix $1 \le i_0 \le \nu$, $1 \le j_0 < \mu_{i_0}$, take

$$V(z) = \prod_{i \neq i_0} (z - c_i)^{\mu_i} (z - c_{i_0})^{j_0}.$$

Then $V^{(j_0)}(c_{i_0}) \neq 0$ and $V^{(j)}(c_i) = 0$ for any other (i, j). Therefore $A_{i_0}^{j_0} = 0$. The proof is completed.   □

3 Transversality Results for Rational Maps

Throughout this section we again consider a map f in the space $\mathbf{Rat}_d^{\mu}$ of rational maps of degree d, with v distinct critical points $c_1, c_2, \ldots, c_v$ with multiplicities $\mu = (\mu_1, \mu_2, \ldots, \mu_v)$ where $\sum_{i=1}^{v} \mu_i = 2d - 2$. For g in a small neighborhood of f in $\mathbf{Rat}_d^{\mu}$, the critical points $c_1(g), c_2(g), \ldots, c_v(g)$ depends holomorphically on g.

We are interested in the smoothness of sets defined by a set of critical relations of the form $g^m(c_i(g)) = g^n(c_j(g))$. A particular case of our main result in this direction is the following:

Theorem 3.1 *Let $f \in \mathbf{Rat}_d^{\mu}$ and assume that there exists $1 \leq i, j \leq v$ and $m > 0$ so that $f^m(c_i) = c_j$. Then the equation*

$$g^m(c_i(g)) = c_j(g)$$

defines an embedded submanifold of $\mathbf{Rat}_d^{\mu}$ of codimension one near f.

In order to state a more general result, we have to prepare some terminology. Let us say that a quadruple $(i, j; m, n)$ is a *(candidate) critical relation* if $1 \leq i, j \leq v$, and m, n are non-negative integer with $m + n > 0$. We say that this critical relation is *realized by f* if $f^m(c_i(f)) = f^n(c_j(f))$.

Given f, let $\zeta = \zeta(f) \geq 0$ be the maximal number of critical points with pairwise disjoint infinite orbits. Note that this number is well-defined, but that one *cannot* say *which* critical points are 'free'. For example, if f has three distinct critical points c_1, c_2, c_3, so that the forward orbits of $f(c_1) = f(c_2)$ and c_3 are disjoint and infinite, then $\zeta(f) = 2$; of course one could consider c_1, c_3 as the free critical points of f, but equally well also c_2, c_3.

In this section we will show

Theorem 3.2 *Assume $f \in \mathbf{Rat}_d^{\mu}$ is not a flexible Lattès map. Then there exists a set*

$$\mathcal{F} = \{(i_k, j_k; m_k, n_k), k = 1, \ldots, N\} \text{ with } N = v - \zeta(f)$$

of critical relations $f^{m_k}(c_{i_k}) = f^{n_k}(c_{j_k})$ which are realised by f, such that the Jacobian of the map

$$\mathcal{R}_{\mathcal{F}}^{\sigma} : g \mapsto (\sigma(g^{m_k}(c_{i_k}(g))) - \sigma(g^{n_k}(c_{j_k}(g))))_{k=1}^{N} \tag{3.1}$$

at $g = f$ has rank N, whenever σ is a Möbius transformation for which $\sigma(f^{m_k}(c_{i_k})) \in \mathbb{C}, k = 1, \ldots, N$.

Remark 3.1 The assumption that $\sigma(f^{m_k}(c_{i_k})) \in \mathbb{C}$ is made to ensure that (3.1) is holomorphic near f. The kernel of the Jacobian of $\mathcal{R}_{\mathcal{F}}^{\sigma}$ at f, hence its rank, does not depend on σ, as long as $\sigma(f^{m_k}(c_{i_k})) \neq \infty$ for all $k = 1, \ldots, N$. Indeed, a tangent vector of $\mathbf{Rat}_d^{\mu}$ at f belongs to the kernel if and only if it has the same image under the tangent map of the maps $g \mapsto (g^{m_k}(c_{i_k}(g)))_{k=1}^{N}$ and $g \mapsto (g^{n_k}(c_{j_k}(g)))_{k=1}^{N}$ at $g = f$ (both are holomorphic maps from a neighborhood of f in $\mathbf{Rat}_d^{\mu}$ into $\overline{\mathbb{C}}^N$).

In particular, to prove Theorem 3.2, we can and will assume that the critical obits of f avoid ∞ and only prove that $\mathcal{R}_{\mathcal{F}} = \mathcal{R}_{\mathcal{F}}^{id}$ has rank N at $g = f$. Indeed, we can always choose z_0 (arbitrarily close to ∞) which avoids the critical orbits of f. Put $\sigma(z) = z_0 z/(z_0 - z)$ and $\tilde{f} = \sigma \circ f \circ \sigma^{-1}$. Then ∞ avoids the critical orbits of $\tilde{f}$. Since $\mathcal{R}_{\mathcal{F}}^{\sigma}(g) = \mathcal{R}_{\mathcal{F}}^{id}(\sigma \circ g \circ \sigma^{-1})$, once we prove that the Jacobian of $g \mapsto \mathcal{R}_{\mathcal{F}}^{id}(g)$ has rank N at $g = \tilde{f}$, it follows that the Jacobian of $\mathcal{R}_{\mathcal{F}}^{\sigma}$ has rank N at $g = f$.

Remark 3.2 There are several ways of assigning a set of critical relations $\mathcal{F}$ to f. As we will prove in Sect. 4.2, for *any* set of critical relations which is *full* in the sense of Definition 3.6, Theorem 3.2 holds.

Remark 3.3 A *flexible Lattès map* is by definition a rational map that is conformally conjugate to a map of the form $L/\sim: T/\sim \to T/\sim$, where $T = \mathbb{C}/(\mathbb{Z} \oplus \gamma\mathbb{Z}), \gamma \in \mathbb{H}$ (where $\mathbb{H}$ is the upper-half plane), $\sim$ is the equivalence relation on $\mathbb{C}$ defined by $z \sim -z$ and $L: \mathbb{C} \to \mathbb{C}$ is of the form $L(z) = az + b$ with $a \in \mathbb{Z}$ and $2b \in \mathbb{Z} \oplus \gamma\mathbb{Z}$, see [27]. Such maps can be of two types: either each critical point is mapped in two iterates into a repelling fixed point or in one iterate into a repelling periodic point of period two, see [27].

Remark 3.4 Theorem 3.2 and the implicit function theorem, imply that manifolds defined by critical relations corresponding to disjoint subsets $\mathcal{F}', \mathcal{F}''$ of $\mathcal{F}$ are smooth and transversal to one another.

For completeness we prove the following corollary of Theorem 3.2:

Corollary 3.5 *If each critical point c_i is eventually mapped to a repelling periodic point p_i with $f^{m_i}(c_i) = p_i$ and $f^j(c_i) \notin \{c_1, \ldots, c_\nu\}$ for all $j = 1, \ldots, m_i$ then the Jacobian of*

$$\mathbf{Rat}_d^{\mu} \ni g \mapsto \{\sigma(g^{m_i}(c_i(g))) - \sigma(p_i(g))\}_{i=1}^{\nu} \tag{3.2}$$

has maximal rank at $g = f$, where σ is a Möbius transformation with $\sigma(p_i) \neq \infty$ for all i.

Proof For the same reason as explained in Remark 3.1, we only need to consider the case where ∞ avoids the critical orbits and $\sigma = id$. Let $\mathcal{R}$ denote the map in (3.2). The corollary follows from the following claim by Theorem 3.2.

Claim. If f_t is a holomorphic curve in $\mathbf{Rat}_d^{\mu}$ passing through f at $t = 0$ which represents a vector in the kernel of $D_f \mathcal{R}$, then for any critical relations $(i, j; m, n)$ realized by f, we have

$$f_t^m(c_i(f_t)) - f_t^n(c_j(f_t)) = o(t) \text{ as } t \to 0.$$

Indeed, the claim implies that the kernel of $D_f \mathcal{R}$ is contained in the kernel of $D_f \mathcal{R}_{\mathcal{F}}$ for any finite collection $\mathcal{F}$ of critical relations. By Theorem 3.2, we can choose $\mathcal{F}$ such that $D_f \mathcal{R}_{\mathcal{F}}$ has maximal rank. Thus $D_f \mathcal{R}$ has maximal rank.

Let us prove the claim. Choose k large enough such that $m + k \geq m_i$ and $n + k \geq m_j$. Since $f^{m+k-m_i}(p_i) = f^{n+k-m_j}(p_j)$, the periodic points p_i and p_j have the same period, denoted by s. Moreover, if $p_i(t)$ (resp. $p_j(t)$) denotes the repelling periodic point of f_t^s near p_i (resp. p_j), then

$$f_t^{m+k-m_i}(p_i(t)) = f_t^{n+k-m_j}(p_j(t)).$$

Since $f_t^{m_i}(c_i(f_t)) - p_i(t) = o(t)$ as $t \to 0$, we have

$$f_t^{m+k}(c_i(f_t)) - f_t^{m+k-m_i} p_i(t) = o(t) \text{ as } t \to 0.$$

Similarly, we have

$$f_t^{n+k}(c_j(f_t)) - f_t^{n+k-m_j}(p_j(t)) = o(t) \text{ as } t \to 0.$$

Therefore,

$$f_t^{m+k}(c_i(f_t)) - f_t^{n+k}(c_j(f_t)) = o(t) \text{ as } t \to 0.$$

Since $f^{m+k'}(c_i)$ is not critical for each $0 \leq k' < k$, it follows that $f_t^m(c_i(f_t)) - f_t^n(c_j(f_t)) = o(t)$ as $t \to 0$. □

3.1 How to Associate Critical Relations to a Rational Map

There are several ways to record the (infinitely many) critical relations of a rational map. In this subsection we will show how one can associate these in an efficient way so that in particular no critical relation is counted twice.

As above, let $c_1, c_2, \ldots, c_v$ be the critical points of a rational map in the class $\mathbf{Rat}_d^\mu$. For an arbitrary collection $\mathcal{F}$ of critical relations realized by f, let $\sim_{\mathcal{F}}$ denote the *smallest* equivalence relation in the set $\Sigma := \{(i, m) : 1 \leq i \leq v, m \geq 0\}$ such that $(i, m + k) \sim_{\mathcal{F}} (j, n + k)$ for each $(i, j; m, n) \in \mathcal{F}$ and each $k \geq 0$.

So $\sim_{\mathcal{F}}$ defines the set of critical relations that can be 'read off' from $\mathcal{F}$. So for example, if $v = 4$ and $\mathcal{F} = \{(1, 2; 1, 1), (1, 3; 1, 1)\}$ then $(i, 1 + k) \sim_{\mathcal{F}} (j, 1 + k)$ for all $i, j \in \{1, 2, 3\}$ and all $k \geq 0$, but $(i, m) \nsim_{\mathcal{F}} (4, n)$ for $i \in \{1, 2, 3\}$ and all $m, n \geq 0$.

Roughly speaking, we say that a collection $\mathcal{F}$ of critical relations is *full* if it 'essentially' explains all critical relations of f and $\mathcal{F}$ is *minimally full* if it does not contain redundant critical relations. More precisely,

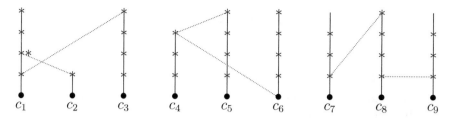

Fig. 1 The orbit diagram of a map with $\zeta(f) = 3$ and $f^2(c_1) = f(c_2)$, $f(c_1) = f^4(c_3)$, $f^3(c_4) = f^4(c_5) = c_6$, $f(c_7) = f^4(c_8)$, $f(c_8) = f(c_9)$. Each of the collections $\{(2, 1; 1, 2), (3, 1; 4, 1),$ $(4, 6; 3, 0), (5, 4; 4, 3), (8, 7; 4, 1), (9, 8; 1, 1)\}$, $\{(2, 1; 1, 2), (3, 1; 4, 1), (4, 6; 3, 0), (5, 6; 4, 0),$ $(9, 7; 4, 1), (9, 8; 1, 1)\}$ and $\{(2, 1; 2, 3), (3, 1; 5, 2), (4, 6; 3, 0), (5, 6; 4, 0), (9, 7; 4, 1), (9, 8;$ $2, 2)\}$ is minimally full

Definition 3.6 We say that a collection $\mathcal{F}$ of critical relations realized by f is *full* if for any critical relation $(i, j; m, n)$ realised by f, i.e. whenever $f^m(c_i) = f^n(c_j)$, there exists $k \geq 0$ such that $(i, m + k) \sim_{\mathcal{F}} (j, n + k)$ and such that $f^{m+k'}(c_i) = f^{n+k'}(c_j) \notin \{c_1, \ldots, c_v\}$ for each $0 \leq k' < k$.

Note that any full collection contains at least $v - \zeta(f)$ relations. A full collection $\mathcal{F}$ is called *minimally full* if $\#\mathcal{F} = v - \zeta(f)$ (Fig. 1).

If $\mathcal{F}$ is minimally full then in particular there exists no $1 \leq i_1, i_2, \cdots, i_k \leq v$, $k \geq 2$, such that

$$(i_1, i_2; 1, 1), (i_2, i_3; 1, 1), \ldots, (i_k, i_1; 1, 1) \in \mathcal{F}. \qquad (3.3)$$

We refer to the last property as the *non-cyclic condition*.

So if f has critical points $c_1, \ldots, c_4$ with critical relations $f^k(c_1) = f^k(c_2) = f^k(c_3)$ for all $k \geq 1$ and there are no other critical relations, then $\zeta(f) = 2$ and

$$\mathcal{F}_1 = \{(1, 2; 1, 1), (1, 3; 1, 1)\} \text{ but also } \mathcal{F}_2 = \{(1, 2; 2, 2), (1, 3; 3, 3)\}$$

are minimally full collections.

Note that if $\mathcal{F}$ is minimally full and $(i, j; m, n) \in \mathcal{F}$ then $(j, i; n, m) \notin \mathcal{F}$. Later on, we will define a convenient choice for a minimally full collection $\mathcal{F}$, see Definition 4.2 and in Lemma 4.4 we will show that such a choice can always be made.

3.2 An Even More General Theorem

Let g be a rational map with critical points $c_1(g), \ldots, c_v(g)$. Associate to each $(i, j; m, n)$ the following rational map

$$Q^g_{i,j;m,n}(z) = \sum_{r=1}^{m} \frac{Dg^{m-r}(g^r(c_i(g)))}{z - g^r(c_i(g))} - \sum_{s=1}^{n} \frac{Dg^{n-s}(g^s(c_j(g)))}{z - g^s(c_j(g))},$$

when $g^r(c_i(g)), g^s(c_j(g)) \neq \infty$ for all $1 \leq r \leq m$, $1 \leq s \leq n$. (Convention: For $m = 0$ or $n = 0$, the corresponding sum is understood as 0.)

Given a meromorphic quadratic differential $Q = q(z)dz^2$, define its push-forward as $f_*Q = \widehat{q}(z)dz^2$, where

$$\widehat{q}(z) = \sum_{w \in f^{-1}(z)} \frac{q(w)}{f'(w)^2}.$$

It is not difficult to check that f_*Q is again a meromorphic quadratic differential. The assignment $Q \mapsto f_*Q$ is often called the Thurston operator, see [14, 23], and was used in Thurston's rigidity theorem, see [6]. M. Tsujii was probably the first to use quadratic differentials in the context of transversality, see [26, 32, 33], but see also [2, 8, 9, 17, 18, 20, 22].

Theorem 3.2 will follow from

Theorem 3.3 *Assume that the critical orbits of $f \in Rat^\mu_d$ avoid ∞. Let $\mathcal{F}$ be a finite set of critical relations $(c_{i_k}, c_{j_k}, m_k, n_k)$, $k = 1, 2, \ldots, N$, which are realized by f and which satisfies the non-cyclic condition (3.3). If the Jacobian of the map*

$$Rat^\mu_d \ni g \mapsto \{g^{m_k}(c_{i_k}(g)) - g^{n_k}(c_{j_k}(g))\}_{k=1}^{N} \tag{3.4}$$

at $g = f$ has rank less than N, then there exist complex numbers $a_1, a_2, \cdots, a_N$, such that

- *for some k, $(m_k, n_k) \neq (1, 1)$ and $a_k \neq 0$;*
- *$f_*(q(z)dz^2) = q(z)dz^2$, where*

$$q(z) = \sum_{\substack{1 \leq k \leq N \\ (m_k, n_k) \neq (1,1)}} a_k Q^f_{i_k, j_k; m_k, n_k}(z). \tag{3.5}$$

If in addition f is not a flexible Lattés example, then $q(z) \equiv 0$.

Remark 3.7 If f has $2d - 2$ distinct critical values, then the converse statement of the theorem also holds. Namely, for any finite set $\mathcal{F}$ as above, if $f_*(q(z)dz^2) = q(z)dz^2$, then the Jacobian of the map (3.1) has rank less than N.

Remark 3.8 Take $f \in Rat^\mu_d$ and a manifold S passing through f of dimension p, that is transverse to the orbit $O(f)$ of f under Möbius conjugacies. Assume that the map defined in (3.1) has maximal rank. Then the restriction of this map to S also has maximal rank. This holds because the value of the map (3.1) is constant on $O(f)$.

4 Theorems 3.1–3.2 Follow from Theorem 3.3

4.1 Proof of Theorem 3.1

If $f^m(c_i) = c_j$ then f is not a Lattés example. We may assume without loss of generality that ∞ avoids the critical orbits of f so that Theorem 3.3 applies. It is clear that

$$Q^f_{i,j;m,0}(z) = \frac{Df^{m-1}(f(c_i))}{z - f(c_i)} + \cdots + \frac{1}{z - f^m(c_i)}$$

has a pole at c_j, so it is not identically zero and thus the conclusion follows from the last sentence of Theorem 3.3.   □

4.2 An Improved Way to Organise Critical Relations and the Proof of Theorem 3.2

In general, one can associate several full collections $\mathcal{F}$ to f each giving rise to a map $\mathcal{R}^\sigma_\mathcal{F}$ as in (3.1). Let us first prove, as claimed in Remark 3.2, that any full collection gives rise to the same rank:

Lemma 4.1 *For any full collections $\mathcal{F}$ and $\mathcal{F}'$ of critical relations for f, the Jacobian matrices of $\mathcal{R}^\sigma_\mathcal{F}$ and $\mathcal{R}^\sigma_{\mathcal{F}'}$ at $g = f$ have the same rank.*

Proof According to Remark 3.1, we may assume the critical orbits avoid ∞ and $\sigma = id$. Consider a holomorphic curve f_t, passing through f at $t = 0$. This curve represents a vector in the kernel of $D\mathcal{R}_\mathcal{F}$ if and only if the derivative of $t \mapsto f^m_t(c_i(f_t)) - f^n_t(c_j(f_t))$ vanishes at $t = 0$ for each $(i, j; m, n) \in \mathcal{F}$, and therefore if and only if $t \mapsto f^m_t(c_i(f_t)) - f^n_t(c_j(f_t))$ vanishes at $t = 0$ for each $(i, m) \sim_\mathcal{F} (j, n)$ where $\sim_\mathcal{F}$ is the equivalence relation associated to $\mathcal{F}$ as defined in the first paragraph of Sect. 3.1.

Assume that $(i, j; m, n)$ is realised by f. Since $\mathcal{F}$ is full, there exists $k \geq 0$ so that $(i, m + k) \sim_\mathcal{F} (j, n + k)$ and so that $Df^k(f^m(c_i)) = Df^k(f^n(c_j)) \neq 0$. So if f_t represents a vector in the kernel of $D\mathcal{R}_\mathcal{F}$ then the derivative of $t \mapsto f^{m+k}_t(c_i(f_t)) - f^{n+k}_t(c_j(f_t))$ vanishes at $t = 0$. Since $f^m(c_i) = f^n(c_j)$ and $Df^k(f^m(c_i)) = Df^k(f^n(c_j)) \neq 0$, this implies that the derivative of $t \mapsto f^m_t(c_i(f_t)) - f^n_t(c_j(f_t))$ vanishes at $t = 0$.

On the other hand, if for each $(i, j; m, n)$ which is realised by f the derivative of $t \mapsto f^m_t(c_i(f_t)) - f^n_t(c_j(f_t))$ vanishes at $t = 0$, then in particular this holds for each $(i, j; m, n) \in \mathcal{F}$ and so the holomorphic curve f_t represents a vector in the kernel of $D\mathcal{R}_\mathcal{F}$.

It follows that f_t represents a vector in the kernel of $D\mathcal{R}_\mathcal{F}$ if and only if for each $(i, j; m, n)$ which is realised by f the derivative of $t \mapsto f^m_t(c_i(f_t)) - f^n_t(c_j(f_t))$ vanishes at $t = 0$. The last condition is independent of the choice of the full collection

$\mathcal{F}$. Since both $\mathcal{F}$ and $\mathcal{F}'$ are full, the rank-nullity theorem implies that the rank of the Jacobian matrices are the same.   □

We will find it convenient to prove Theorem 3.2 for a conveniently chosen minimal collection $\mathcal{F}$, namely one which satisfies the following stronger minimality assumption.

Definition 4.2 We say that a *collection* $\mathcal{F}$ is *proper* for f if it is minimally full and satisfies the following extra properties:

(1) $(i, j; m, n) \in \mathcal{F}$ implies $m > 0$ and either $i \geq j$ or $n = 0$. If $i = j$ then $m > n$.
(2) if $(i, j; m, n) \in \mathcal{F}$ then the collection of points $f^k(c_i)$, $k = 1, \ldots, m - 1$ is pairwise disjoint and does not intersect $c_1, \ldots, c_v$ nor the forward orbits of $c_1, \ldots, c_{i-1}$.
(3) For each $1 \leq i \leq v$ there exists at most one critical relation of the form $(i, j; m, n) \in \mathcal{F}$.
(4) For each $1 \leq j \leq v$ there exists at most one critical relation of the form $(i, j; m, 0) \in \mathcal{F}$;
(5) For each $1 \leq j \leq v$ and each $n > 1$ there exists at most one critical relation of the form $(i, j; 1, n) \in \mathcal{F}$;
(6) If $(i, j; m, n) \in \mathcal{F}$ with $m > 1$ and $n > 0$, and $(k, i; 1, l) \in \mathcal{F}$ for some k and l, then $l < m$.

Remark 4.3 If the collection $\mathcal{F}$ is proper then it satisfies the non-cyclic condition (3.3).

Lemma 4.4 *There exists a proper collection of critical relations which are realised by f.*

Proof For each $i = 1, \ldots, v$, inductively define $m_i > 0$ maximal so that $f(c_i), \ldots, f^{m_i-1}(c_i)$ are distinct and also distinct from

$$\{f^k(c_j); 0 \leq k < m_j, j = 1, \ldots, i - 1\} \cup \{c_1, \ldots, c_v\}.$$

(When $i = 1$ we take this union to be $\{c_1, \ldots, c_v\}$.) If m_i is finite, then there are two possibilities:

(a) $f^{m_i}(c_i) = f^{n_{j_i}}(c_{j_i})$ for some $1 \leq j_i \leq i$ and some finite n_{j_i} with $0 < n_{j_i} < m_{j_i}$. In this case associate to c_i the critical relation (i, j_i, m_i, n_{j_i}).
(b) $f^{m_i}(c_i) = c_j$ with $1 \leq j \leq v$ and in this case associate to c_i the critical relation $(i, j, n_i, 0)$.

These choices ensure that properties (1) and (2) in the above definition hold. To take care that properties (3)–(6) hold we also make the following requirement:
   If both (a) and (b) hold, then only assign to c_i the critical relation as in (a). If (a) holds for several $j_i \leq i$, then choose the smallest possible j_i with $n_{j_i} = 1$ and if there is no j_i with $n_{j_i} = 1$ then simply choose the smallest possible j_i. Assign to i

only the corresponding critical relation. Once we have done this for i then repeat this construction for $i + 1$.

In this way we define no new critical relation for each $1 \leq i \leq \nu$ whose orbit is infinite and disjoint from forward orbits of $c_1, \ldots, c_{i-1}$ and from $c_1, \ldots, c_\nu$, but a unique critical relation for each of the other i's. Thus we get $N = \nu - \zeta(f)$ critical relations.

The resulting set of critical relations is realised by f. By construction $\mathcal{F}$ is proper.   □

Proof of Theorem 3.2. By Remark 3.1, it is enough to consider the case that σ is the identity and the critical orbits avoid ∞. Let $\mathcal{F}$ be a proper collection of critical relations realized by f. Note that if $m, n \geq 1$ then $Q^f_{i,k;m,n}(z)$ is equal to

$$Q^f_{i,j;m,n}(z) = \sum_{r=1}^{m-1} \frac{Df^{m-r}(f^r(c_i))}{z - f^r(c_i)} - \sum_{s=1}^{n-1} \frac{Df^{n-s}(f^s(c_j))}{z - f^s(c_j)} \tag{4.1}$$

and if $m = n = 1$ then $Q^f_{i,j;m,n}(z) = 0$. By property (2) of Definition 4.2, $(i, j; m, n) \in \mathcal{F}$ and $m, n \geq 1$ imply

$$Df^{m-1}(f(c_i)) \neq 0, \; Df^{n-1}(f(c_j)) \neq 0 \tag{4.2}$$

and if $i \neq j$ then

$$f(c_i), \ldots, f^{m-1}(c_i), f(c_j), \ldots, f^{n-1}(c_j) \tag{4.3}$$

are all distinct and distinct from $c_1, \ldots, c_\nu$. Similarly, if $i = j$ then by properties (1), (2) of Definition 4.2, $m > n$ and $f(c_i), \ldots, f^{m-1}(c_i), c_1, \ldots, c_\nu$ are all distinct. Hence, if $m, n \geq 1$ and $i \neq j$ then $Q^f_{i,j;m,n}(z)$ has a non-removable pole in each of the points from the collection (4.3) and nowhere else. In particular, $c_1, \ldots, c_\nu$ is not a pole for any $Q_{i,j;m,n}(z)$ when $m, n \geq 1$ (this holds even when $i = j$). On the other hand, $Q^f_{i,j;m,0}(z)$ does have a pole at c_j and only critical relations of this form in $\mathcal{F}$ have a pole at c_j.

Suppose that the Jacobian does not have full rank. By Theorem 3.3 this implies

$$\sum_{\substack{1 \leq k \leq N \\ (m_k, n_k) \neq (1,1)}} a_k Q^f_{i_k, j_k; m_k, n_k}(z) = 0. \tag{4.4}$$

Let $\mathcal{F}_0$ be the set of relations $(i_k, j_k; m_k, n_k)$ in $\mathcal{F}$ in this sum for which $a_k \neq 0$ and with $(m_k, n_k) \neq (1, 1)$. So (4.4) is equal to the sum over the set $\mathcal{F}_0$. By Theorem 3.3, $\mathcal{F}_0$ consists of at least one critical relation, and obviously the properties stated in Definition 4.2 are also satisfied for $\mathcal{F}_0$.

Suppose first that there exists a critical relation $(i, j; m, 0) \in \mathcal{F}_0$. In this case by property (4) in Definition 4.2 there exists no $i' \neq i, m' > 0$ so that $(i', j; m', 0) \in \mathcal{F}_0$.

It follows from this that $(i, j; m, 0)$ is the only term in the sum (4.4) which leads to a pole at $z = c_j$. So the corresponding coefficient $a_k = 0$, a contradiction.

From now on, let us assume that for any $(i, j; m, n) \in \mathcal{F}_0, n > 0$. Then by property (1) of Definition 4.2, we have $i \geq j$. Because of property (3) of Definition 4.2 we can rearrange, if necessary, the critical relations in $\mathcal{F}_0$ so that they are of the form $(i_k, j_k; m_k, n_k)$, $1 \leq k \leq N_0$, with $i_1 < i_2 < \cdots < i_{N_0}$. If $m_k = 1$ holds for all $1 \leq k \leq N_0$, then by property (5), (j_k, n_k) are pairwise distinct. Since $Q^f_{i_k, j_k; 1, n_k}$ has poles precisely at the points $f(c_{j_k})$, $f^2(c_{j_k})$, $\cdots$, $f^{n_k - 1}(c_{j_k})$, $\sum_{k=1}^{N_0} a_k Q^f_{i_k, j_k; m_k, n_k}$ has a pole, a contradiction! So let us assume that there is a maximal $N_1 \leq N_0$ such that $m_{N_1} \geq 2$. By property (6) of Definition 4.2, for each $N_0 \geq k > N_1$, either $j_k \neq i_{N_1}$, or $j_k = i_{N_1}$ and $n_k < m_{N_1}$. Together with property (2) of Definition 4.2, this implies that $Q^f_{i_k, j_k; m_k, n_k} = Q^f_{i_k, j_k; 1, n_k}$ does not have a pole at $f^{m_{N_1} - 1}(c_{i_{N_1}})$. For each $k < N_1$, since $i_{N_1} > i_k \geq j_k$, by property (2) of Definition 4.2, $Q^f_{i_k, j_k; m_k, n_k}$ does not have a pole at $f^{m_{N_1} - 1}(c_{N_1})$ either. Therefore $\sum_{k=1}^{N_0} a_k Q^f_{i_k, j_k; m_k, n_k}$ has a pole at $f^{m_{N-1} - 1}(c_{N_1})$, a contradiction! $\qquad\square$

5 A Proof of Theorem 3.3

Remark 5.1 A proof of Theorem 3.3 is contained essentially in [20]. We only outline it here (and then present another proof). Denote $v_j(f) = f(c_j(f))$ for $j = 1, \cdots, v$. Conjugating f by a Möbius transformation, one can assume that $f(\infty) = \infty$, $Df(\infty) \neq 0$. We label the critical values so that for some $0 \leq v' \leq v$ the following holds: $v_j(f) \in \mathbb{C}$ for $1 \leq j \leq v'$ and $v_j(f) = \infty$ for $v' < j \leq v$. Consider a subset $\Lambda_{f,v} \subset \mathrm{Rat}^\mu_d$ of maps g such that there exists $\sigma(g), b(g) \in \mathbb{C}$ so that $g(z) = \sigma(g)z + b(g) + O(1/z)$ as $z \to \infty$. By [19], $\Lambda_{f,v}$ has a structure of $v + 2$ dimensional complex manifold and $(\sigma(g), b(g), v_1(g), \cdots, v_{v'}(g), v_{v'+1}(g)^{-1}, \cdots, v_v(g)^{-1})$ is a holomorphic coordinate of $g \in \Lambda_{f,v}$. Proposition 10 of [20] implies that for any $(i, j; m, n)$, if $(i, j; m, n)$ is realized by f and $f^m(c_i(f))$, $f^n(c_j(f)) \neq \infty$, then

$$Q^f_{i,j;m,n}(x) - \widehat{Q}^f_{i,j;m,n}(x) =$$

$$\sum_{k=1}^{v'} \frac{1}{v_k(f) - x} \frac{\partial(g^m(c_i(g)) - f^n(c_j(g)))}{\partial v_k}\Big|_{g=f},$$

(5.1)

where $\widehat{Q}^f_{i,j;m,n}(x)dx^2 = f_*(Q^f_{i,j;m,n}(x)dx^2)$. Now Theorem 3.3 can be proved by repeating the proof of the main result of [20] after replacing Proposition 13 of that paper by (5.1). Instead of going into more details we give here a direct and short proof of the theorem.

5.1   Proof of Theorem 3.3.

Let us first apply Thurston's pull back argument to obtain a relation of partial derivatives of $g \mapsto g^m(c_i(g)) - g^n(c_j(g))$ with the quadratic differential $Q_{i,j;m,n}(z)dz^2$. Let $L_\infty(\mathbb{C})$ denote the space of all Borel measurable functions μ with $\|\mu\|_\infty < \infty$. Note that $f^*\mu(z) = \mu(f(z))\overline{f'(z)}/f'(z)$ also belongs to the class $L_\infty(\mathbb{C})$.

Lemma 5.2   *Given $\mu \in L_\infty(\mathbb{C})$ which vanishes in a neighborhood of ∞ and $f(\infty)$, there exists a holomorphic family f_t of rational maps of degree d, $t \in \mathbb{D}_\varepsilon$, with $f_0 = f$, and such that the following holds: For any $(i, j; m, n)$,*

$$-\frac{1}{\pi} \int_\mathbb{C} (\mu - f^*\mu) Q_{i,j;m,n} |dz|^2 = \frac{d(f_t^m(c_i(f_t)) - f_t^n(c_j(f_t)))}{dt}\bigg|_{t=0}.$$

Proof Assume without loss of generality that $\|\mu\|_\infty \leq 1$. Then for each $t \in \mathbb{D}$, there are qc maps $\varphi_t, \psi_t : \mathbb{C} \to \mathbb{C}$ with complex dilatations $t\mu$ and $tf^*\mu$ respectively such that (see e.g. [1])

- $\varphi_t(z) = z + o(1)$, $\psi_t(z) = z + o(1)$ as $z \to \infty$ for each t;
- f_t defined by $f_t \circ \psi_t = \varphi_t \circ f$ is a family of rational maps.

Then φ_t and ψ_t depends on t holomorphically [1] and thus $\partial f_t/\bar{\partial}t = 0$ in the sense of distribution, which implies that f_t depends holomorphically on t. Let

$$\mathcal{L}_n(z) = \frac{df_t^n(z)}{dt}\bigg|_{t=0}$$

and $L(z) = \mathcal{L}_1(z)$. Then

$$L(x) + Df(z)\frac{d}{dt}\psi_t(z) = \frac{d}{dt}\phi_t(f(z)) \tag{5.2}$$

and therefore

$$L(z) + Df(z)\widehat{X}(z) = X(f(z)), \tag{5.3}$$

where

$$X(z) = -\frac{1}{\pi} \int_\mathbb{C} \frac{\mu(\zeta)}{\zeta - z}|d\zeta|^2, \ \widehat{X}(z) = -\frac{1}{\pi} \int_\mathbb{C} \frac{f^*\mu(\zeta)}{\zeta - z}|d\zeta|^2.$$

The latter formulas come from the following fact. Let $\nu \in L_\infty(\mathbb{C})$ have a compact support, $\|\nu\|_\infty \leq 1$ and h_t ($|t| < 1$) is the (unique) qc map with complex dilatation $t\nu$ such that $h_t(z) = z + o(1)$ as $z \to \infty$. Then

$$\frac{dh_t(z)}{dt}\bigg|_{t=0} = -\frac{1}{\pi} \int_\mathbb{C} \frac{\nu(\zeta)}{\zeta - z}|d\zeta|^2. \tag{5.4}$$

The formula (5.4) is well-known and follows for example from the formula (6) in the proof of Theorem 1 of [1], Chap. V (noting that a different normalisation for h_t is chosen there) or by differentiating the formula on the 2nd line of page 25 in [4].

For any $h \in \{1, 2, \ldots, v\}$ and non-negative integer l, define

$$S_l(c_h) = \sum_{r=1}^{l} Df^{l-r}(f^r(c_h))X(f^r(c_h))$$

and

$$\widehat{S}_l(c_h) = \sum_{r=1}^{l} Df^{l-r}(f^r(c_h))\widehat{X}(f^r(c_h))$$

It suffices to show that for any l, h as above,

$$S_l(c_h) - \widehat{S}_l(c_h) = \left.\frac{df_t^l(c_h(f_t))}{dt}\right|_{t=0} - \widehat{X}(f^l(c_h)). \tag{5.5}$$

If $l = 0$, then the left hand is equal to zero, and the right hand side is also equal to zero, since $\widehat{X}(c_h) = \left.\frac{d\psi_t(c_h)}{dt}\right|_{t=0}$ and $c_h(f_t) = \psi_t(c_h)$. For $l \geq 1$, we use (5.3):

$$S_l(c_h) = \sum_{r=1}^{l} Df^{l-r}(f^r(c_h))X(f^r(c_h))$$

$$= \sum_{r=1}^{l} Df^{l-r}(f^r(c_h))L(f^{r-1}(c_h)) + \sum_{r=2}^{l} Df^{l-r+1}(f^{r-1}(c_h))\widehat{X}(f^{r-1}(c_h))$$

$$= \mathcal{L}_l(c_h) + \sum_{r=1}^{l-1} Df^{l-r}(f^r(c_h))\widehat{X}(f^r(c_h))$$

$$= \mathcal{L}_l(c_h) + \widehat{S}_l(c_h) - \widehat{X}(f^l(c_h)).$$

Since

$$\left.\frac{df_t^l(c_h(f_t))}{dt}\right|_{t=0} = \left.\frac{df_t^l(c_h)}{dt}\right|_{t=0},$$

Equation (5.5) follows. □

Proof (Proof of Theorem 3.3) Assume that the Jacobian matrix has rank less than N. Then there exist complex numbers $a_1, a_2, \cdots, a_N$ such that all the partial derivatives of the map

$$g \mapsto \sum_{k=1}^{N} a_k \left(g^{m_k}(c_{i_k}(g)) - g^{n_k}(c_{j_k}(g))\right) \tag{5.6}$$

is equal to 0 at $g = f$. Since $\mathcal{F}$ satisfies the non-cyclic condition (3.3), by Theorem 2.1, there exists k such that $(m_k, n_k) \neq (1, 1)$ and $a_k \neq 0$.

Given $\mu \in L_\infty(\mathbb{C})$ which vanishes in a neighbourhood of ∞ and $f(\infty)$, let f_t be given by the previous lemma. Then for each $k = 1, 2, \ldots, N$, we have

$$-\frac{1}{\pi} \int_{\mathbb{C}} (\mu - f^* \mu) Q_{i_k, j_k; m_k, n_k} |dz|^2 = \left. \frac{d(f_t^m(c_i(f_t)) - f_t^n(c_j(f_t)))}{dt} \right|_{t=0}.$$

Thus for q defined as in (3.5), and $\widehat{q}(z)dz^2 = f_*(q(z)dz^2)$, we have

$$\int_{\mathbb{C}} \mu(\widehat{q} - q)|dz|^2$$
$$= \int_{\mathbb{C}} (\mu - f^*\mu)q(z)|dz|^2$$
$$= -\pi \sum_{k=1}^{N} a_k \left. \frac{d(f_t^m(c_i(f_t)) - f_t^n(c_j(f_t)))}{dt} \right|_{t=0} = 0,$$

where the last equality follows from the argument in the previous paragraph. It follows that $\widehat{q} = q$.

Assume now that f is not a flexible Lattés example. Let us prove that $q = 0$. To this end, first assume $f(\infty) \neq \infty$. Let φ_i be the local inverse diffeomorphic branches of f near ∞. Then $\widehat{q} = q$ implies that

$$q(z) = \sum_{i=1}^{d} q(\varphi_i(z))\varphi_i'(z)^2$$

holds near ∞. Since $\varphi_i(\infty) \in \mathbb{C}$ (and is not equal to one of the finitely many poles of q) and $\varphi_i'(z) = O(1/z^2)$ as $z \to \infty$, it follows from the displayed formula $q(z) = O(1/z^4)$ at infinity. Thus $q(z)dz^2$ is an integrable meromorphic quadratic differential. By a well-known argument, this implies that $q(z) = 0$, see for example Sect. 3.5 of [6, 24].

If $f(\infty) = \infty$, then we can find a sequence of Möbius transformations $\sigma_l, l = 1, 2, \ldots$, converging to the identity uniformly, such that $f_{(l)} = \sigma_l \circ f \circ \sigma_l^{-1}$ satisfies $f_{(l)}(\infty) \neq \infty$ and ∞ avoids the critical orbits of $f_{(l)}$. Putting $g_{(l)} = \sigma_l \circ g \circ \sigma_l^{-1}$, by (5.6), all partial derivatives of the map

$$g \mapsto \sum_{k=1}^{N} a_k \left(\sigma_l^{-1} \left(g_{(l)}^{m_k} \left(c_{i_k}(g_{(l)}) \right) \right) - \sigma_l^{-1} \left(g_{(l)}^{n_k} \left(c_{j_k}(g_{(l)}) \right) \right) \right)$$

are equal to zero at $g = f$, hence all partial derivatives of the map

$$g \mapsto \sum_{k=1}^{N} \frac{a_k}{\sigma'_{(l)}(f^{m_k}(c_{i_k}))}((g^{m_k}(c_{i_k}(g)) - g^{n_k}(c_{j_k}(g)))$$

are equal to zero at $g = f_{(l)}$. Since $f_{(k)}$ is not a Lattés example, as above we obtain that $q_{(l)} := \sum_{k=1}^{N} \frac{a_k}{\sigma'_{(l)}(f^{m_k}(c_{i_k}(f)))} Q_{i_k,j_k;m_k,n_k}^{f_{(l)}} \equiv 0$. By continuity we conclude that $q = 0$.   □

6 The Polynomial Case

The previous theorems also hold in the space of polynomials of degree d. In that case, let $\boldsymbol{\mu} = (\mu_1, \ldots, \mu_\nu)$ so that $\sum_{i=1}^{\nu} \mu_i = d - 1$ and let $\mathbf{Pol}_d^\mu$ be the set of maps with critical points $c_1, \ldots, c_\nu \in \mathbb{C}$ of orders $\mu_1, \ldots, \mu_\nu$. The space $\mathbf{Pol}_d^\mu$ is clearly an embedded submanifold of $\mathbf{Rat}_d^{\hat{\mu}}$ of codimension one, where $\hat{\mu} = (\mu_1, \mu_2, \cdots, \mu_\nu, d - 1)$.

Theorem 6.1 *Assume $f \in \mathbf{Pol}_d^\mu$. Then there exists a set*

$$\mathcal{F} = \{(i_k, j_k; m_k, n_k), k = 1, \ldots, N\} \text{ with } N = \nu - \zeta(f)$$

of critical relations $f^{m_k}(c_{i_k}) = f^{n_k}(c_{j_k})$ which are realised by f, such that the Jacobian of the map

$$\mathbf{Pol}_d^\mu \ni g \mapsto (g^{m_k}(c_{i_k}(g)) - g^{n_k}(c_{j_k}(g)))_{k=1}^{N} \tag{6.1}$$

at $g = f$ has rank N.

Proof Let $c_{\nu+1} = \infty$. For maps g in $\mathbf{Rat}_d^{\hat{\mu}}$ close to f, let $c_j(g)$ denote the critical point of g close to c_i, $1 \leq i \leq \nu + 1$. By Theorem 3.2, there is a set $\widehat{\mathcal{F}} = \{(i_k, j_k; m_k, n_k)\}_{j=1}^{N+1}$ of critical relations of f so that the Jacobian of the map

$$\mathcal{R}_{\widehat{\mathcal{F}}}^\sigma : \mathbf{Rat}_d^{\hat{\mu}} \ni g \mapsto (\sigma(g^{m_k}(c_{i_k}(g))) - \sigma(g^{n_k}(c_{j_k}(g))))_{k=1}^{N+1}$$

has rank $N + 1$ at $g = f$, where σ is a Möbius tansformation such that $\sigma(f^{m_k}(c_{i_k})) \neq \infty$ for all k. Since $\widehat{\mathcal{F}}$ is full, there is k_0 such that $(i_{k_0}, j_{k_0}; m_{k_0}, n_{k_0}) = (\nu + 1, \nu + 1; 1, 0)$ (or $(\nu + 1, \nu + 1; 0, 1)$). Assume without loss of generality $k_0 = N + 1$. Let $\mathcal{F} = \{(i_k, j_k; m_k, n_k) : 1 \leq k \leq \nu\}$ and let $\mathcal{R}$ denote the map defined by (6.1). Note that the kernel of $D_f \mathcal{R}$ is contained in the kernel of $D_f \mathcal{R}_{\widehat{\mathcal{F}}}^\sigma$, so its dimension is at most $\dim(\mathbf{Rat}_d^{\hat{\mu}}) - (N + 1) = \dim(\mathbf{Pol}_d^\mu) - N$. Thus the rank of $D_f \mathcal{R}$ is at least N. The rank is not more than N, so it is equal to N.   □

Acknowledgements The authors thank Alex Eremenko for a discussion about Theorem 2.1, Adam Epstein for pointing out an error in the last paragraph of Sect. 5.1 in a previous version, Xavier Buff and Lasse Rempe-Gillen for helpful comments on the introduction of the final version of this paper and the referee for carefully reading the paper. This project was partly supported by the ISF grant no: 1226/17, ERC AdG grant no: 339523 RGDD and the NSFC grant no: 11731003.

References

1. L.V. Ahlfors, *Lectures on quasiconformal mappings*, vol. 38, 2nd edn. Van Nostrand, Princeton 1966; A.M.S. University Lecture Series (2006)
2. M. Astorg, Summability condition and rigidity for finite type maps, arXiv:1602.05172v1
3. X. Buff, A. Epstein, Bifurcation measure and postcritically finite rational maps. Complex Dyn. **491**, 512 (2009)
4. L. Carleson, T.W. Gamelin, *Complex Dynamics* (Springer, 1992)
5. A. Douady, J.H. Hubbard, Étude dynamique des polynômes complexes. Publications Mathématiques d'Orsay, 84–2. Université de Paris-Sud, Département de Mathématiques, Orsay (1984), 75 pp
6. A. Douady, J.H. Hubbard, A proof of Thurston's topological characterization of rational functions. Acta Math. **171**(2), 263–297 (1993)
7. A. Epstein, Infinitesmimal Thurston rigidity and the Fatou-Shishikura inequality, Stony Brook IMS preprint 1999#1
8. A. Epstein, Transversality in holomorphic dynamics, http://homepages.warwick.ac.uk/~mases/Transversality.pdf
9. A. Epstein, Slides of talk available in https://icerm.brown.edu/materials/Slides/sp-s12-w1/Transversality_Principles_in_Holomorphic_Dynamics_%5D_Adam_Epstein,_University_of_Warwick.pdf
10. A. Eremenko, A. Gabrielov, Rational functions with real critical points and the B. and M. Shapiro conjecture in real enumerative geometry. Ann Math. **155**, 105–129 (2002)
11. A. Eremenko, M. Lyubich, Dynamical properties of some classes of entire functions. Ann. Inst. Fourier **42**(4), 1–32 (1992)
12. A. Eremenko, A Markov-type inequality for arbitrary plane continua. Proc. AMS **135**, 1505–1510 (2007)
13. C. Favre, T. Gauthier, Distribution of postcritically finite polynomials. Isr. J. Math. **209**, 235–292 (2015)
14. F. Gardiner, *Teichmuller theory and quadratic differentials* (Wiley, 1987)
15. G. Levin, M.L. Sodin, P.M. Yuditski, A Ruelle operator for a real Julia set. Comm. Math. Phys. **141**(1), 119–132 (1991)
16. G.M. Levin, *On the theory of iterations of polynomial families in the complex plane*. Translation from: Toeriya Funkzii, Funkzionalnyi Analiz i Ih Prilozheniya, no. 51, 94–106 (1989)
17. G.M. Levin, Polynomial Julia sets and Pade's approximations (in Russian), in *Proceedings of XIII Workshop on Operator's Theory in Functional Spaces, Kyubishev, 6–13 October 1988* (Kyubishev State University, Kyubishev, 1988), pp. 113–114
18. G. Levin, On an analytic approach to the Fatou conjecture. Fund. Math. **171**(2), 177–196 (2002)
19. G. Levin, Multipliers of periodic orbits in spaces of rational maps. Ergod. Theory Dynam. Syst. **31**, 197–243 (2011)
20. G. Levin, Perturbations of weakly expanding critical orbits. Front. Complex Dyn., 163–196 (2014). Princeton Math. Ser., 51, Princeton University Press, Princeton, NJ
21. G. Levin, W. Shen, S. van Strien, Monotonicity of entropy for one-parameter families of interval maps. Preprint Oct 2016
22. P. Makienko, Remarks on the Ruelle operator and the invariant line fields problem. II. Ergod. Theory Dynam. Syst. **25**(5), 1561–1581 (2005)

23. C. McMullen, Amenability, Poincaré series and quasiconformal maps. Invent. Math **97**, 95–127 (1989)
24. C. McMullen, *Complex Renormalisation and Renormalisation* (Princeton University Press, Princeton, 1994)
25. W. de Melo, S. van Strien, *One-dimensional dynamics, Ergebnisse der Mathematik und ihrer Grenzgebiete* (Springer, Berlin, 1993)
26. J. Milnor, Tsujii's monotonicity proof for real quadratic maps, unpublished 2000
27. J. Milnor, On Lattès maps, in *Dynamics on the Riemann Sphere*, pp. 9–43 (2006). Eur. Math. Soc, Zürich
28. J. Milnor, W. Thurston, On iterated maps of the interval, in *Dynamical Systems*, College Park, MD, 1986–87. Lecture Notes in Math., 1342 (Springer, Berlin, 1988), pp. 465–563
29. J. Milnor, C. Tresser, On entropy and monotonicity for real cubic maps. Commun. Math. Phys. **209**, 123–178 (2000)
30. S. van Strien, Misiurewicz maps unfold generically (even if they are critically non-finite). Fund. Math. **163**(1), 39–54 (2000)
31. O. Teichmüller, Eine Anwendung, quasikonformer Abbildungen auf das Typenproblem. Deutsche Math. **2**, 321–327 (1937). Gesammelte Abhandlungen, Springer, Berlin, 1982, 171–177
32. M. Tsujii, A note on Milnor and Thurston's monotonicity theorem, in *Geometry and Analysis in Dynamical Systems* (Kyoto, 1993), pp. 60–62, Adv. Ser. Dynam. Systems, 14, World Sci. Publ., River Edge, NJ, 1994
33. M. Tsujii, A simple proof of monotonicity of entropy in the quadratic family. Ergod. Theory Dynam. Syst. **20**, 925–933 (2000)

Teichmüller Space of Fibonacci Maps

Mikhail Lyubich

Dedicated to memory of Welington de Melo.

Abstract We prove that all smooth Fibonacci maps with quadratic critical point are quasisymmetrically conjugate. The proof is based upon an idea of asymptotically conformal extension, which provides a link between smooth and holomorphic dynamics.

Keywords Fibonacci map · Asymptotically conformal extension · Teichmüller metric

Preamble This paper was written 25 year ago, but appeared only as a Stony Brook Preprint (# 12, 1993). Its main result states that all smooth Fibonacci maps with quadratic critical point are quasisymmetrically conjugate. The proof is based upon an idea of *asymptotically conformal extension*. It provides a link between smooth and holomorphic dynamics that allowed us to apply some holomorphic techniques to the smooth case. Today this paper can serve as an illustration of this idea in a simple setting which is not overshadowed by combinatorial and geometric complications. Further applications of the asymptotically conformal extension to smooth dynamics have been carried in the work of Avila and Krikoryan [24], Graczek et al. [26], and Guarino and de Melo [27]. A strategy of how our result can be generalized to arbitrary smooth unimodal maps was outlined in [28], §12.2. Very recently, Clark and van Strien have treated the general polymodal case [25]. We reproduce this article in the original form, with a few typos fixed, a couple of footnotes added, and the bibliography updated.

M. Lyubich (✉)
Stony Brook University, New York, USA
e-mail: mlyubich@math.stonybrook.edu

© Springer Nature Switzerland AG 2019
M. J. Pacifico and P. Guarino (eds.), *New Trends in One-Dimensional Dynamics*, Springer Proceedings in Mathematics & Statistics 285,
https://doi.org/10.1007/978-3-030-16833-9_12

1 Introduction

According to Sullivan, a space $\mathcal{E}$ of unimodal maps with the same combinatorics (modulo smooth conjugacy) should be treated as an infinite-dimensional Teichmüller space. This is a basic idea in Sullivan's approach to the Renormalization Conjecture [14, 15]. One of its principle ingredients is to supply $\mathcal{E}$ with the Teichmüller metric. To have such a metric one has to know, first of all, that all maps of $\mathcal{E}$ are quasi-symmetrically conjugate. This was proved in [6, 7] for some classes of non-renormalizable maps (when the critical point is not too recurrent). Here we consider a space of non-renormalizable unimodal maps with in a sense fastest possible recurrence of the critical point (called Fibonacci). Our goal is to supply this space with the Teichmüller metric.

Let f be a unimodal map with critical point c. A Fibonacci unimodal map f can be defined by saying that the closest returns of the critical point occur at the Fibonacci moments. This combinatorial type was suggested by Hofbauer and Keller [5] as extremal among non-renormalizable types (see [11] for more detailed history). Its combinatorial, geometric and measure-theoretical properties were studied in [11] under the assumption that f is *quasi-quadratic*, i.e., it is C^2-smooth and has the quadratic-like critical point (see also [8]). We will assume this regularity throughout the paper.

A principal object of our combinatorial considerations is a nested sequence of intervals $I^0 \supset I^1 \supset \ldots$ obtained subsequently by pulling back along the critical orbit. Our proof is based upon the geometric result of [11] which says that the scaling factors $\mu_n = |I^n|/|I^{n-1}|$ characterizing the geometry of the Fibonacci map decay exponentially. It follows that appropriately defined renormalizations $R^n f$ are becoming purely quadratic near the critical point. This reduces the renormalization process to compositions of quadratic maps.

The next idea is to consider a quasi-conformal continuation of f to the complex plane which is asymptotically conformal on the real line.[1] Then we consider complex generalized renormalizations, and prove that the renormalized maps are becoming purely quadratic in the complex plane as well. Hence the geometric patterns of renormalized maps are subsequently obtained by the Thurston pull-back transformation (up to an exponentially small error) in an appropriate Teichmüller space. It follows that these patterns converge (after rescaling) to the corresponding pattern of the quadratic map $p : z \mapsto z^2 - 1$. In particular, the shape of the complex puzzle-pieces converges to the Julia set of p, see Fig. 1 (this is perhaps the most surprising outcome of our analysis).

To each renormalization we then associate a pair of pants Q_n by removing from the critical puzzle-piece of level n two puzzle-pieces of the next level. Using the same type of argument as above, we show that the pairs of pants Q^n and $\tilde{Q}^n$ stay on bounded distance. This yields the quasi-conformal equivalence of the critical sets of f and $\tilde{f}$.

---

[1] This idea was also used in [7] with a reference to D. Sullivan's lectures.

Fig. 1 A Fibonacci puzzle-piece (made by S. Sutherland and B. Yarrington)

To complete the construction of the quasi-symmetric conjugacy, we apply a Sullivan-like pull-back argument. However, this is not quite straightforward since there is no dilatation control away from the real line.

In the last section we prove that two Fibonacci maps that stay on zero Teichmüller distance are smoothly conjugate. So this pseudo-metric is non-degenerate on the smooth equivalence classes.

We will use abbreviations qc and qs for "quasi-conformal" and "quasi-symmetric" respectively.

Remark 1 The dilatation of the conjugacy we construct depends only on the geometry of the maps in question.

Remark 2 It is proved in [12] that, as in the Fibonacci case, the scaling factors of any non-renormalizable quasi-quadratic map decay exponentially. This allows us to generalize the above result to all combinatorial classes of quasi-quadratic maps. The exposition of this result is more technical, and it will be the subject of forthcoming notes.[2] Note that for polynomial-like maps this result follows from the Yoccoz Theorem (see [4] for the exposition of this theorem, and [10] for an alternative proof based upon a pull-back argument).

Remark 3 In this paper we concentrate on the dynamical constructions, and do not touch the issue of the sharp regularity for which the theory can be built up. Compare [2] and [15].

---

[2]See [28], §12.2.

2 Asymptotically Conformal Continuation and Generalized Renormalization

Real renormalization (see [11]). Given a Fibonacci map f, there is a sequence of maps

$$g_n : I_0^n \cup I_1^n \to I_0^{n-1}, \quad n = 1, 2, \ldots$$

constructed in the following way. Let $I^0 \equiv I_0^0$ be a c-symmetric interval[3] satisfying the property $f^n(\partial I^0) \cap I^0 = \emptyset, n = 1, 2, \ldots$. Now given $I^{n-1} \equiv I_0^{n-1} \ni c$ by induction, let us consider the first return map $f_n : \cup I_j^n \to I^{n-1}$. Its domain of definition generally consists of infinitely many intervals $I_j^n \subset I^{n-1}$. However, for the Fibonacci map, only two of them, $I^n \equiv I_0^n \ni c$ (the "central" one) and I_1^n, intersect the critical set $\omega(c)$. Let us define g_n as the restriction of f_n to these two intervals. These maps satisfy the following properties:

(i) $g_n : I_1^n \to I_0^{n-1}$ is a diffeomorphism and $g_n(\partial I_0^n) \subset \partial I_0^{n-1}$;

(ii) $g_n I_0^n \supset I_0^n$ (*high return*);

(iii) $g_n c \in I_1^n$ and $g_n^2 c \in I_0^n$.

By rescaling I^n to some definite size T (e.g., $T = [0, 1]$), we obtain the generalized n-fold renormalization

$$R^n f : T_0^n \cup T_1^n \to T$$

of f. The asymptotic properties of the renormalized maps express the small scale information about the critical set $\omega(c)$.

Let us now introduce the principle geometric parameters, the scaling factors

$$\mu_n = \frac{|I^n|}{|I^{n-1}|} = \frac{|T^n|}{|T|}.$$

The main result of [11] says that they decrease to 0 exponentially at the following rate:

$$\mu_n \sim a \left(\frac{1}{2}\right)^{n/3}. \tag{1}$$

It follows by the Koebe principle that up to an exponentially small error the restriction of $R^n f$ to the central interval T_0^n is purely quadratic, while the restriction to T_1^n is linear. This is all we need to know for the comprehensive study of f.

Asymptotically conformal continuation. Let us represent f as $h \circ \phi$ where $\phi(z) = (z - c)^2$ is the quadratic map, while h is a C^2-diffeomorphism of appropriate intervals. Let us continue h to a diffeomorphism of a bounded C^2 norm on the whole real line, and then consider the Ahlfors-Beurling continuation of h to the complex plane:

---

[3]We assume for simplicity that f is even.

$$\hat{h}(x+iy) = \frac{1}{2y}\int_{x-y}^{x+y} h(t)dt + \frac{1}{y}\left(\int_x^{x+y} h(t)dt - \int_{x-y}^x h(t)dt\right).$$

This is clearly a C^2-map, and one can check by calculation that $\bar{\partial}\hat{h} = 0$ on the real line. Hence $\bar{\partial}\hat{h}/\partial\hat{h} = O(|y|)$ as $|y| \to 0$. This provides us with a C^2 extension of f which is asymptotically conformal on the real line in the sense that

$$\mu(z) \equiv \bar{\partial}\hat{f}/\partial\hat{f} = O(|y|) \tag{2}$$

as well. In what follows we denote the extended h and f by the same letters.

Complex pull-back. Given an interval $I \subset \mathbf{R}$ and $\theta \in (0, \pi/2)$, let $D_\theta(I)$ denote the domain bounded by the union of two $\mathbf{R}$-symmetric arcs of the circles which touch the real line at angle θ. In particular, $D_{\pi/2}(I) \equiv D(I)$ is the Euclidean disk with diameter I. Observe that I is a hyperbolic geodesic in the domain $\mathbf{C} \setminus (\mathbf{R} \setminus I)$ and $D_\theta(I)$ is its hyperbolic neighborhood of radius depending only on θ.

We say than an interval $\tilde{I}$ is obtained from the I by α-scaling if these intervals are cocentric and $|\tilde{I}| = (1 + \alpha)|I|$.

Lemma 1 *Let $\alpha < 1$, n be sufficiently big. Let us consider the α-scaled interval $\tilde{I}^n \supset I^n$. Let $\Delta = D(\tilde{I}^n)$, and Δ' be the pull-back of Δ by $g_{n+1}|I^{n+1}$. Then $\Delta' \subset D(\tilde{I}^{n+1})$ where $\tilde{I}^{n+1}$ is obtained from I^{n+1} by β-scaling with $\beta = \alpha + O(\mu_n)$.*

Proof Let us skip the index n in the notations of objects of level n and mark the objects of level $n + 1$ with a prime. Set $g|I' = f^p$, and let us consider the pull-back $I, I_{-1}, ..., I_{-p} \equiv I'$ of I along the orbit $\{f^k c\}_{k=0}^p$. Then

$$\sum_{k=0}^p |I_{-k}| = O(\mu). \tag{3}$$

Since the map $f^k : I_{-k} \to I$ has the Koebe space covering I^{n-1}, the pull-back $\tilde{I}_{-k}$ of $\tilde{I}$ along the same orbit also has the total length $O(\mu)$.

Let us now take the disk Δ and pull it back along the same orbit. We obtain a sequence of pieces Δ_{-k} based upon the intervals $\tilde{I}_{-k}$. Assume by induction that $\Delta_{-l} \subset D_{\theta(k)}(\tilde{I}_{-l})$, $l = 0, ..., k < p$, with

$$\theta_l = \alpha + O(\sum_{j=0}^{l-1} |\tilde{I}_{-j}|). \tag{4}$$

Represent f as $h \circ \phi$ and carry out the next pull-back in two steps: first by the diffeomorphism h and then by the quadratic map ϕ. Let $h^{-1}\tilde{I}_{-k} = L_{-k}$. If we rescale the intervals $\tilde{I}_{-k}$ and L_{-k} to the unit size, the C^1-distance from the rescaled map $H^{-1} : [0, 1] \to [0, 1]$ to id is $O(|I_{-k}|)$. It follows that

$$h^{-1}\Delta_{-k} \subset D_{\theta(k+1)}(L_{-k}) \tag{5}$$

with $\theta(k+1)$ as in (4).

Consider now two cases. Let first $k < p - 1$. Then $\phi : \tilde{I}_{-(k+1)} \to L_{-k}$ is a diffeomorphism and by the Schwarz lemma (see the above hyperbolic interpretation of the $D_\theta(I)$)

$$\Delta_{-(k+1)} \subset D_{\theta(k+1)}(\tilde{I}_{-(k+1)}).$$

Let us now carry out the last pull-back corresponding to $k = p - 1$. Then $\phi|I_{-(k+1)} = \phi|I'$ is the quadratic folding map into $L \equiv L_{-(p-1)}$. Moreover, what is important is that $\phi I'$ covers at least half (up to an error of order $O(\mu)$) of the interval L (It follows from the high return property of g and the estimate of its non-linearity). Hence we can find an interval $K \supset L$ centered at the critical value $f(c)$ such that

$$D_{\theta(p-1)}(L) \subset D(K)$$

and

$$|K| = 2|\phi I'|(1 + O(\mu)).$$

Two last equations together with (4) yield the required.   □

Let us now take the Euclidean disk $\Delta = D(I^m)$ and pull it back by the maps g_n continued to the complex plane. Denote the corresponding domains by Δ_0^n and Δ_1^n, $n > m$.

Corollary 2 *If m is sufficiently big then the $\mathrm{diam}\Delta_j^n$ is commensurable with the $\mathrm{diam}I_j^n$.*

Proof Applying the previous lemma $n - m$ times, we see that diam Δ_j^n is $|I_j^n|\left(1 + O\left(\sum_{k=m}^n \mu_k\right)\right)$. Since μ_k decay exponentially, we are done.   □

3 Thurston's Transformation and the Shape of the Complex Puzzle-Pieces

Let us consider the quadratic map $p : z \mapsto z^2 - 1$ and mark on $\mathbf{C}$ a set A of three points $-1, 0$, and $a = (1 + \sqrt{5})/2$. The first two form a cycle, while the last one is fixed. Taking a conformal structure ν on the thrice punctured plane $S = \mathbf{C} \smallsetminus A$, we can pull it back by p. This induces a "Thurston's transformation" L of the Teichmüller space T_S of thrice punctured planes into itself (compare [1] or [13]). The main property of L is that it strictly contracts the Teichmüller metric, and hence all trajectories $L^n \tau$ exponentially converge to the single fixed point $\tau_0 \in T_S$ represented by the standard conformal structure.

Let us consider the involution $\rho : T_S \to T_S$ induced by the reflection of the conformal structure about the real line. This involution commutes with L, and so the subspace T_S^* of **R**-symmetric structures is L-invariant. This subspace can be identified with the set of triples on the real line up to affine transformations. We can normalize the triples, say, as follows: $\{\gamma, 0, a\}, \gamma < 0$. To pull back such a triple, we should take the quadratic polynomial p_γ which fixes a and carries 0 to γ, and take the negative preimage of 0.

Let us rescale both intervals I^n and I^{n-1} to the size $T = [-a, a]$ with a as above. Let $G_n : T \to T$ be the rescaled $g_n : I^n \to I^{n-1}$ (observe that this is a non-dynamical procedure, compare [9]). Let us select the orientation in such a way that 0 is the minimum point of G_n.

Lemma 3 *The maps G_n converge to the polynomial $p(z) = z^2 - 1$ in C^1 norm on compact subsets of* **C**.

Proof If we pull back the Euclidean disk $\Delta = D(I^n)$, we obtain a sequence of puzzle-pieces whose diameter is commensurable with their traces on the real line (Corollary 2). By the Denjoy distortion argument,

$$Dh_n^{-1}(z) = Dh_n^{-1}(0)(1 + O(\sqrt{\mu_n})), \quad z \in \Delta,$$

so that h_n^{-1} in Δ is an exponentially small perturbation of a linear map. Rescaling, we conclude that $G_n = H_n \circ p_{\gamma(n)}$ where H_n are diffeomorpisms converging exponentially to id in C^1 on compact sets, and $p_{\gamma(n)}$ are quadratic polynomials introduced above.

Let us now consider a sequence $\tau_n \in T_S^*$ represented by triples $(G_n(0), 0, a)$. It was shown in [11] that $|G_n(0)|/a$ stays away from 0 and 1. Hence $\tau_{n+1} = L \circ Q_n(\tau_n)$ where L is the Thurston transformation, while Q_n is exponentially close to id in the Teichmüller metric. Since L is strictly contracting, τ_n must converge to its fixed point τ_0.

We conclude that $G_n(0) \to -1$, hence $p_{\gamma(n)} \to p$ and $G_n \to p$. $\qquad\square$

Let us consider the following topology on the space $\mathcal{K}$ of connected full compact subsets K of **C**. Let $\psi_K : \{z : |z| > 1\} \to \mathbf{C} \setminus K$ be the Riemann map normalized at ∞ by $\psi(z) \sim qz$ with $q > 0$. Then the topology on $\mathcal{K}$ is induced by the compact device open topology on the space of univalent functions.[4]

Let us now consider the complex pieces Δ^n based upon the intervals I^n. Here Δ^n is the g_n-pull-back of Δ^{n-1}. Rescaling of I^n to T leads to the corresponding rescaled pieces P_n.

Lemma 4 *The pieces P_n converge to the filled-in Julia set of $p(z) = z^2 - 1$.*

Proof The piece P_n is the G_n-pull-back of P_{n-1}. By Lemma 1, $\text{diam} P_n$ is bounded. Hence $G_n|P_n$ is an exponentially small perturbation of p which yields the desired. $\qquad\square$

---

[4]It is called the *Carathéodory topology*.

4   Qc Conjugacy on the Critical Sets

Let us consider the complex renormalizations of f,

$$F_n = R^n f : V_0^n \cup V_1^n \to P^n,$$

where V_i^n are the rescaled puzzle-pieces based upon the intervals T_i^n. We use the same letters for the complex extensions of different maps. In particular, let $G_n : P^n \to P^{n-1}$ be the rescaled $g_n : \Delta^n \to \Delta^{n-1}$ (see[5] Fig. 2).

Let us parametrize smoothly the boundary of the piece P^0, $\gamma : \mathbf{T} \to \partial P^0$. This parametrization can be naturally lifted to the parametrization $\gamma_1 : \mathbf{T} \to \partial P^1$, namely $G_1 \circ \gamma_1 = \gamma(z^2)$, then to the parametrization of ∂P^2 etc. We refer to these parametrizations the *boundary markings*.

Let us also consider another Fibonacci map $\tilde{f}$ whose data will be labeled by tilde. The *Teichmüller distance* between two marked puzzle-pieces is the best dilatation of qc maps between the pieces respecting the boundary marking.

Lemma 5 *The marked puzzle-pieces P^n and $\tilde{P}^n$ stay bounded Teichmüller distance apart.*

Proof Let us have a K-qc map $H_{n-1} : P^{n-1} \to \tilde{P}^{n-1}$ of the marked pieces respecting the positions of the critical points and the critical values, that is, $H_{n-1}(0) = 0$ and $H_{n-1}(\gamma_{n-1}) = \tilde{\gamma}_{n-1}$. It can be lifted to the $K(1 + O(\mu_n))$-qc map $h_n : P^n \to \tilde{P}^n$. This map respects boundary marking and 0-points but it does not respect γ-points. However, it respects these points up to exponentially small error, namely $h_n(\gamma_n)$ and $\tilde{\gamma}_n$ are exponentially close.

Indeed, let $q_n \in T_1^n$ be the G_n-preimage of 0. As the length of T_n is exponentially small, the points q_n and γ_n are exponentially close. Moreover, by Lemma 4 the distance from these points to the boundary ∂P^n is bounded from below. By the Hölder continuity of qc maps we conclude that $h_n(q_n)$ and $h_n(\gamma_n)$ are also exponentially close. As $h_n(q_n) = \tilde{q}_n$, the points $h_n(\gamma_n)$ and $\tilde{\gamma}_n$ are exponentially close as well.

As the distance from these points to the boundary $\partial \tilde{P}^n$ and from 0 is bounded from below, they are exponentially close with respect to the Poincaré metric of $\tilde{P}^n$. Hence there is a diffeomorphism $\psi : \tilde{P}^n \to \tilde{P}^n$ with exp small dilatation keeping $\partial \tilde{P}^n$ and 0 fixed, and pushing $h_n(\gamma_n)$ to $\tilde{\gamma}_n$. Then $H_n = \psi \circ h_n$ is a $(K+$exp small)-qc map between the marked puzzle-pieces P_n and $\tilde{P}_n$ respecting the positions of the critical points and the critical values.

Proceeding in a such a way we construct uniformly qc maps between P^n and $\tilde{P}^n$ on all levels (as the exponentially small addings to dilatation sum up to a finite value).   □

Let us now consider the pairs of pants $Q^n = P^n \setminus (V_0^n \cup V_1^n)$, where $V_0^n \equiv P^{n+1}$, with the naturally marked boundary.

---

[5] According to our convention, V_1^n on this picture should actually be placed to the left of V_0^n.

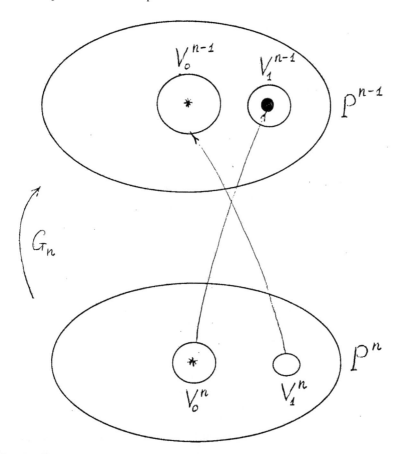

Fig. 2 The Fibonacci scheme

Lemma 6 *The pairs of pants Q^n and $\tilde{Q}^n$ stay bounded Teicmüller distance apart.*

Proof Let us consider a K-qc homeomorphism $H_{n-1} : Q^{n-1} \to \tilde{Q}^{n-1}$ of marked pairs of pants. It follows from the previous lemma that we can extend these maps across V_j^{n-1}. Indeed, the previous lemma provides us with the continuation to V_0^{n-1}. Moreover, it provides us with a map $P^{n-1} \to \tilde{P}^{n-1}$ which then can be pulled back to V_1^{n-1}. Let us keep the notation H_{n-1} for this extension.

Let us now consider the pull-back $W^{n-1} \subset V_1^{n-1}$ of V_0^{n-1} by F_{n-1}. Its boundary is also naturally marked. By one more pull-back of H_{n-1} we can reconstruct it in such a way that it will respect this marking. Let us consider the annulus $A^{n-1} = P^{n-1} \smallsetminus W^{n-1}$ with marked boundary.

The annulus $L^n = P^n \smallsetminus V_0^n$ double covers A^{n-1} under G_n. So we can pull H^{n-1} back to a K-qc map $H^n : L^n \to \tilde{L}^n$. Moreover, this map respects the parametrization of ∂V_1^n, and hence can be restricted to the K-qc map of marked pairs of pants of level n. $\qquad\square$

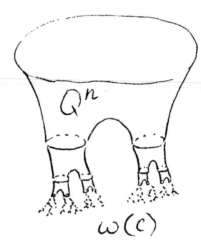

Fig. 3 Pairs of pants decomposition

We are prepared to obtain the desired result of this section.

Lemma 7 *There is an **R**-symmetric qc map which conjugates f and $\tilde{f}$ on their critical sets.*

Proof The critical set can be represented as

$$\omega(c) = \cap_{n=1}^{\infty} \cup Q_i^n,$$

where Q_i^n are dynamically constructed disjoint pairs of pants (see Fig. 3). They are obtained by univalent pull-backs of appropriate central pairs of pants. As these pull-backs have bounded dilatations, Lemma 6 implies that Q_i^n stay a bounded Teichmüller distance from $\tilde{Q}_i^n$. Gluing together all these pairs of pants, we obtain the desired result.   □

5   Pull-Back Argument

Sullivan's pull-back argument allows us to construct a qc conjugacy between two polynomial-like maps as long as there is a qc conjugacy on their critical sets. In this paper we deal with asymptotically conformal maps, so that we need the dilatation control of pull-backs. Lemma 1 will provide us with such a control along the real line. However, away from the real line the dilatation can grow, so that we should stop the construction at an appropriate moment. Let us show how it works. First we need some extra analysis on the real line.

Let $f_n : \cup I_j^n \to I^{n-1}$ be the full return map to the interval I^{n-1}.

Lemma 8 *Let $I^n \equiv J_0, J_{-1}, \ldots$ be any pull-back (finite or infinite) of the interval I^n. Then*

$$\sum |J_{-k}| = O(\mu_n).$$

Proof Denote by $\mathcal{J}$ the union of the intervals in the pull-back. Let us first assume that the intervals J_{-k} do not intersect I^n. Let $K_0 \equiv J_0, K_1, \dots$ be the piece of the pull-back which belongs to $I^{n-1}, \mathcal{K} = \mathcal{J} \cap I^{n-1}$ be the union of these intervals. This is actually the pull-back under the map f_n. This map is expanding with a bounded distortion on the I_j^n (actually, is very strongly expanding and is almost linear on the I_j^n). Hence

$$\sum |K_j| = O(\mu_n). \tag{6}$$

Let us now consider all intervals L_i obtained by pulling I^{n-1} back which are maximal in the sense that they do not belong to another pull-back interval. In other words, there is an $m = m(i)$ such that $f^m L_i = I^{n-1}$ but $f^l L_i \cap I^{n-1} = \emptyset$ for $l < m$. These intervals are mutually disjoint (and cover almost everything).

Let $\mathcal{K}_i = \mathcal{J} \cap L_i$. Then $f^{m(i)}$ maps $\mathcal{K}_i$ with bounded distortion (actually almost linearly) onto $\mathcal{K}$. Hence $\text{dens}(\mathcal{K}_i | L_i) = O(\mu_n)$. Summing up over i we get the claim.

Assume now that there are intervals in I^n but there are no ones in I^{n+1}. Let J_{-l} be the first interval belonging to I^n. Then for the further pull-backs we can repeat the same argument on level n instead of $n - 1$ (taking into account that the Poincaré lengths of the I_j^{n+1} in I^n are $O(\mu_n)$).

In general, let us divide the pull-back into the pieces $\mathcal{J}_l$ between the first landing at I^l and the first landing at I^{l+1}. Let us pull I^l along the corresponding piece. This pull-back does not intersect I^{l+1} either, and according to the previous considerations its total length is $O(\mu_l)$. All the more this is true for the total length of $\mathcal{J}_l$.

Hence the total length of $\mathcal{J}$ is $O\left(\sum_{l \geq n} \mu_l\right) = O(\mu_n)$. $\qquad\square$

Let us now state the complex version of the above lemma.

Lemma 9 *Let $\Omega = D(I^n), \Omega_{-1}, \dots$ be any pull-back of the disk Ω along the real line. Then*

$$\sum \text{diam}\,\Omega_{-k} = O(\mu_n).$$

Proof Let $\mathcal{W}$ denote the union of the disks in this pull-back. As in the above argument, let us decompose it into the strings $\mathcal{W}_j$ in between levels j and $j + 1$. Let Ω^j be the first puzzle-piece in the jth string.

On the other hand, let Δ^j denote the pull-backs of Ω based upon the intervals I^j. Then by the Markov property of the whole family of pull-backs, $\Omega^j \subset \Delta^j$. Hence the pull-back $\mathcal{W}_j$ can be inscribed into the corresponding pull-back $\mathcal{D}_j$ of the puzzle-piece Δ^j.

It follows from Lemma 1 that the sum of the diameters of pieces in $\mathcal{D}_j$ is commensurable with the total length of its trace on the real line. By the previous lemma, the latter is $O(\mu_n)$, and we are done. $\qquad\square$

Let us now select a high level n and consider the complex renormalization F_n : $V_0^n \cup V_1^n \to P^n$. Let us re-denote all these objects as $F : U_0^1 \cup U_1^1 \to U^0$. As above, the corresponding objects for another Fibonacci map $\tilde{f}$ will be labeled with the tilde. The following statement shows that two renormalizations of sufficiently high order are qc-conjugate.

Proposition 10 *There is a qc map $U^0 \to \tilde{U}^0$ which conjugates F and $\tilde{F}$ on the real line.*

Proof By Lemma 7, there is a qc map $h_0 : U^0 \to \tilde{U}^0$ which conjugates F to $\tilde{F}$ on the critical sets and on the $\partial(U_0^1 \cup U_1^1)$. Let us start to pull it back.

Let U_j^n denote the family of puzzle-pieces of depth n (that is, the components of $F^{-n}U^0$) *which meet the real line.* Let us assume by induction that we have already constructed a qc map $h_n : U^0 \to \tilde{U}^0$ which conjugates F to $\tilde{F}$ on their critical sets and on $(U_0^1 \cup U_1^1) \setminus \text{int}(\cup U_j^n)$. Then construct h_{n+1} as the lift of h_n to all puzzle-pieces U_j^n.

Since the puzzle-pieces U_j^n shrink to points, the sequence h_n has the continuous pointwise limit h which conjugates F and $\tilde{F}$ on the real line. Moreover, by (2) and Lemma 9, the h_n have uniformly bounded dilatations. Hence h is qc. $\square$

Let us re-denote I^n by $J \equiv J^0$, and let $\Delta = D(J)$. Let us now consider the full first return map f_1 to Δ. Its domain intersects the real line by the union of intervals $J_j^1 \equiv I_j^{n+1}$. Let Δ_j^1 be the pull-back of Δ intersecting the real line by I_j^{n+1}, $\mathcal{D}^1 = \cup \Delta_j^1$ (see Fig. 4).

The goal of the next three lemmas is to construct a qc map $h : \Delta \to \tilde{\Delta}$ which conjugates $f_1 | \partial \mathcal{D}^1$ to $\tilde{f}_1 | \partial \tilde{\mathcal{D}}^1$ (as well as $f_1 | \omega(c)$ to $\tilde{f}_1 | \omega(\tilde{c})$). This will be the start-

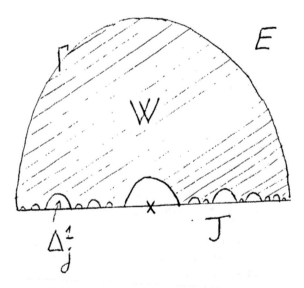

Fig. 4 The initial quasi-disk

ing data for the pull-back argument. The problem is that the boundary $\partial \mathcal{D}$ is not piecewise-smooth.

Given a set U, denote by U^+ the intersection of U with the upper half-plane.

Lemma 11 *The topological discs Δ_j^1 are pairwise disjoint. The set $W = (\Delta \setminus \mathcal{D})^+$ is a quasi-disk.*

Proof The map $f_n : \Delta_j^1 \to \Delta$ has exponentially small non-linearity. Hence Δ_j^1 is a minor distorted round disk. On the other hand, the intervals J_i^1 and J_j^1 are exponentially small as compared with the gap G_{ij} in between. It follows that the disks Δ_i^1 and Δ_j^1 are disjoint.

Let $\Gamma = \partial W$. It follows from the previous discussion that this curve is rectifiable. Take two close points $z, \zeta \in \Gamma$. Let δ be the shortest path connecting z and ζ in $\Gamma \cup \mathbf{R}$ (it is "typically" the union of an interval of the real line and two almost circle arcs), and γ be the shortest arc in Γ connecting z and ζ. Then the length of δ is commensurable with both the length of γ and the dist(z, ζ). $\square$

For the further discussion it is convenient to make a more special choice of the interval J (compare [3, 4, 7]). Namely, let α be the fixed point of f with negative multiplier $\sigma \equiv f'(\alpha)$. Let $\mathcal{Y}^{(0)}$ be the partition of T by α into two intervals. Pulling this partition back, we obtain partitions $\mathcal{Y}^{(n)}$ by n-fold preimages of α. Let us call the elements of this partition *the puzzle-pieces of depth* n. The element containing c is called *critical*. We select $J = [\beta, \beta']$ as the critical puzzle-piece of sufficiently high depth N.

Set $\tau = \log |\tilde{\sigma}| / \log |\sigma|$.

Let us now start with a qc $\mathbf{R}$-symmetric map $H : \Delta \to \tilde{\Delta}$ which carries the critical set of f_1 to the critical set of $\tilde{f}_1$ and such that

$$|H(z) - \tilde{\beta}| \asymp |z - \beta|^\tau. \tag{7}$$

Moreover, let H commute with the symmetry around c induced by f and $\tilde{f}$.

Pull H back to a map $h : \mathcal{D} \to \tilde{\mathcal{D}}$. Since the union $\cup J_j^1$ is dense in J, this map can be continued to a homeomorphism $h : J \to \tilde{J}$. Let also $h|\partial \Delta = H$. This defines h on the topological semi-circle $S = \partial \Delta^+$. Since S and $\tilde{S}$ are piecewise smooth curves, we can naturally define the notion of a quasi-symmetric map between them.

Lemma 12 *The map $h : S \to \tilde{S}$ is quasi-symmetric.*

Proof Let us consider an extension $H : T \to \tilde{T}$ of $H : J \to \tilde{J}$ which carries the puzzle-pieces of depth N to the corresponding puzzle-pieces, and has the asymptotics (7) near the boundary points of these puzzle-pieces.

Let K be the expanding Cantor set of points which never land in int$J$. Each component L of $T \setminus K$ (a "gap") is a monotone pull-back of J with bounded distortion. So we can lift the map H to qs maps on all the gaps L. These maps clearly glue together to a homeomorphism $\phi : T \to \tilde{T}$ which respect the dynamics on the Cantor sets K and $\tilde{K}$. Moreover, if we rescale the corresponding gaps L and $\tilde{L}$ to the unit size then the rescaled ϕ near the boundary points will have asymptotics (7) uniformly in L.

Furthermore, it easily follows from the bounded distortion properties of expanding dynamics that $\phi|K$ can be extended to a qs conjugacy ψ in a neighborhood of K. This conjugacy must have the same asymtotics (7) on the rescaled gaps (since the conjugacy near the fixed points has such asymptotics). It follows that ϕ and ψ are comparable on the gaps, and hence ϕ is qs on the whole interval.

Observe now that $h : J \to \tilde{J}$ is the lift of ϕ by the almost quadratic maps $f|J$ and $\tilde{f}|\tilde{J}$. Hence $h|J$ is qs and has asymptotics (7) near the boundary. Since it has the same asymptotics on the opposite side of β, β' on the arc $S \setminus J$, it is qs on S. $\qquad\square$

Lemma 13 *The map* $h : \partial W \to \partial \tilde{W}$ *allows a qc extension to* $W \to \tilde{W}$.

Proof Let E be the exterior component of $\mathbf{C} \setminus S$. By the previous lemma, there is a qc extension of h from S to $h_0 : E \to \tilde{E}$ (which changes the original values of h below the real line).

We can now glue $h : \mathcal{D}^+ \to \tilde{\mathcal{D}}^+$ with h_0 to a qc map $h_* : \mathbf{C} \setminus W \to \tilde{\mathbf{C}} \setminus W$ (since they agree on the real line). Since W is a quasi-disk (by Lemma 11), h_* can be reflected to the interior of W, and this is a desired extension. $\qquad\square$

Corollary 14 *There is an* $\mathbf{R}$-*symmetric qc map* $h : \Delta \to \tilde{\Delta}$ *which conjugates* f_1 *to* $\tilde{f}_1$ *on the critical sets and on the boundary of* $\mathcal{D}$.

Proof Lemma 13 gives us a desired qc extension of the original h from $\mathcal{D} \cup \partial \Delta$ to Δ. $\qquad\square$

Now we are ready to prove the main result.

Theorem I *Any two Fibonacci quasi-quadratic maps are qc conjugate.*

Proof Starting with the qc map h given by Corollary 14, we can go through the pull-back argument in the same way as in Proposition 10. This provides us with a qs conjugacy between the return maps f_1 and $\tilde{f}_1$. Then we can spread it around the whole interval T as in the proof of Lemma 12. $\qquad\square$

6   Teichmüller Metric

Let K_h denote the dilatation of a qc map h. Given two Fibonacci maps f and g and the qs conjugacy between them, the Teichmüller pseudo-distance $\mathrm{dist}_T(f, g)$ is defined as the infimum of $\log K_h$ for all qc extensions of h.

Theorem II *If* $\mathrm{dist}_T(f, g) = 0$ *then* f *and* g *are smoothly conjugate.*[6]

---

[6]This result was generalized by M. Martens and W. de Melo to arbitrary unimodal maps which are not infinitely renormalizable [29].

Proof Our first step is the same as Sullivan's [14]: If $\mathrm{dist}_T(f, g) = 0$ then the multipliers of the corresponding periodic orbits of the maps are equal. However, as we do not have yet a proper thermodynamical formalism for unimodal maps, we will proceed by a concrete geometric analysis.

The next observation is that the parameter a in (1) must be the same for f and g. Indeed, it can be explicitly expressed via the multipliers of the fixed points of the return maps $g_n : I^n \to I^{n-1}$ (since the g_n are asymptotically quadratic). By [11] this already yields the smoothness of the conjugacy on the critical sets.

Let us now take a point $x \in I^n \setminus I^{n-1}$ and push it forward by iterates of g_n until the first moment it lands in I^n (if any), then apply the iterates of g_{n+1} until the first moment it lands in I^{n+1}, etc. This provides us with a nested sequence of intervals around x whose lengths can be expressed (up to a bounded error) through the scaling factors and the multipliers of appropriate periodic points (by shadowing). This implies that h is Lipschitz continuous. Moreover, when we approach the critical point, then the errors in the above argument exponentially decrease. Hence h is smooth at the critical point.

Given now any pair of intervals $I \supset J$, let us show that

$$\left| \frac{|hJ|}{|J|} : \frac{|hI|}{|I|} - 1 \right| = O(|I|). \tag{8}$$

This is enough to prove locally at any point b. By the previous considerations, this is true at the critical point. Since the critical set $\omega(c)$ is minimal, this is also true for any $b \in \omega(c)$.

Let now $b \notin \omega(c)$, and I be a tiny interval around b. Remark that almost all points $x \in I$ eventually return back to I. Let us take the pull-back of I corresponding to this return. This provides us with the covering of almost all of I by intervals L_k. The distortion of the return map g is $O(|I|)$ on all $L'_k s$. Let σ_k be the multiplier of the g-fixed point in L_k. Then we conclude that

$$\left| \frac{|I|}{|L_k|} : \sigma_k - 1 \right| = O(|I|), \tag{9}$$

and the analogous estimate holds for the second map. Since the corresponding multipliers of these maps are equal, we obtain (8) with $J = L_k$. Repeating now this procedure for returns of higher order, we obtain an arbitrarily fine covering of almost the whole of I by intervals for which (8) hold. This implies (8) for any $J \subset I$.

Let $\epsilon_n = 1/2^n$, and let us consider the sequence of functions

$$\rho_n(x) = \frac{h(x + \epsilon_n) - h(x - \epsilon_n)}{2\epsilon_n}.$$

According to (8) and Lipschitz continuity,

$$|\rho_n(x) - \rho_{n+1}(x)| = O(\epsilon_n) \tag{10}$$

uniformly in x. Hence the ρ_n uniformly converge to the derivative of h.

References

1. A. Douady, J.H. Hubbard, A proof of Thurston's topological characterization of rational functions. Preprint Institut Mittag-Leffler, 1986
2. F.P. Gardiner. Lacunary series as quadratic differentials, Preprint 1992
3. J. Guckenheimer, S. Johnson, Distortion of S-unimodal maps. Ann. Math. **132**, 71–130 (1990)
4. J.H. Hubbard, Local connectivity of Julia sets and bifurcation loci: three theorems of J.-C. Yoccoz, in *Topological Methods in Modern Mathematics, A Symposium in Honor of John Milnor's 60th Birthday* (Publish or Perish, 1993)
5. F. Hofbauer, G. Keller, Some remarks on recent results about S-unimodal maps. Ann. Institut Henri Poincaré **53**, 413–425 (1990)
6. Y. Jiang, Generalized Ulam-von Neumann transformations. Thesis, 1990
7. M. Jakobson, G. Swiatek, Quasisymmetric conjugacies between unimodal maps. Preprint IMS at Stony Brook, #1991/16
8. G. Keller, T. Nowicki, Fibonacci maps revisited. Preprint 1992
9. J. Ketoja, O. Piirila, On the abnormality of the period doubling bifurcation. Phys. Lett. A **138**, 488–492 (1989)
10. J. Kahn, Holomorphic Removability of Julia Sets. Manuscript in preparation
11. M. Lyubich, J. Milnor, The unimodal Fibonacci map. Preprint IMS at Stony Brook #1991/15. To appear in the Journal of AMS
12. M. Lyubich, Combinatorics, geometry and attractors of quasi-quadratic maps. Preprint IMS at Stony Brook #1992/18
13. J. Milnor, W. Thurston, On iterated maps of the interval, Preprint of 1977 and pp. 465–563 of Dynamical Systems, Proc. U. Md., 1986–87, ed. J. Alexander, Lecture Notes in Mathematics, vol. 1342 (Springer, Heidelberg, 1988)
14. D. Sullivan, Quasiconformal homeomorphisms in dynamics, topology and geometry, in *Proceedings of the ICM*, Berkeley, vol. 2 (1986), p. 1216
15. D. Sullivan, Bounds, quadratic differentials, and renormalization conjectures, 1990. To appear in AMS Centennial Publications. **2**: Mathematics into Twenty-first Century

Addendum 1: Bibliography Update (2018)

16. Appeared in Acta Math., v. 171 (1993), 263–297
17. Appeared in Contemp Math., v. 169 (1994), 307–330 (joint with D. Sullivan)
18. Appeared in Erg. Th. & Dyn. Syst., v. 14 (1994), 721–755
19. Appeared in Erg. Th. & Dyn. Syst., v. 15 (1995), 99–120
20. Is available as a Stony Brook Preprint, 1998, # 11, http://www.math.stonybrook.edu/cgi-bin/preprint.pl?ims98-11
21. Appeared in JAMS, v. 6 (1993), 425–457
22. Appeared in Annals of Math., v. 140 (1994), 347–404
23. Appeared in AMS Centennial Publications, v. 2: Mathematics into Twenty-first Century (1992)

Addendum 2: Additional References (Added in 2018)

24. A. Avila, R. Krikoryan, Monotonic cocycles. Invent. Math. **202**, 271–331 (2015)
25. T. Clark, S. van Strien, Quasisymmetric rigidity in one-dimensional dynamics, arXiv:1804.16122 (2018)
26. J. Graczek, G. Swiatek, D. Sands, Decay of geometry for unimodal maps: negative Schwarzian case. Ann. Math. **161**, 613–677 (2005)
27. P. Guarino, W. de Melo, Rigidity of smooth critical circle maps. J. Eur. Math. Soc. **19**, 1729–1783 (2017)
28. M. Lyubich, Dynamics of quadratic polynomials, I–II. Acta Math. **178**, 185–297 (1997)
29. M. Martens, W. de Melo, The multipilers of peridic points in one-dimensional dynamics. Non-linearity **12**, 217–227 (1999)

On the Three-Legged Accessibility Property

Jana Rodriguez Hertz and Raúl Ures

Abstract We show that certain types of the three-legged accessibility property of a partially hyperbolic diffeomorphism imply the existence of a unique minimal set for one strong foliation and the transitivity of the other one. In case the center dimension is one, we also give a criterion to obtain three-legged accessibility in a robust way. We show some applications of our results to the time-one map of Anosov flows, skew products and certain Anosov diffeomorphisms with partially hyperbolic splitting.

Keywords Accessibility · Partial hyperbolicity · Minimal set

1 Introduction

A diffeomorphism f of a closed manifold M is *partially hyperbolic* if the tangent bundle TM of M, splits into three invariant sub-bundles: $TM = E^s \oplus E^c \oplus E^u$ such that all unit vectors $v^\sigma \in E^\sigma_x$ ($\sigma = s, c, u$) with $x \in M$ satisfy:

$$\|T_x f v^s\| < \|T_x f v^c\| < \|T_x f v^u\| \tag{1.1}$$

for some suitable Riemannian metric. The *stable bundle* E^s must also satisfy $\|Tf|_{E^s}\| < 1$ and the *unstable bundle*, $\|Tf^{-1}|_{E^u}\| < 1$. The bundle E^c is called *center bundle*.

It is a well-known fact that the *strong* bundles, E^s and E^u, are uniquely integrable [3, 13]. That is, there are invariant strong foliations $\mathcal{W}^s$ and $\mathcal{W}^u$ tangent, respectively to the invariant bundles E^s and E^u (However, the integrability of E^c is a more delicate matter).

J. Rodriguez Hertz · R. Ures (✉)
Department of Mathematics, Southern University of Science and Technology, 1088 Xueyuan Rd., Xili, Nanshan District, Shenzhen 518055, Guangdong, China
e-mail: ures@sustc.edu.cn

J. Rodriguez Hertz
e-mail: rhertz@sustc.edu.cn

© Springer Nature Switzerland AG 2019
M. J. Pacifico and P. Guarino (eds.), *New Trends in One-Dimensional Dynamics*, Springer Proceedings in Mathematics & Statistics 285,
https://doi.org/10.1007/978-3-030-16833-9_13

In this paper we will deal with partially hyperbolic diffeomorphisms that satisfy a certain type of accessibility property. Recall that such a diffeomorphism f satisfies the accessibility property (we will also say that f is accessible) if any pair of points can be joined by a curve that is piecewise tangent to either E^s or E^u. We will say that f has the *three-legged accessibility property* if in the definition of accessibility you can choose the curve joining each pair of points consisting of three arcs tangent to either E^s or E^u and with uniformly bounded length. Moreover, we will say that f is *sus-accessible* if it satisfies the three-legged accessibility property and the three-legged curve of the definition can be chosen, for all pair of points, in such a way that the first arc is stable, the second unstable and the last one stable. The *usu*-accessibility property is defined in a analogous way.

These more restrictive accessibility properties impose some limitations to the strong foliations of a partially hyperbolic diffeomorphism. Our first result is in this direction and states the following:

Theorem 1.1 [1] *Let $f \in C^1$ be an usu-accessible partially hyperbolic diffeomorphism. Then,*

(1) $\mathcal{W}^u$ has a unique minimal set.
(2) $\mathcal{W}^s$ is transitive, that is, it has a dense leaf.

The importance of the uniqueness of the minimal sets of $\mathcal{W}^u$ lies in the fact that it imposes important obstructions to the presence of attractors, in fact, under certain conditions they are unique. We will establish this result in the following theorem.

Theorem 1.2 *Let f be a $C^{1+\alpha}$ partially hyperbolic diffeomorphism and assume that $\mathcal{W}^u(f)$ has a unique minimal set Λ. If there is an u-Gibbs measure μ supported on Λ for which all center Lyapunov exponents are negative then, μ is the unique u-Gibbs measure for f and, as a consequence, it is SRB.*

Before showing some applications of Theorem 1.1 we give some necessary conditions to obtain *sus*-accessibility.

Theorem 1.3 *Let $f \in C^1$ be an accessible partially hyperbolic diffeomorphism with one-dimensional center such that $\mathcal{W}^u$ is minimal. Then, there is an open C^1-neighborhood $\mathcal{U}$ of f such that every $g \in \mathcal{U}$ is usu-accesible.*

Of course we have the analogous result if $\mathcal{W}^s$ is minimal. In fact, in their pioneer paper about stable ergodicity [11], Grayson, Pugh and Shub showed that the time-one map φ of the geodesic flow of a closed surface of constant negative curvature is both *sus*- and *usu*-accessible. Dolgopyat [8] used this property to get some consequences about the uniqueness of SRB measures in a neighborhood of φ. In Sect. 4 we show that most time-one maps of Anosov flows, including the case of geodesic flows, are both *sus*- and *usu*-accessible. This generalizes Burns, Pugh and Wilkinson [5]

---

[1] The authors jointly with Federico Rodriguez Hertz already had a proof of this result more than ten years ago.

result about the accessibility of Anosov flows. Also, as an application of Theorem 1.1 we obtain a generalization of the main result of Bonatti and Guelman [1] about the approximation of time-one maps of Anosov flows by Axiom A diffeomorphisms.

We also give some applications to skew products (Sect. 5) and to Anosov diffeomorphisms (Sect. 6). In the case of skew products we show that our results can be applied to an open and dense set of isometric circle extensions of volume-preserving partially hyperbolic diffeomorphisms. In the case of Anosov diffeomorphisms, we answer positively the following conjecture of Gogolev, Maimon and Kolgomorov [10]. Let A be a hyperbolic automorphism of the 3-torus with three real eigenvalues $|\lambda_1| < 1 < |\lambda_2| < |\lambda_3|$.

Conjecture 1.4 *For all analytic diffeomorphisms f in a sufficiently small neighborhood of A the strong unstable foliation $\mathcal{W}^{uu}$ is transitive, i.e., it has a dense leaf.*

In fact, we get a stronger result. We obtain the transitivity of $\mathcal{W}^{uu}$ for any 3-dimensional C^1 Anosov diffeomorphism f with a partially hyperbolic splitting.

2 Preliminaries

2.1 Partial Hyperbolicity

Definition 2.1 A diffeomorphism f of a closed manifold M is *partially hyperbolic* if the tangent bundle TM of M, splits into three invariant sub-bundles: $TM = E^s \oplus E^c \oplus E^u$ such that all unit vectors $v^\sigma \in E^\sigma_x$ ($\sigma = s, c, u$) with $x \in M$ satisfy:

$$\|T_x f v^s\| < \|T_x f v^c\| < \|T_x f v^u\| \tag{2.1}$$

for some suitable Riemannian metric. The *stable bundle* E^s must also satisfy $\|Tf|_{E^s}\| < 1$ and the *unstable bundle*, $\|Tf^{-1}|_{E^u}\| < 1$. The bundle E^c is called *center bundle*.

We also call $E^{cu} = E^c \oplus E^u$ and $E^{cs} = E^c \oplus E^s$.

It is a well-known fact that the *strong* bundles, E^s and E^u, are uniquely integrable [3, 13]. That is, there are invariant strong foliations $\mathcal{W}^s(f)$ and $\mathcal{W}^u(f)$ tangent, respectively, to the invariant bundles E^s and E^u. (However, the integrability of E^c is a more delicate matter) In general, we will call $\mathcal{W}^\sigma(f)$ any foliation tangent to E^σ, $\sigma = s, u, c, cs, cu$, whenever it exists and $W^\sigma_f(x)$ the leaf of $\mathcal{W}^\sigma(f)$ passing through x. A subset Λ is σ-*saturated* if $\subset \Lambda$ for every $x \in \Lambda$. A closed σ-saturated subset K is *minimal* if $\overline{W^\sigma_f(x)} = \Lambda$ for every $x \in \Lambda$. We say that a foliation is minimal if M is a minimal set for it. A foliation is *transitive* if it has a dense leaf.

2.2   Skew Products

In this subsection we consider skew products for which the base is a volume-preserving Anosov diffeomorphism and the fibers are circles (also known as isometric circle extensions) That means that we have $F_\varphi : M \times \mathbb{S}^1 \to M \times \mathbb{S}^1$ such that $F_\varphi(x, \theta) = (f(x), R_{\varphi(x)}\theta)$ where $f : M \to M$ is a C^r volume-preserving Anosov diffeomorphisms, R_α is the rotation of angle α and $\varphi : M \to \mathbb{S}^1$ is a C^r map, $r \geq 2$. These diffeomorphisms are partially hyperbolic, see [3]. Since f is volume preserving we have that F_φ preserves the measure given by the product of the volume of M and the Lebesgue measure on $\mathbb{S}^1$ (the Haar measure of $\mathbb{S}^1$ seen as a Lie group).

The following is proved in [3] (see also Proposition 2.1 in [4]) The center bundle of F_φ is tangent to the circle fibers and the strong stable and strong unstable bundles have the same dimension as the corresponding bundles of the base Anosov diffeomorphism. In particular, it has a center foliation and its leaves are the circle fibers. F_φ is dynamically coherent, that means that there are invariant foliations $\mathcal{W}^{c\sigma}(F_\varphi)$ tangent tangent to $E^{c\sigma}, \sigma = s, u$. Each leaf of $\mathcal{W}^{c\sigma}(F_\varphi)$ is the preimage under π of a leaf of $\mathcal{W}^\sigma(f), \sigma = s, u$ ($\pi : M \times \mathbb{S}^1 \to M$ is the projection on the first coordinate) In particular, any leaf of $\mathcal{W}^{c\sigma}(F_\varphi)$ is dense and it is the product of a leaf of $\mathcal{W}^\sigma(f)$ and $\mathbb{S}^1$ again $\sigma = s, u$. Each leaf of a strong foliation is is the graph of a C^r function from the corresponding leaf of the corresponding strong foliation of f to $\mathbb{S}^1$.

2.3   u-Gibbs Measures

We will only give a very brief introduction to the u-Gibbs measures with the purpose of doing this paper as self-contained as possible. For a more complete presentation we refer to [2, 7].

Definition 2.2 Let f be a $C^{1+\alpha}$ partially hyperbolic diffeomorphism. An f-invariant probability measure μ is u-Gibbs if it admits conditional measures along local strong unstable leaves which are absolutely continuous with respect to Lebesgue.

The condition that f is $C^{1+\alpha}$ is needed in order to get the absolute continuity of the strong unstable foliation.

The densities of the conditional measures are bounded away from zero and infinity and then, the support of a u-Gibbs measure consists of entire strong unstable leaves. The set of u-Gibbs measures is a convex compact subset of the set of probability measures.

Recall that an *SRB measure* is an invariant probability measure such that admits conditional measures along unstable manifolds which are absolutely continuous with respect to Lebesgue. Observe the difference with u-Gibbs measures for which these conditional measures are taken along **strong** unstable leaves. SRB measures of partially hyperbolic diffeomorphisms are always u-Gibbs measures. In general the converse is not true but if f admits a unique u-Gibbs measure μ then, μ is SRB (see, for instance, [2, Theorem 11.16]).

3 Proofs of the Main Results

3.1 Proof of Theorem 1.1

Proof of (a). Let A, $B \subset M$ be two closed u-saturated sets, $a \in A$, $b \in B$ and K a bound of the length of the curves given by the definition of usu-accessibility. Let $n \in \mathbb{N}$ and γ a curve of length less than K joining $f^{-n}(a)$ and $f^{-n}(b)$ consisting of three arcs γ_i $i = 1, 2, 3$ such that γ_1 and γ_3 are tangent to E^u and γ_2 is tangent to E^s. Observe that, since A and B are u-saturated, $\gamma_1 \subset f^{-n}(A)$ and $\gamma_3 \subset f^{-n}(B)$. On the one hand, the previous observation gives that $f^n(\gamma_2)$ is an arc joining A and B. On the other hand, since γ_2 is tangent to E^s, we have that the length of $f^n(\gamma_2)$ is less than $K\lambda^n$ for some uniform $\lambda < 1$. Thus, the distance between A and B is less than $K\lambda^n$ for all $n \in \mathbb{N}$. This implies that $A \cap B \neq \emptyset$. We have that any pair of closed u-saturated subsets has nonempty intersection and since the intersection of u-saturated sets is an u-saturated set we obtain that the family of all closed u-saturated sets satisfies the FIP. This implies that $S = \bigcap\{A \subset M;\ A \text{ is closed and } u\text{-saturated}\} \neq \emptyset$. It is not difficult to see that S is the unique minimal set of $\mathcal{W}^u$. This ends the proof of the first part of Theorem 1.1. $\qquad\square$

Remark 3.1 Observe that since f is a diffeomorphism and $\mathcal{W}^u$ is f-invariant we have that the minimal set given by Theorem 1.1 is f-invariant.

Proof of (b). Let U, $V \subset M$ be to open sets. It is enough to show that there is a leaf of $\mathcal{W}^s$ that has nonempty intersection with both U and V.

Take $x \in U$, $y \in V$ and $n \in \mathbb{N}$. Then, there is curve γ of length less than K joining $f^n(x)$ and $f^n(y)$ consisting of three arcs γ_i $i = 1, 2, 3$ such that γ_1 and γ_3 are tangent to E^u and γ_2 is tangent to E^s. Since γ_1 and γ_3 are tangent to E^u we have that the length of $f^{-n}(\gamma_i)$ $i = 1, 3$ is less than $K\lambda^n$ for for some uniform $\lambda < 1$. Observe that x and y are extreme points of the arcs $f^{-n}(\gamma_1)$ and $f^{-n}(\gamma_3)$ respectively and the other two extremes are in the same stable leaf. Since $K\lambda^n \to 0$ we have that for n large enough $f^{-n}(\gamma_1) \subset U$ and $f^{-n}(\gamma_3) \subset V$. Then, there is a stable leaf intersecting simultaneously U and V. This finishes the proof of the second part of Theorem 1.1. $\qquad\square$

3.2 Proof of Theorem 1.3

In order to proof Theorem 1.3 we need to introduce a new definition. In this subsection we will assume that $\dim(E^c) = 1$.

Definition 3.2 We say that two unstable disks U_1 and U_2 are skew iff

(1) There is a dimension $\dim(E^u) + 1$ disk D containing U_2.
(2) U_2 separates D into two connected components.
(3) U_1 and D define the holonomy map $h_s : U_1 \to D$ as $h_s(x) = W_\varepsilon^s(x) \cap D$ for ε small.

(4) $h_s(U_1)$ intersects both connected components of $D \setminus U_2$

Remark 3.3 (1) Being skew is an open condition in the following sense: if we have two disks U_1' and U_2' close enough to U_1 and U_2 they are also skew and this remains true if U_1' and U_2' are unstable disks of a diffeomorphism g close enough to f.

(2) If U_1 and U_2 are skew there exist $x_i \in U_i$ $i = 1, 2$ such that $x_2 \in W_\varepsilon^s(x_1)$.

Proof of Theorem 1.3. It is already known that accessibility implies the existence of two unstable disks U_i $i = 1, 2$ that are skew (see [6, 16]) Given neighborhoods V_i of U_i $i = 1, 2$ there exists $C > 0$ such that $W_C^u(x)$ intersect both V_1 and V_2. This implies that given any pair of points $x_i \in M$ $i = 1, 2$, $W_C^u(x_i)$ contains a disk in V_i for $i = 1, 2$. The continuos dependence of the unstable foliation on the diffeomorphisms imply that the same is true, with the same constant C, for any g close enough to f (maybe we have to take V_i $i = 1, 2$ a little bit larger) So, the two disks are skew (see Remark 3.3) and have two points y_i $i = 1, 2$ that are joined by a stable curve of length less than ε. The previous considerations imply that y_i con be joined to x_i by an unstable arc of length less than C, $i = 1, 2$. That means that g is *usu*-accessible for all g C^1 close enough to f with $K = 2C + \varepsilon$ and finishes the proof of the theorem. $\square$

3.3 Proof of Theorem 1.2

Proof Let F be a leaf of $\mathcal{W}^u(f)$. Since Λ is the unique minimal set of $\mathcal{W}^u(f)$ we have that $\Lambda \subset \overline{F}$. Since μ is supported on Λ and all its center Lyapunov exponents are negative, the absolute continuity of the stable partition implies that F has a positive Lebesgue measure set of points that are in the basin of μ. This clearly implies that μ is the unique u-Gibbs measure for f. $\square$

4 Anosov Flows

In this section we will apply our results to the time-one maps of Anosov flows. On the one hand, Burns, Pugh and Wilkinson [5] (see also [6]) proved that if φ is the time-one map of a transitive Anosov flow, accessibility is equivalent to the fact of $E^s \oplus E^u$ not being integrable. On the other hand, Plante [14] showed that in case $E^s \oplus E^u$ is integrable the flow is topologically equivalent to a suspension of a hyperbolic diffeomorphism. Moreover, in dimension three, joint integrability implies that the flow **is** a suspension. These comments lead to the following result.

Theorem 4.1 *Let φ_t be a transitive Anosov flow and assume that $E^s \oplus E^u$ is not integrable. Then, there is a C^1-neighborhood $\mathcal{U}$ of $\varphi = \varphi_1$ such that the strong* *unstable and the strong stable foliations of any* $g \in \mathcal{U}$ *are transitive and have a unique minimal set.*

Proof Since the Anosov flow is transitive and $E^s \oplus E^u$ is not integrable we have that φ is accessible and both strong foliations are minimal. Since φ is partially hyperbolic with one dimensional center we can apply Theorem 1.3 that immediately implies the thesis. □

In particular, any $g \in \mathcal{U}$ can have at most one transitive hyperbolic attractor (in fact it would be topologically mixing) and one transitive hyperbolic repeller. Then, we obtain as corollary the following generalization of the main result of [1].

Corollary 4.2 *If the time-one map of a transitive Anosov flow* φ_t *is* C^1*-approximated by diffeomorphisms having more than one transitive hyperbolic attractor, then* $E^s \oplus E^u$ *is integrable. In particular,* φ_t *is topologically equivalent to the suspension of a hyperbolic diffeomorphism. Moreover, if the dimension of the ambient manifold is three then,* φ_t *is the suspension of an Anosov diffeomorphisms.*

Proof Since an attractor is a compact u-saturated set it necessarily contains the unique minimal set given by Theorem 4.1. Since transitive hyperbolic attractors are disjoint $g \in \mathcal{U}$ can have only one hyperbolic attractor. This proves the corollary. □

In fact, to obtain the thesis of this corollary the only thing we need is the robustness of the uniqueness of the minimal set of the strong unstable foliation. Then, the same proof yields the following theorem.

Theorem 4.3 *Let* f *be a stably usu-accessible partially hyperbolic diffeomorphism. Then, there is a* C^1 *neighborhood* $\mathcal{U}$ *of* f *such that every* $g \in \mathcal{U}$ *has at most one transitive hyperbolic attractor.*

Observe that the uniqueness of the minimal set of $\mathcal{W}^u(f)$ implies that the transitive hyperbolic attractor, if it exists, is in fact topologically mixing.

In case the diffeomorphisms are $C^{1+\alpha}$ the previous results are consequence of Theorem 1.2.

5 Skew Products

In this section we consider skew products (also known as isometric extensions) for which the base is a volume-preserving Anosov diffeomorphism and the fibers are circles. That means that we have $F_\varphi : M \times \mathbb{S}^1 \to M \times \mathbb{S}^1$ such that $F_\varphi(x, \theta) = (f(x), R_{\varphi(x)}\theta)$ where $f : M \to M$ is a C^r volume-preserving Anosov diffeomorphisms, R_α is the rotation of angle α and $\varphi : M \to \mathbb{S}^1$ is a C^r map. $r \geq 2$. These diffeomorphisms are partially hyperbolic and their ergodic theory is well known. Burns and Wilkinson [4] proved the following (in fact they prove a much stronger result).

Theorem 5.1 ([4]) *The set of accessible isometric $\mathbb{S}^1$-extensions of a volume-preserving Anosov diffeomorphism is open and dense in the C^r-topology.*

We will show that if a skew product as above satisfies the accessibility property, both the strong stable and the strong unstable foliation are minimal. The arguments to show this fact are inspired in similar arguments of Plante [14].

Observe that F_φ commutes with $G_\alpha : M \times \mathbb{S}^1 \to M \times \mathbb{S}^1, G_\alpha(x, \theta) = (x, R_\alpha \theta)$ $\forall \alpha \in \mathbb{S}^1$. This implies that $\mathcal{W}^\sigma(F_\varphi)$ is G_α-invariant for $\sigma = s, u$.

Theorem 5.2 *Let F_φ be as above and suppose that it satisfies the accessibility property. Then, $\mathcal{W}^\sigma(F_\varphi)$ is minimal, $\sigma = s, u$.*

In fact, the previous theorem is a consequence of the following stronger fact.

Proposition 5.3 *Let F_φ be as above and suppose that $\mathcal{W}^\sigma(F_\varphi)$ is not minimal. Then, $E^s \oplus E^u$ is integrable.*

Proof Suppose that there is $x \in M \times \mathbb{S}^1$ such that $W^u_{F_\varphi}(x)$ is not dense. Since $\overline{W^u_{F_\varphi}(x)}$ is u-saturated we can take $K \subset \overline{W^u_{F_\varphi}(x)}$ a minimal subset. Call $\pi : M \times \mathbb{S}^1 \to M$ to the projection onto M, that is $\pi(y, \theta) = y$. It is not difficult to see that $\pi(W^u_{F_\varphi}(z)) = W^u_f(\pi(z))$. In particular, since f is transitive and K is compact and u-saturated we have that $\pi(K) = M$. The strategy is to prove that K is also s-saturated. This will immediately imply that F_φ is not accessible (moreover, the invariance of the strong foliations under $G_\alpha \forall \alpha \in \mathbb{S}^1$ will imply that $E^s \oplus E^u$ is integrable).

Our first claim is that $\#(K \cap \{y\} \times \mathbb{S}^1)$ is finite for every $y \in M$. Suppose that it is false, then there is $y_0 \in M$ such that for every $\varepsilon > 0$, $S = \{y_0\} \times \mathbb{S}^1$ has two points (y_0, θ_1) and (y_0, θ_2) at distance less than ε. This means that $\alpha = \theta_2 - \theta_1 < \varepsilon$. Observe that $G_\alpha(y_0, \theta_1) = (y_0, \theta_2)$ and then, $G_\alpha(K) \cap K \neq \emptyset$. Since K is minimal and $\mathcal{W}^u(F_\varphi)$ is G_α-invariant we have that $G_\alpha(K) = K$. Then, $G_{n\alpha}(y_0, \theta_1) \in K$ $\forall n \in \mathbb{Z}$. In particular, K is ε-dense in S. Since ε is arbitrary we obtain that $S \subset K$. This implies that $W^u_{F_\varphi}(S) \subset K$ and it is not difficult to see that $W^u_{F_\varphi}(S)$ is dense, then $K = M \times \mathbb{S}^1$ contradicting that $W^u_{F_\varphi}(x)$ is not dense.

The previous considerations allow us to define $\Psi : M \to \mathbb{N}$ as $\Psi(y) = \#(K \cap \{y\} \times \mathbb{S}^1)$. Our second claim is that this function is upper semicontinuous. Suppose it is false. Then, there is a sequence $(y_n)_n \subset M$ such that $y_n \to y$ and $\Psi(y_n) \to \eta > \Psi(y)$. Since K is compact, we also have that $\lim(K \cap \{y_n\} \times \mathbb{S}^1) \subset K \cap \{y\} \times \mathbb{S}^1$. In particular, we have that for any $\varepsilon > 0$ and n large enough there two points in $\{y_n\} \times \mathbb{S}^1$ at distance less than ε. An argument similar to the one used in the proof of our first claim gives that $\{y\} \times \mathbb{S}^1 \subset K$ and then, $K = M \times \mathbb{S}^1$, a contradiction.

Now we want to proof that Ψ is constant. On the one hand, observe that if $y' \in W^u_f(y), \Psi(y') = \Psi(y)$. On the other hand, since Ψ is semicontinuous it has continuity points and it is locally constant at this kind of points. Then, the minimality of the unstable foliation of f implies that Ψ is equal to a constant h.

The previous considerations imply that (K, π) is an h-fold covering of M and it is not difficult to see that $\mathcal{K} = \cup_{\alpha \in \mathbb{S}^1} G_\alpha(K)$ is a C^0 foliation of $M \times \mathbb{S}^1$ with compact leaves homeomorphism to K. Since F_φ is an isometric extension $\mathrm{dist}_c(K_1, K_2) =$

$\min\{dist(k_1, k_2); \ k_1 \in K_1, \ k_2 \in K_2, \ \pi(k_1) = \pi(k_2)\} = \text{dist}_c(F_\varphi(K_1), F_\varphi(K_2))$ where $K_1, K_2 \in \mathcal{K}$. In particular, $\text{dist}(F_\varphi^n(K_1), F_\varphi^n(K_2)$ does not go to zero. Then, any stable manifold intersects only one leaf of $\mathcal{K}$. That means that the leaves of $\mathcal{K}$ are s-saturated and then, $\mathcal{K}$ is a foliation tangent to $E^s \oplus E^u$. □

We get the following corollary of Theorems 5.1 and 1.3 for an open and dense set of skew products over volume-preserving Anosov diffeomorphisms.

Corollary 5.4 *Let F be an accessible isometric circle extension of a volume-preserving Anosov diffeomorphism. Then, there is a C^1 neighborhood $\mathcal{U}$ of F such that:*

- *$\mathcal{W}^\sigma(F)$ has a unique minimal set, $\sigma = s, u$.*
- *$\mathcal{W}^\sigma(F)$ is transitive, $\sigma = s, u$.*

Moreover, we can apply Theorems 1.2 and 4.3 to these diffeomorphisms and obtain the corresponding conclusions.

Remark 5.5 Hammerlindl and Potrie [12] have shown that partially hyperbolic diffeomorphisms on 3-nilmanifolds (different than the torus) always are sus- and usu-accessible. In particular we can also apply Theorems 1.2 and 4.3 to them. On the other hand, Shi [17] has announced that, on 3-nilmanifolds, there are partially hyperbolic diffeomorphism satisfying Axiom A. Of course, they have only one attractor and one repeller.

6 Anosov Diffeomorphisms

In this section we will suppose that A is a hyperbolic automorphism of $\mathbb{T}^3$ with three real eigenvalues $|\lambda_1| < 1 < |\lambda_2| < |\lambda_3|$. Along this section we will assume that f is an Anosov diffeomorphism isotopic to A and such that it admits a partially hyperbolic splitting $E^s \oplus E^c \oplus E^u$. Observe that $E^c \oplus E^u$ corresponds to the unstable bundle of the hyperbolic splitting of f. We obtain the following result that answers Conjecture 1.4 by Gogolev, Maimon and Kolmorgorov.

Theorem 6.1 *Let f be as above. Then, the strong unstable foliation of f is transitive.*

Proof Call h to the conjugacy between f and A. Observe that since the strong stable manifolds of f are its stable manifolds we have that h sends strong stable manifolds of f into strong stable manifolds of A. Thus, the strong stable foliations of f is minimal.

Ren, Gan and Zhang [15] have proved that, in our setting, the following statements are equivalent:

- f is not accessible.
- $E^s \oplus E^u$ is integrable.
- h sends strong unstable manifolds of f into strong unstable manifolds of A.

Then, on the one hand, if f is not accessible we have that $\mathcal{W}^u(f)$ is minimal and, in particular, it is transitive. On the other hand, if f is accessible, since we already know that $\mathcal{W}^s(f)$ is minimal, we can apply Theorem 1.3 (with the roles of u and s reversed) We get the transitivity of $\mathcal{W}^u(f)$ by applying Theorem 1.1. This ends the proof of the theorem. $\square$

References

1. C. Bonatti, N. Guelman, Transitive Anosov flows and Axiom-A diffeomorphisms. Ergod. Theory Dyn. Syst. **29**, 817–848 (2009)
2. C. Bonatti, L. Díaz, M. Viana, *Dynamics Beyond Uniform Hyperbolicity. A Global Geometric and Probabilistic Perspective*. Encyclopaedia of Mathematical Sciences, Mathematical Physics, III, vol. 102 (Springer, Berlin, 2005)
3. M. Brin, Y. Pesin, Partially hyperbolic dynamical systems. Izv. Ross. Akad. Nauk. Seriya Mat. **38**, 170–212 (1974)
4. K. Burns, A. Wilkinson, Stable ergodicity of skew products. Ann. Sci. l'École Norm. Supérieure **32**, 859–889 (1999)
5. K. Burns, C. Pugh, A. Wilkinson, Stable ergodicity and Anosov flows. Topology **39**, 149–159 (2000)
6. P. Didier, Stability of accessibility. Ergod. Theory Dyn. Syst. **23**, 1717–1731 (2003)
7. D. Dolgopyat, *Lectures on u-Gibbs States* (2001), www.math.psu.edu/dolgop/papers.html
8. D. Dolgopyat, On differentiability of SRB states for partially hyperbolic systems. Invent. Math. **155**, 389–449 (2004)
9. J. Franks, Anosov diffeomorphisms, in *Global Analysis. Proceedings of Symposia in Pure Mathematics*, Berkeley, California,1968, vol. XIV (1970), pp. 61–93
10. A. Gogolev, I. Maimon, A. Kolgomorov, *A Numerical Study of Gibbs u-Measures for Partially Hyperbolic Diffeomorphisms on* $\mathbb{T}^3$ (2017)
11. M. Grayson, C. Pugh, M. Shub, Stably ergodic diffeomorphisms. Ann. Math. **140**, 295–329 (1994)
12. A. Hammerlindl, R. Potrie, Pointwise partial hyperbolicity in three-dimensional nilmanifolds. J. Lond. Math. Soc. **89**, 853–875 (2014)
13. M. Hirsch, C. Pugh, M. Shub, *Invariant manifolds*, Lecture Notes in Mathematics, vol. 583 (Springer, Berlin, 1977)
14. J. Plante, Anosov flows. Am. J. Math. **94**, 729–754 (1972)
15. Y. Ren, S. Gan, P. Zhang, Accessibility and homology bounded strong unstable foliation for Anosov diffeomorphisms on 3-torus. Acta Math. Sin. **33**, 71–76 (2017)
16. F. Rodriguez Hertz, J. Rodriguez Hertz, R. Ures, Accessibility and stable ergodicity for partially hyperbolic diffeomorphisms with 1D-center bundle. Invent. Math. **172**, 353–381 (2008)
17. Y. Shi, Partially hyperbolic diffeomorphisms on Heisenberg nilmanifolds and holonomy maps. Comptes Rendus Math. **352**, 743–747 (2014)

Adapted Metrics for Codimension One Singular Hyperbolic Flows

Luciana Salgado and Vinicius Coelho

Abstract For a partially hyperbolic splitting $T_\Gamma M = E \oplus F$ of Γ, a C^1 vector field X on a m-manifold, we obtain singular-hyperbolicity using only the tangent map DX of X and its derivative DX_t whether E is one-dimensional subspace. We show the existence of adapted metrics for singular hyperbolic set Γ for C^1 vector fields if Γ has a partially hyperbolic splitting $T_\Gamma M = E \oplus F$ where F is volume expanding, E is uniformly contracted and a one-dimensional subspace.

Keywords Dominated splitting · Partial hyperbolicity · Sectional hyperbolicity · Lyapunov function

2000 Mathematics Subject Classification Primary: 37D30 · Secondary: 37D25

1 Introduction

Let M be a connected compact finite m-dimensional manifold, $m \geq 3$, with or without boundary. We consider a vector field X, such that X is inwardly transverse to the boundary ∂M, if $\partial M \neq \emptyset$. The flow generated by X is denoted by X_t.

A hyperbolic set for a flow X_t on a finite dimensional Riemannian manifold M is a compact invariant set Γ with a continuous splitting of the tangent bundle, $T_\Gamma M = E^s \oplus E^X \oplus E^u$, where E^X is the direction of the vector field, for which the

---

L.S. is partially supported by Fapesb-JCB0053/2013, PRODOC-UFBA/2014 and CNPq 2017 post-doctoral schoolarship at Universidade Federal do Rio de Janeiro, V.C. is supported by CAPES.

L. Salgado (✉)
Universidade Federal do Rio de Janeiro, Av. Athos da Silveira Ramos, 149, CT-Bloco C,
P.O. Box 68530, Cidade Universitaria, Rio de Janeiro-RJ 21941909, Brazil
e-mail: lsalgado@im.ufrj.br

V. Coelho
Universidade Federal da Bahia, Instituto de Matemática, Av. Adhemar de Barros, S/N Ondina,
Salvador, BA 40170-110, Brazil
e-mail: vinicius.coelho@ufba.br

© Springer Nature Switzerland AG 2019
M. J. Pacifico and P. Guarino (eds.), *New Trends in One-Dimensional Dynamics*, Springer Proceedings in Mathematics & Statistics 285,
https://doi.org/10.1007/978-3-030-16833-9_14

subbundles are invariant under the derivative DX_t of the flow X_t

$$DX_t \cdot E_x^* = E_{X_t(x)}^*, \quad x \in \Gamma, \quad t \in \mathbb{R}, \quad * = s, X, u; \tag{1}$$

and E^s is uniformly contracted by DX_t and E^u is likewise expanded: there are $K, \lambda > 0$ so that

$$\|DX_t\mid_{E_x^s}\| \leq Ke^{-\lambda t}, \quad \|(DX_t\mid_{E_x^u})^{-1}\| \leq Ke^{-\lambda t}, \quad x \in \Gamma, \quad t \in \mathbb{R}. \tag{2}$$

Very strong properties can be deduced from the existence of such hyperbolic structure; see for instance [9, 10, 15, 22, 24].

An important feature of hyperbolic structures is that it does not depends on the metric on the ambient manifold (see [13]). We recall that a metric is said to be *adapted* to the hyperbolic structure if we can take $K = 1$ in Eq. (2).

Weaker notions of hyperbolicity (e.g. dominated splitting, partial hyperbolicity, volume hyperbolicity, sectional hyperbolicity, singular hyperbolicity) have been developed to encompass larger classes of systems beyond the uniformly hyperbolic ones; see [8] and specifically [2, 6, 26] for singular hyperbolicity and Lorenz-like attractors.

In the same work [13], Hirsch, Pugh and Shub asked about adapted metrics for dominated splittings. The positive answer was given by Gourmelon [12] in 2007, where it is given adapted metrics to dominated splittings for both diffeomorphisms and flows, and he also gives an adapted metric for partially hyperbolic splittings as well.

In fact, in [29], Wojtkowski proved that if a diffeomorphism f is strictly $\mathcal{J}$-separated then it has a dominated splitting and affirmed that the continuous riemannian metric induced by $\mathcal{J}$ is an adapted one. In other words, we can use the quadratic form to produce an adapted metric. Here, in Lemma 3.1, we give a proof of this affirmation in our setting.

Proving the existence of some hyperbolic structure is, in general, a non-trivial matter, even in its weaker forms.

In [16], Lewowicz stated that a diffeomorphism on a compact riemannian manifold is Anosov if and only if its derivative admits a nondegenerate Lyapunov quadratic function.

An example of application of the adapted metric from [12] is contained in [3], where the first author jointly with V. Araújo, following the spirit of Lewowicz's result, construct quadratic forms which characterize partially hyperbolic and singular hyperbolic structures on a trapping region for flows.

In [4], the first author and V. Araújo provided an alternative way to obtain singular hyperbolicity for three-dimensional flows using the same expression as in Proposition 2.3 applied to the infinitesimal generator of the exterior square $\wedge^2 DX_t$ of the cocycle DX_t. This infinitesimal generator can be explicitly calculated through the infinitesimal generator DX of the linear multiplicative cocycle DX_t associated to the vector field X.

Here, we provide a similar result as above for m-dimensional flows if this admits a partially hyperbolic splitting for which one of the invariant subbundles is one-dimensional.

Moreover, we show the existence of adapted metrics for a singular hyperbolic set Γ for C^1 vector fields if Γ has a partially hyperbolic splitting $T_\Gamma M = E \oplus F$, where F is volume expanding, E is uniformly contracted and one-dimensional subbundle.

1.1 Statements of Main Results

In the sequel, we write $\tilde{J}(v) = < \tilde{J}_x v, v >$, where $\tilde{J}_x$ is given in Proposition 2.3, that is, $\tilde{J}(v)$ is the time derivative of a quadratic form J under the action of the flow.

The absolute value of the *cross product* (also called *vector product*) on a 3-dimensional vector space V, denote by $w = u \times v$, provides the length of the vector w. It is very useful to calculate the area expansion of the parallelogram generated by u, v, under the action of a linear operator.

Following this way, in [4], the first author and V. Araújo proved the result below.

Theorem 1.1 ([4, Theorem B]) *Suppose that X is 3-dimensional vector field on M which is non-negative strictly J-separated over a non-trivial subset Γ, where J has index 1. Then*

(1) $\wedge^2 DX_t$ is strictly $(-J)$-separated;
(2) Γ is a singular hyperbolic set if either one of the following properties is true

    (a) $\tilde{\Delta}_0^t(x) \xrightarrow[t \to +\infty]{} -\infty$ for all $x \in \Gamma$.
    (b) $\tilde{J} - 2\operatorname{tr}(DX)J > 0$ on Γ.

Here, we generalized this result to m and $k = m - 1$, as follows.

If $\wedge^k DX_t$ is strictly separated with respect to some family J of quadratic forms, then there exists the function δ_k as stated in Proposition 2.3 with respect to the cocyle $\wedge^k DX_t$. We set

$$\tilde{\Delta}_a^b(x) := \int_a^b \delta_k(X_s(x)) \, ds$$

the area under the function $\delta_k : U \to \mathbb{R}$ given by Proposition 2.3 with respect to $\wedge^k DX_t$ and its infinitesimal generator.

If $k = m - 1$, it is not difficult to see that this function is related to X and δ as follows: let $\delta : \Gamma \to \mathbb{R}$ be the function associated to J and DX_t, as given by Proposition 2.3, then $\delta_k = 2\operatorname{tr}(DX) - \delta$, where $\operatorname{tr}(DX)$ represents the trace of the linear operator $DX_x : T_x M \circlearrowleft, x \in M$.

We recall that $\tilde{J} = \partial_t J$ is the time derivative of J along the flow; see Remark 2.4.

Our first main result is the following.

Theorem A *Suppose that X is m-dimensional vector field on M which is non-negative strictly $\mathfrak{J}$-separated over a non-trivial subset Γ, where $\mathfrak{J}$ has index 1. Then*

(1) $\wedge^{(m-1)} DX_t$ *is strictly* $(-\mathfrak{J})$*-separated;*
(2) Γ *is a singular hyperbolic set if either one of the following properties is true*

    (a) $\tilde{\Delta}_0^t(x) \xrightarrow[t \to +\infty]{} -\infty$ *for all* $x \in \Gamma$.
    (b) $\tilde{\mathfrak{J}} - 2 \operatorname{tr}(DX)\mathfrak{J} > 0$ *on* Γ.

We work here with exterior products of codimension one. See [11] for more details on this subject.

This result provides useful sufficient conditions for a m-dimensional vector field to be singular hyperbolic if $k = m - 1$, using only one family of quadratic forms $\mathfrak{J}$ and its space derivative DX, avoiding the need to check cone invariance and contraction/expansion conditions for the flow X_t generated by X on a neighborhood of Γ.

Now we recall the definition of adapted metrics in the singular hyperbolic setting.

Definition 1 We say a Riemannian metric $\langle \cdot, \cdot \rangle$ *adapted to a singular hyperbolic splitting* $T\Gamma = E \oplus F$ if it induces a norm $| \cdot |$ such that there exists $\lambda > 0$ satisfying for all $x \in \Gamma$ and $t > 0$ simultaneously

$$|DX_t|_{E_x}| \cdot |(DX_t|_{F_x})^{-1}| \le e^{-\lambda t}, \quad |DX_t|_{E_x}| \le e^{-\lambda t} \quad \text{and} \quad |\det(DX_t|_{F_x})| \ge e^{\lambda t}.$$

We call it *singular adapted metric*, for simplicity.

In [4], the first author and V. Araújo proved the next result.

Theorem 1.2 ([4, Theorem C]) *Let Γ be a singular-hyperbolic set for a C^1 three-dimensional vector field X. Then Γ admits a singular adapted metric.*

Here, we generalize this result for any codimension one singular hyperbolic flow in higher dimensional manifolds. Consider a partially hyperbolic splitting $T_\Gamma M = E \oplus F$ where E is uniformly contracted and F is volume expanding. We show that for C^1 flows having a singular-hyperbolic set Γ such that E is one-dimensional subspace there exists a metric adapted to the partial hyperbolicity and the area expansion, as follows.

Theorem B *Let Γ be a singular-hyperbolic set of codimension one for a C^1 m-dimensional vector field X. Then Γ admits a singular adapted metric.*

We present the relevant definitions and auxiliary results in the next section.

The paper is organized as follow. In the present Section we provide an introduction and statement of main results. In Sect. 2 we give the main definitions and useful properties of quadratic forms. In Sect. 3 we provide some auxiliary results. Finally, in Sect. 4 are given the proofs of our theorems.

2 Preliminary Definitions and Results

We now present preliminary definitions and results.

We recall that a *trapping region* U for a flow X_t is an open subset of the manifold M which satisfies: $X_t(U)$ is contained in U for all $t > 0$, and there exists $T > 0$ such that $\overline{X_t(U)}$ is contained in the interior of U for all $t > T$. We define $\Gamma(U) = \Gamma_X(U) := \cap_{t>0}\overline{X_t(U)}$ to be the *maximal positive invariant subset in the trapping region U*.

A *singularity* for the vector field X is a point $\sigma \in M$ such that $X(\sigma) = \mathbf{0}$ or, equivalently, $X_t(\sigma) = \sigma$ for all $t \in \mathbb{R}$. The set formed by singularities is the *singular set of X* denoted $\mathrm{Sing}(X)$. We say that a singularity is hyperbolic if the eigenvalues of the derivative $DX(\sigma)$ of the vector field at the singularity σ have nonzero real part.

Definition 2 A *dominated splitting* over a compact invariant set Λ of X is a continuous DX_t-invariant splitting $T_\Lambda M = E \oplus F$ with $E_x \neq \{0\}$, $F_x \neq \{0\}$ for every $x \in \Lambda$ and such that there are positive constants K, λ satisfying

$$\|DX_t|_{E_x}\| \cdot \|DX_{-t}|_{F_{X_t(x)}}\| < Ke^{-\lambda t}, \text{ for all } x \in \Lambda, \text{ and all } t > 0. \qquad (3)$$

A compact invariant set Λ is said to be *partially hyperbolic* if it exhibits a dominated splitting $T_\Lambda M = E \oplus F$ such that subbundle E is *uniformly contracted*, i.e., there exists $C > 0$ and $\lambda > 0$ such that $\|DX_t|_{E_x}\| \leq Ce^{-\lambda t}$ for $t \geq 0$. In this case F is the *central subbundle* of Λ. Or else, we may replace uniform contraction along E by uniform expansion along F (the right hand side condition in (2)).

We say that a DX_t-invariant subbundle $F \subset T_\Lambda M$ is a *sectionally expanding* subbundle if $\dim F_x \geq 2$ is constant for $x \in \Lambda$ and there are positive constants C, λ such that for every $x \in \Lambda$ and every two-dimensional linear subspace $L_x \subset F_x$ one has

$$|\det(DX_t|_{L_x})| > Ce^{\lambda t}, \text{ for all } t > 0. \qquad (4)$$

Definition 3 ([17, Definition 2.7]) A *sectional-hyperbolic set* is a partially hyperbolic set whose central subbundle is sectionally expanding.

This is a particular case of the so called *singular hyperbolicity* whose definition we recall now. A DX_t-invariant subbundle $F \subset T_\Lambda M$ is said to be a *volume expanding* if in the above condition 4, we may write

$$|\det(DX_t|_{F_x})| > Ce^{\lambda t}, \text{ for all } t > 0. \qquad (5)$$

Definition 4 ([18, Definition 1]) A *singular hyperbolic set* is a partially hyperbolic set whose central subbundle is volume expanding.

Clearly, in the three-dimensional case, these notions are equivalent.

This is a feature of the Lorenz attractor as proved in [25] and also a notion that extends hyperbolicity for singular flows, because sectional hyperbolic sets without singularities are hyperbolic; see [2, 19].

2.1 Linear Multiplicative Cocycles Over Flows

Let $A : G \times \mathbb{R} \to G$ be a smooth map given by a collection of linear bijections

$$A_t(x) : G_x \to G_{X_t(x)}, \quad x \in \Gamma, t \in \mathbb{R},$$

where Γ is the base space of the finite dimensional vector bundle G, satisfying the cocycle property

$$A_0(x) = Id, \quad A_{t+s}(x) = A_t(X_s(x)) \circ A_s(x), \quad x \in \Gamma, t, s \in \mathbb{R},$$

with $\{X_t\}_{t \in \mathbb{R}}$ a complete smooth flow over $M \supset \Gamma$. We note that for each fixed $t > 0$ the map $A_t : G \to G, v_x \in G_x \mapsto A_t(x) \cdot v_x \in G_{X_t(x)}$ is an automorphism of the vector bundle G.

The natural example of a linear multiplicative cocycle over a smooth flow X_t on a manifold is the derivative cocycle $A_t(x) = DX_t(x)$ on the tangent bundle $G = TM$ of a finite dimensional compact manifold M. Another example is given by the exterior power $A_t(x) = \wedge^k DX_t$ of DX_t acting on $G = \wedge^k TM$, the family of all k-vectors on the tangent spaces of M, for some fixed $1 \le k \le \dim G$.

It is well-known that the exterior power of a inner product space has a naturally induced inner product and thus a norm. Thus $G = \wedge^k TM$ has an induced norm from the Riemannian metric of M. For more details see e.g. [7].

In what follows we assume that the vector bundle G has a smoothly defined inner product in each fiber G_x which induces a corresponding norm $\| \cdot \|_x, x \in \Gamma$.

Definition 5 A continuous splitting $G = E \oplus F$ of the vector bundle G into a pair of subbundles is *dominated* (with respect to the automorphism A over Γ) if

- the splitting is *invariant*: $A_t(x) \cdot E_x = E_{X_t(x)}$ and $A_t(x) \cdot F_x = F_{X_t(x)}$ for all $x \in \Gamma$ and $t \in \mathbb{R}$; and
- there are positive constants K, λ satisfying

$$\|A_t|_{E_x}\| \cdot \|A_{-t}|_{F_{X_t(x)}}\| < Ke^{-\lambda t}, \text{ for all } x \in \Gamma, \text{ and all } t > 0. \tag{6}$$

We say that the splitting $G = E \oplus F$ is *partially hyperbolic* if it is dominated and the subbundle E is uniformly contracted: $\|A_t \mid E_x\| \le Ce^{-\mu t}$ for all $t > 0$ and suitable constants $C, \mu > 0$.

2.2   Fields of Quadratic Forms, Positive and Negative Cones

Let E_U be a finite dimensional vector bundle with inner product $\langle \cdot, \cdot \rangle$ and base given by the trapping region $U \subset M$. Let $\mathcal{J} : E_U \to \mathbb{R}$ be a continuous family of quadratic forms $\mathcal{J}_x : E_x \to \mathbb{R}$ which are non-degenerate and have index $0 < q < \dim(E) = n$. The index q of $\mathcal{J}$ means that the maximal dimension of subspaces of non-positive vectors is q. Using the inner product, we can represent $\mathcal{J}$ by a family of self-adjoint operators $J_x : E_x \circlearrowleft$ as $\mathcal{J}_x(v) = \langle J_x(v), v \rangle, v \in E_x, x \in U$.

We also assume that $(\mathcal{J}_x)_{x \in U}$ is continuously differentiable along the flow. The continuity assumption on $\mathcal{J}$ means that for every continuous section Z of E_U the map $U \ni x \mapsto \mathcal{J}(Z(x)) \in \mathbb{R}$ is continuous. The C^1 assumption on $\mathcal{J}$ along the flow means that the map $\mathbb{R} \ni t \mapsto \mathcal{J}_{X_t(x)}(Z(X_t(x))) \in \mathbb{R}$ is continuously differentiable for all $x \in U$ and each C^1 section Z of E_U.

Using Lagrange diagonalization of a quadratic form, it is easy to see that the choice of basis to diagonalize $\mathcal{J}_y$ depends smoothly on y if the family $(\mathcal{J}_x)_{x \in U}$ is smooth, for all y close enough to a given x. Therefore, choosing a basis for T_x adapted to $\mathcal{J}_x$ at each $x \in U$, we can assume that locally our forms are given by $\langle J_x(v), v \rangle$ with J_x a diagonal matrix whose entries belong to $\{\pm 1\}$, $J_x^* = J_x$, $J_x^2 = I$ and the basis vectors depend as smooth on x as the family of forms $(\mathcal{J}_x)_x$.

We let $\mathcal{C}_\pm = \{C_\pm(x)\}_{x \in U}$ be the family of positive and negative cones associated to $\mathcal{J}$

$$C_\pm(x) := \{0\} \cup \{v \in E_x : \pm\mathcal{J}_x(v) > 0\}, \quad x \in U,$$

and also let $\mathcal{C}_0 = \{C_0(x)\}_{x \in U}$ be the corresponding family of zero vectors $C_0(x) = \mathcal{J}_x^{-1}(\{0\})$ for all $x \in U$.

2.3   Strict $\mathcal{J}$-Separation for Linear Multiplicative Cocycles

Let $A : E \times \mathbb{R} \to E$ be a linear multiplicative cocycle on the vector bundle E over the flow X_t. The following definitions are fundamental to state our results.

Definition 6 Given a continuous field of non-degenerate quadratic forms $\mathcal{J}$ with constant index on the positively invariant open subset U for the flow X_t, we say that the cocycle $A_t(x)$ over X_t is

- $\mathcal{J}$-*separated* if $A_t(x)(C_+(x)) \subset C_+(X_t(x))$, for all $t > 0$ and $x \in U$ (simple cone invariance);
- *strictly* $\mathcal{J}$-*separated* if $A_t(x)(C_+(x) \cup C_0(x)) \subset C_+(X_t(x))$, for all $t > 0$ and $x \in U$ (strict cone invariance).
- $\mathcal{J}$-*monotone* if $\mathcal{J}_{X_t(x)}(DX_t(x)v) \geq \mathcal{J}_x(v)$, for each $v \in T_x M \setminus \{0\}$ and $t > 0$;
- *strictly* $\mathcal{J}$-*monotone* if $\partial_t \big(\mathcal{J}_{X_t(x)}(DX_t(x)v)\big) \big|_{t=0} > 0$, for all $v \in T_x M \setminus \{0\}, t > 0$ and $x \in U$;

- $\mathcal{J}$-*isometry* if $\mathcal{J}_{X_t(x)}(DX_t(x)v) = \mathcal{J}_x(v)$, for each $v \in T_x M$ and $x \in U$.

We say that the flow X_t is (strictly) $\mathcal{J}$-*separated* on U if $DX_t(x)$ is (strictly) $\mathcal{J}$-*separated* on $T_U M$. Analogously, the flow of X on U is (strictly) $\mathcal{J}$-*monotone* if $DX_t(x)$ is (strictly) $\mathcal{J}$-*monotone*.

Remark 2.1 If a flow is strictly $\mathcal{J}$-separated, then for $v \in T_x M$ such that $\mathcal{J}_x(v) \leq 0$ we have $\mathcal{J}_{X_{-t}(x)}(DX_{-t}(v)) < 0$, for all $t > 0$, and x such that $X_{-s}(x) \in U$ for every $s \in [-t, 0]$. Indeed, otherwise $\mathcal{J}_{X_{-t}(x)}(DX_{-t}(v)) \geq 0$ would imply $\mathcal{J}_x(v) = \mathcal{J}_x(DX_t(DX_{-t}(v))) > 0$, contradicting the assumption that v was a non-positive vector.

This means that a flow X_t is strictly $\mathcal{J}$-separated if, and only if, its time reversal X_{-t} is strictly $(-\mathcal{J})$-separated.

Remark 2.2 Let V be a real finite dimensional vector space, and $L : V \to V$ be a $\mathcal{J}$-separated linear operator. Then L can be uniquely represented by $L = RU$, where U is a $\mathcal{J}$-isometry(i.e. $\mathcal{J}(U(v)) = \mathcal{J}(v)$, $v \in V$) and R is $\mathcal{J}$-symmetric with positive spectrum; the operator operator R can be diagonalized by a $\mathcal{J}$-isometry, and there exist constants r_- and r_+ such that the operator L is (strictly) $\mathcal{J}$-monotonous if, and only, if $r_- \leq (<) 1$ and $r_+ \geq (>) 1$. For more details see [3, Proposition 2.4] and comments below of the Theorem 1.2 in [29].

A vector field X is $\mathcal{J}$-*non-negative* on U if $\mathcal{J}(X(x)) \geq 0$ for all $x \in U$, and $\mathcal{J}$-*non-positive* on U if $\mathcal{J}(X(x)) \leq 0$ for all $x \in U$. When the quadratic form used in the context is clear, we will simply say that X is non-negative or non-positive.

We say that a C^1 family $\mathcal{J}$ of indefinite and non-degenerate quadratic forms is *compatible* with a continuous splitting $E_\Gamma \oplus F_\Gamma = E_\Gamma$ of a vector bundle over some compact subset Γ if E_x is a $\mathcal{J}$-negative subspace and F_x is a $\mathcal{J}$-positive subspace for all $x \in \Gamma$.

Proposition 2.3 ([3, Proposition 1.3]) *A $\mathcal{J}$-non-negative vector field X on U is strictly $\mathcal{J}$-separated if, and only if, there exists a compatible family $\mathcal{J}_0$ of forms and there exists a function $\delta : U \to \mathbb{R}$ such that the operator $\tilde{J}_{0,x} := J_0 \cdot DX(x) + DX(x)^* \cdot J_0$ satisfies*

$$\tilde{J}_{0,x} - \delta(x) J_0 \text{ is positive definite, } x \in U,$$

where $DX(x)^*$ is the adjoint of $DX(x)$ with respect to the adapted inner product.

Remark 2.4 The expression for $\tilde{J}_{0,x}$ in terms of J_0 and the infinitesimal generator of DX_t is, in fact, the time derivative of $\mathcal{J}_0$ along the flow direction at the point x, which we denote $\partial_t J_0$; see item 1 of Proposition 2.10. We keep this notation in what follows.

A characterization of dominated splittings, via quadratic forms is given in [3] (see also [29]) as follow.

Theorem 2.5 [3, Theorem 2.13] *The cocycle $A_t(x)$ is strictly $\mathcal{J}$-separated if, and only if, E_U admits a dominated splitting $F_- \oplus F_+$ with respect to $A_t(x)$ on the maximal invariant subset Λ of U, with constant dimensions $\dim F_- = q$, $\dim F_+ = p$, $\dim M = p + q$.*

This is an algebraic/geometrical way to prove the existence of dominated splittings. As we have said in the introduction, proving existence of some hyperbolic structure is not an easy work to do, in general. One of the most habitual way is to use cone field techniques, see for instance [14, 20, 21].

In [4, Example 5], the first author and V. Araújo checked out the singular hyperbolicity of geometric Lorenz attractor, in a most simple way, by using Theorem 1.1. It was proved by Tucker [25], under computer assistance, that the Lorenz attractor exist for the classical parameters. It is expected, in a work in progress, that Theorem 1.1 may be used to prove the same result without computer assistance or at least simplify the proof given by Tucker.

In fact, we have an analogous result about partial hyperbolic splittings, as follow.

We say that a compact invariant subset Λ is *non-trivial* if

- either Λ does not contain singularities;
- or Λ contains at most finitely many singularities, Λ contains some regular orbit and is connected.

Theorem 2.6 ([3, Theorem A]) *A non-trivial compact invariant subset Γ is a partially hyperbolic set for a flow X_t if, and only if, there is a C^1 field $\mathcal{J}$ of non-degenerate and indefinite quadratic forms with constant index, equal to the dimension of the stable subspace of Γ, such that X_t is a non-negative strictly $\mathcal{J}$-separated flow on a neighborhood U of Γ.*

Moreover E is a negative subspace, F a positive subspace and the splitting can be made almost orthogonal.

Here strict $\mathcal{J}$-separation corresponds to strict cone invariance under the action of DX_t and $\langle \cdot, \cdot \rangle$ is a Riemannian inner product in the ambient manifold. We recall that the index of a field quadratic forms $\mathcal{J}$ on a set Γ is the dimension of the $\mathcal{J}$-negative space at every tangent space $T_x M$ for $x \in U$. Moreover, we say that the splitting $T_\Gamma M = E \oplus F$ is *almost orthogonal* if, given $\epsilon > 0$, there exists a smooth inner product $\langle \cdot, \cdot \rangle$ on $T_\Gamma M$ so that $|\langle u, v \rangle| < \epsilon$, for all $u \in E$, $v \in F$, with $\|u\| = 1 = \|v\|$.

We note that the condition stated in Theorem 2.6 allows us to obtain partial hyperbolicity checking a condition at every point of the compact invariant set that depends only on the tangent map DX to the vector field X together with a family $\mathcal{J}$ of quadratic forms without using the flow X_t or its derivative DX_t. This is akin to checking the stability of singularity of a vector field using a Lyapunov function. For example, it is well known by Lyapunov's Stability Theorem that if a singularity σ of a C^1 vector field $Y : U \subset \mathbb{R}^n \to \mathbb{R}^n$, defined over an open set U, admits a strict Lyapunov function on σ, then this is a asymptotically stable singularity. Lewowicz, in [16], used this idea replacing stability of a singularity by topological stability of Anosov diffeomorphisms.

2.4 Exterior Powers

We note that if $E \oplus F$ is a DX_t-invariant splitting of $T_\Gamma M$, with $\{e_1, \ldots, e_\ell\}$ a family of basis for E and $\{f_1, \ldots, f_h\}$ a family of basis for F, then $\tilde{F} = \wedge^k F$ generated by $\{f_{i_1} \wedge \cdots \wedge f_{i_k}\}_{1 \le i_1 < \cdots < i_k \le h}$ is naturally $\wedge^k DX_t$-invariant by construction. In addition, $\tilde{E}$ generated by $\{e_{i_1} \wedge \cdots \wedge e_{i_k}\}_{1 \le i_1 < \cdots < i_k \le \ell}$ together with all the exterior products of i basis elements of E with j basis elements of F, where $i + j = k$ and $i, j \ge 1$, is also $\wedge^k DX_t$-invariant and, moreover, $\tilde{E} \oplus \tilde{F}$ gives a splitting of the kth exterior power $\wedge^k T_\Gamma M$ of the subbundle $T_\Gamma M$. Let $T_\Gamma M = E_\Gamma \oplus F_\Gamma$ be a DX_t-invariant splitting over the compact X_t-invariant subset Γ such that dim $F = k \ge 2$. Let $\tilde{F} = \wedge^k F$ be the $\wedge^k DX_t$-invariant subspace generated by the vectors of F and $\tilde{E}$ be the $\wedge^k DX_t$-invariant subspace such that $\tilde{E} \oplus \tilde{F}$ is a splitting of the kth exterior power $\wedge^k T_\Gamma M$ of the subbundle $T_\Gamma M$.

We consider the action of the cocycle $DX_t(x)$ on k-vector that is the k-exterior $\wedge^k DX_t$ of the cocycle acting on $\wedge^k T_\Gamma M$.

We denote by $\|\cdot\|$ the standard norm on k-vectors induced by the Riemannian norm of M, see [7].

Remark 2.7 Let V to be a vector space of dimension N.

(i) The dimension of space $\wedge^r V$ is dim $\wedge^r V = \binom{N}{r}$. If $\{e_1, \cdots, e_N\}$ is a basis of V, so the set $\{e_{k_1} \wedge \cdots \wedge e_{k_r} : 1 \le k_1 < \cdots < k_r \le N\}$ is a basis in $\wedge^r V$ with $\binom{N}{r}$ elements.

(ii) If V has the inner product $\langle, \rangle$, then the bilinear extension of

$$\langle u_1 \wedge \cdots \wedge u_r, v_1 \wedge \cdots \wedge v_r \rangle := \det(\langle u_i, v_j \rangle)_{r \times r}$$

defines a inner product in $\wedge^r V$. In particular, $\|u_1 \wedge \cdots \wedge u_r\| = \sqrt{\det(\langle u_i, u_j \rangle)_{r \times r}}$ is the volume of r-dimensional parallelepiped H spanned by $u_1, \cdots, u_r$, we write $\text{vol}(u_1, \cdots, u_r) = \text{vol}(H) = \det(H) = |\det(u_1, \cdots, u_r)|$.

(iii) If $A : V \to V$ is a linear operator then the linear extension of $\wedge^r A(u_1 \wedge \cdots \wedge u_r) = A(u_1) \wedge \cdots \wedge A(u_r)$ defines a linear operator $\wedge^r A$ on $\wedge^r V$.

(iv) Let $A : V \to V$, and $\wedge^r A : \wedge^r V \to \wedge^r V$ linear operators with G spanned by $v_1, \cdots, v_s \in V$. Define $H := A|_G$, then H is spanned by $A(v_1), \cdots, A(v_s)$. So $|\det A|_G| = \text{vol}(A|_G) = \text{vol}(H) = \text{vol}(A(v_1), \cdots, A(v_s)) = \|A(v_1) \wedge \cdots \wedge A(v_s)\| = \|\wedge^s A(v_1 \wedge \cdots \wedge v_s)\|$.

When $DX_t(u_i) = v_i(t) = v_i$, where G is spanned by $u_1, \cdots, u_r \in T_\Gamma M$, and H is spanned by $v_1, \cdots, v_r$, we have $H = DX_t(G) = DX_t|_G$. Thus,

$$|\det(DX_t|_G)| = \text{vol}(DX_t(u_1), \cdots, DX_t(u_r)) =$$
$$\|DX_t(u_1) \wedge \cdots \wedge DX_t(u_r)\| = \|\wedge^r DX_t(u_1 \wedge \cdots \wedge u_r)\|.$$

It is natural to consider the linear multiplicative cocyle $\wedge^k DX_t$ over the flow X_t of X on U, that is, for any k choice, $u_1, u_2, \cdots, u_k$ of vectors in $T_x M$, $x \in U$ and $t \in \mathbb{R}$ such that $X_t(x) \in U$ we set

$$(\wedge^k DX_t) \cdot (u_1 \wedge u_2 \wedge \cdots \wedge u_k) = (DX_t \cdot u_1) \wedge (DX_t \cdot u_2) \wedge \cdots \wedge (DX_t \cdot u_k)$$

see [7, Chap. 3, Sect. 2.3] or [27] for more details and standard results on exterior algebra and exterior products of linear operator.

In [4], the authors proved the following relation between a dominated splitting and its exterior power.

Theorem 2.8 ([4, Theorem A]) *The splitting $T_\Gamma M = E \oplus F$ is dominated for DX_t if, and only if, $\wedge^k T_\Gamma M = \widetilde{E} \oplus \widetilde{F}$ is a dominated splitting for $\wedge^k DX_t$.*

Hence, the existence of a dominated splitting $T_\Gamma M = E_\Gamma \oplus F_\Gamma$ over the compact X_t-invariant subset Γ, is equivalent to the bundle $\wedge^k T_\Gamma M$ admits a dominated splitting with respect to $\wedge^k DX_t : \wedge^k T_\Gamma M \to \wedge^k T_\Gamma M$.

As a consequence, they obtain the next characterization of three-dimensional singular sets.

Corollary 2.9 ([4, Corollary 1.5]) *Assume that M has dimension 3, E is uniformly contracted by DX_t, and that $k = 2$. Then $E \oplus F$ is a singular-hyperbolic splitting for DX_t if, and only if, $\widetilde{E} \oplus \widetilde{F}$ is partially hyperbolic splitting for $\wedge^2 DX_t$ such that $\widetilde{F}$ is uniformly expanded by $\wedge^2 DX_t$.*

2.5 Properties of $\mathcal{J}$-Separated Linear Multiplicative Cocycles

We present some useful properties about $\mathcal{J}$-separated linear cocycles whose proofs can be found in [3].

Let $A_t(x)$ be a linear multiplicative cocycle over X_t. We define the infinitesimal generator of $A_t(x)$ by

$$D(x) := \lim_{t \to 0} \frac{A_t(x) - Id}{t}. \tag{7}$$

The following is the basis for arguments given by the first author and V. Araújo in [3] to prove the Theorem 2.6.

Proposition 2.10 ([3, Proposition 2.7]) *Let $A_t(x)$ be a cocycle over X_t defined on an open subset U and $D(x)$ its infinitesimal generator. Then*

(1) $\widetilde{\mathcal{J}}(v) = \partial_t \mathcal{J}(A_t(x)v) = \langle \widetilde{J}_{X_t(x)} A_t(x)v, A_t(x)v \rangle$ for all $v \in E_x$ and $x \in U$, where

$$\widetilde{J}_x := J \cdot D(x) + D(x)^* \cdot J \tag{8}$$

and $D(x)^*$ denotes the adjoint of the linear map $D(x) : E_x \to E_x$ with respect to the adapted inner product at x;

(2) the cocycle $A_t(x)$ is $\mathcal{J}$-separated if, and only if, there exists a neighborhood V of Λ, $V \subset U$ and a function $\delta : V \to \mathbb{R}$ such that

$$\tilde{\mathcal{J}}_x \geq \delta(x)\mathcal{J}_x \quad for\ all \quad x \in V. \tag{9}$$

In particular we get $\partial_t \log |\mathcal{J}(A_t(x)v)| \geq \delta(X_t(x))$, $v \in E_x$, $x \in V$, $t \geq 0$;

(3) if the inequalities in the previous item are strict, then the cocycle $A_t(x)$ is strictly $\mathcal{J}$-separated. Reciprocally, if $A_t(x)$ is strictly $\mathcal{J}$-separated, then there exists a compatible family $\mathcal{J}_0$ of forms on V satisfying the strict inequalities of item (2).

(4) For a $\mathcal{J}$-separated cocycle $A_t(x)$, we have $\frac{|\mathcal{J}(A_{t_2}(x)v)|}{|\mathcal{J}(A_{t_1}(x)v)|} \geq \exp \Delta_{t_1}^{t_2}(x)$ for all $v \in E_x$ and reals $t_1 < t_2$ so that $\mathcal{J}(A_t(x)v) \neq 0$ for all $t_1 \leq t \leq t_2$, where $\Delta_{t_1}^{t_2}(x)$ was defined in (10).

(5) we can bound δ at every $x \in \Gamma$ by $\inf_{v \in C_+(x)} \frac{\tilde{\mathcal{J}}(v)}{\mathcal{J}(v)} \leq \delta(x) \leq \sup_{v \in C_-(x)} \frac{\tilde{\mathcal{J}}(v)}{\mathcal{J}(v)}$.

Remark 2.11 We stress that the necessary and sufficient condition in items (2–3) of Proposition 2.10, for (strict) $\mathcal{J}$-separation, shows that a cocycle $A_t(x)$ is (strictly) $\mathcal{J}$-separated if, and only if, its inverse $A_{-t}(x)$ is (strictly) $(-\mathcal{J})$-separated.

Remark 2.12 Item (2) above of Proposition 2.10 shows that δ is a measure of the "minimal instantaneous expansion rate" of $|\mathcal{J} \circ A_t(x)|$.

The area under the function δ provided by Proposition 2.10 allows us to detect different dominated splittings with respect to linear multiplicative cocycles on vector bundles (Proposition 2.13). For this, define the function

$$\Delta_a^b(x) := \int_a^b \delta(X_s(x))\, ds, \quad x \in \Gamma, a, b \in \mathbb{R}. \tag{10}$$

Proposition 2.13 ([3, Theorem 2.23]) *Let Γ be a compact invariant set for X_t admitting a dominated splitting $E_\Gamma = F_- \oplus F_+$ for $A_t(x)$, a linear multiplicative cocycle over Γ with values in E. Let $\mathcal{J}$ be a C^1 family of indefinite quadratic forms such that $A_t(x)$ is strictly $\mathcal{J}$-separated. Then*

(1) $F_- \oplus F_+$ is partially hyperbolic with F_+ uniformly expanding if $\Delta_0^t(x) \xrightarrow[t \to +\infty]{}$ $+\infty$ for all $x \in \Gamma$.

(2) $F_- \oplus F_+$ is partially hyperbolic with F_- uniformly contracting if $\Delta_0^t(x) \xrightarrow[t \to +\infty]{}$ $-\infty$ for all $x \in \Gamma$.

(3) $F_- \oplus F_+$ is uniformly hyperbolic if, and only if, there exists a compatible family $\mathcal{J}_0$ of quadratic forms in a neighborhood of Γ such that $\mathcal{J}_0'(v) > 0$ for all $v \in E_x$ and all $x \in \Gamma$.

For the proof and more details about the Proposition 2.13, see [3].

3   Auxiliary Results

3.1   Adapted Metric for Dominated Splittings from Quadratic Forms

Let $A : G \times \mathbb{R} \to G$ be a linear multiplicative cocycle on the vector bundle G over the flow X_t. We assume from now on that the family $A_t(x)$ of linear multiplicative cocycles on a vector bundle G_U over the flow X_t on trapping region $U \subseteq M$ has been given, together with a field of non-degenerate quadratic forms $\mathfrak{J}$ on G_U.

Let us consider a C^1 field of non-degenerate quadratic forms $\mathfrak{J}$ with constant index on the positively invariant open set U for the flow X_t, such that the cocycle $A_t(x)$ over X_t is strictly $\mathfrak{J}$-separated. In [3, Theorem 2.13], the authors showed that G_U admits a dominated splitting $F_- \oplus F_+$ with to respect to $A_t(x)$, on the maximal invariant subset Γ of U. In fact, in [29, Proposition 4.1], Wojtkowski have made this to diffeomorphisms and affirmed that the $\mathfrak{J}$-metric is an adapted one. In other words, we can obtain adapted metrics from the quadratic forms.

Following the arguments into the proofs from [3, Theorem 2.13] and [29, Proposition 4.1], we are going to show, in the next lemma, that the metric induced by the quadratic forms $\mathfrak{J}$ over G_U is indeed an adapted one to the dominated splitting $F_- \oplus F_+$.

Lemma 3.1 *Consider a C^1 field of non-degenerate quadratic forms $\mathfrak{J}$ with constant index on the positively invariant open subset U for the flow X_t, such that the cocycle $A_t(x)$ over X_t is strictly $\mathfrak{J}$-separated. Then the induced $\mathfrak{J}$-metric on G_U is adapted to the dominated splitting $F_- \oplus F_+$.*

Proof Following [3, Theorem 2.13], we know that for each x in Γ there exist subspaces $F_-(x)$ and $F_+(x)$ such that $G_x = F_-(x) \oplus F_+(x)$ and $A_t(x)(F_a(x)) = F_a(x)$ for $a \in \{+, -\}$. It is also proved there that, for every $x \in \overline{X_t(U)}$ and every pair of unit vectors $u \in F_-$ and $v \in F_+$, we have

$$\frac{|A_t(x)u|}{|A_t(x)v|} \leq \frac{r_-^t(x)}{r_+^t(x)} \leq \omega_t := \sup_{x \in \Gamma} \frac{r_-^t(x)}{r_+^t(x)} < 1,$$

where r_-^t and r_+^t represent the values r_- and r_+ in Remark 2.2 with respect to the strictly $\mathfrak{J}$-separated cocycle $A_t(x)$ with a fixed t.

Then, $\lim\limits_{t \to \infty} \frac{|A_t(x)|_{F_-}|}{|A_t(x)|_{F_+}|} = 0$ for $x \in \Gamma$.

We claim that $\sup\limits_t \omega_t < 1$.

Suppose, in contrary, that there exists a sequence $(\omega_{t_n})_n$ converging to 1.

Then there exist sequences $(x_n)_n$ in Γ, $(u_n)_n$ in F_-, and $(v_n)_n$ in F_+ such that $\frac{|A_{t_n}(x_n)u_n|}{|A_{t_n}(x_n)v_n|}$ converge to 1. By compactness of Γ, we can suppose that $(x_n)_n$ converges to x in Γ.

If $(t_n)_n$ goes to infinity, we have a contradiction with $\lim\limits_{t \to \infty} \frac{|A_t(x)|_{F_-}|}{|A_t(x)|_{F_+}|} = 0$.

Now, suppose that $(t_n)_n$ is a bounded sequence, we can assume that $(t_n)_n$ converges for some t in $\mathbb{R}$, but $\frac{|A_t(x)u|}{|A_t(x)v|} < \omega_t < 1$, this contradiction proves the claim and completes the proof of the Lemma. $\square$

3.2 Exterior Products and Main Lemma

From now, we present some properties about exterior products and the main lemma to prove the Theorem A. Next, we are going to use Proposition 2.13 to obtain sufficient conditions for a flow X_t on a m-manifold M to have a $\wedge^{m-1} DX_t$-invariant one-dimensional uniformly expanding direction orthogonal to the $(m-1)$-dimensional center-unstable bundle.

Let V a m-dimensional vector space, we denote V by V^m, consider $\wedge^k V^m$ where $2 \leq k \leq m$. Let $\mathcal{B} = \{e_1, \cdots, e_m\}$ a basis of V^m. So $\{e_{j_1} \wedge \cdots \wedge e_{j_k} : 1 \leq j_1 < \cdots < j_k \leq m\}$ is a basis of $\wedge^k V^m$, and $J := \{(j_1, \cdots, j_k) \in \mathbb{N}^k : 1 \leq j_1 < \cdots < j_k \leq m\}$. Let $l = \binom{m}{k}$, so we have l combination of k vectors in $\{e_1, \cdots, e_m\}$, and $|J| = l$.

Take $u_1, u_2, \cdots, u_k \in V^m$ where $u_j = (u_j^1, u_j^2, \cdots, u_j^m)_{\mathcal{B}}$ for all $j \in \{1, \cdots, k\}$. Define

$$\mathcal{C} := \begin{pmatrix} u_1^1 & \cdots & u_k^1 \\ \cdots & \cdots & \cdots \\ u_1^m & \cdots & u_k^m \end{pmatrix}_{m \times k} . \tag{11}$$

For $(j_1, \cdots, j_k) \in J$, consider

$$\mathcal{C}^{j_1, \cdots, j_k} := \begin{pmatrix} u_1^{j_1} & \cdots & u_k^{j_1} \\ \cdots & \cdots & \cdots \\ u_1^{j_k} & \cdots & u_k^{j_k} \end{pmatrix}_{k \times k} \tag{12}$$

The following result holds

$$u_1 \wedge \cdots \wedge u_k = \sum_{(j_1, \ldots, j_k) \in J} \det(\mathcal{C}^{j_1, \ldots, j_k})(e_{j_1} \wedge \cdots \wedge e_{j_k}). \tag{13}$$

Let $A : V^m \to V^m$ a linear operator with matrix in basis $\mathcal{B}$ given by

$$\begin{pmatrix} a_{11} & a_{12} & \cdots & a_{1m} \\ \cdots & \cdots & \cdots & \cdots \\ a_{m1} & a_{m2} & \cdots & a_{mm} \end{pmatrix}_{(m \times m)} . \tag{14}$$

We will denote this matrix by A too.

Consider $\wedge^k A : \wedge^k V^m \to \wedge^k V^m$, note that $A(u_1) \wedge \cdots \wedge A(u_k) = \wedge^k A(u_1 \wedge \cdots \wedge u_k)$, by (13) and the linearity of $\wedge^k A$, we have that

$$A(u_1) \wedge \cdots \wedge A(u_k) = \sum_{(j_1, \cdots, j_k) \in J} \det(C^{j_1, \cdots, j_k}) \wedge^k A(e_{j_1} \wedge \cdots \wedge e_{j_k}) \qquad (15)$$

Define $A_j := A(e_j)$, so A_j is the j-th column of A, i.e., $A(e_j) = A_j = (a_{1j}, \cdots, a_{mj})^T$, so $A(e_j) = [a_{ij}]_{m \times 1}$. Let $A_{j_1 \cdots j_k} := (A_{j_1} \cdots A_{j_k})_{m \times k}$ where $(j_1, \cdots, j_k) \in J$. For each $(i_1 \cdots i_k), (j_1 \cdots j_k) \in J$ consider

$$A^{i_1 \cdots i_k}_{j_1 \cdots j_k} := \begin{pmatrix} a_{i_1 j_1} & \cdots & a_{i_1 j_k} \\ \cdots & \cdots & \cdots \\ a_{i_k j_1} & \cdots & a_{i_k j_k} \end{pmatrix}_{k \times k} \qquad (16)$$

Using that $\wedge^k A(e_{j_1} \wedge \cdots \wedge e_{j_k}) = A(e_{j_1}) \wedge \cdots \wedge A(e_{j_k})$ with matrix

$$A_{j_1 \cdots j_k} := (A_{j_1} \cdots A_{j_k})_{m \times k},$$

by (13) we obtain that

$$A(e_{j_1}) \wedge \cdots \wedge A(e_{j_k}) = \sum_{(i_1, \cdots, i_k) \in J} \det(A^{i_1 \cdots i_k}_{j_1 \cdots j_k})(e_{i_1} \wedge \cdots \wedge e_{i_k}). \qquad (17)$$

Lemma 3.2 *Let V to be vector space and $A : V \to V$ to be a linear operator then $\wedge^{(m-1)} A = \det(A) \cdot (A^{-1})^*$.*

Proof Consider $k = m - 1$. We use the following identification between $\wedge^{(m-1)} V$ and V. For each $(j_1, \cdots, j_{m-1}) \in J$, we identify $e_{j_1} \wedge \cdots \wedge e_{j_k}$ in $\wedge^{(m-1)} V$ by $\delta_p e_p$ in V, where $p \notin \{j_1, \cdots, j_{m-1}\}$, $\delta_p = 1$ if p is odd, and $\delta_p = -1$ if p is even.

We must show that for each $(j_1, \cdots, j_{m-1}) \in J$ the exterior product $\wedge^{(m-1)} A(e_{j_1} \wedge \cdots \wedge e_{j_k})$ corresponds to the $\det(A) \cdot (A^{-1})^*(\delta_p e_p)$, where $\delta_p e_p$ is given as above.

Define $S := \det(A) \cdot (A^{-1})^*$, using that $A^{-1} = \frac{1}{\det(A)} \mathrm{Adj}(A)$, we obtain that $S = \mathrm{cof}(A)$ where $\mathrm{cof}(A) = [(-1)^{i+j} M_{ij}]_{m \times m}$ and M_{ij} is the determinant of the submatrix formed by deleting the i-th row and j-th column. We have that $M_{ij} = \det(A^{r_1 \cdots r_k}_{s_1 \cdots s_k})$ where $i \notin r_1, \cdots, r_k$ and $j \notin s_1, \cdots, s_k$.

Note that

$$\mathrm{cof}(A)(\delta_p e_p) = \delta_p \mathrm{cof}(A)(e_p) = \delta_p((-1)^{1+p} M_{1p}, (-1)^{2+p} M_{2p}, \cdots, (-1)^{m+p} M_{mp})_{\mathcal{B}}.$$

In case p is odd, $\delta_p = 1$ and $\mathrm{cof}(A)(\delta_p e_p) = (M_{1p}, -M_{2p}, \cdots, (-1)^{m+p} M_{mp})_{\mathcal{B}}$.
We obtain that

$$\mathrm{cof}(A)(\delta_p e_p) = M_{1p} e_1 + M_{2p}(-e_2) + \cdots + M_{mp}(-1)^{m+p} =$$
$$M_{1p}(e_1 \delta_1) + M_{2p}(e_2 \delta_2) + \cdots + M_{mp}(e_{mp} \delta_{mp}).$$

Using that

$$A(e_{j_1}) \wedge \cdots \wedge A(e_{j_k}) = \sum_{(i_1,\cdots,i_k)\in J} \det(A^{i_1\cdots i_k}_{j_1\cdots j_k})(e_{i_1} \wedge \cdots \wedge e_{i_k})$$

and $M_{ij} = \det(A^{r_1\cdots r_k}_{s_1\cdots s_k})$ where $i \notin r_1,\cdots,r_k$ and $j \notin s_1,\cdots,s_k$, we have that $\mathrm{cof}(A)(\delta_p e_p) \cong A(e_{j_1}) \wedge \cdots \wedge A(e_{j_k})$. $\square$

This concludes the proof.

Remark 3.3 Under suitable identification, the last formula holds for differential of a diffeomorphism of a compact finite dimensional manifold.

The result below generalizes Corollary 2.9 to arbitrary n and k. The main difficulty here is working on the dimensions of the subbundles and its exterior powers.

Lemma 3.4 *The subbundle F_Γ is volume expanding by DX_t if, and only if, $\widetilde{F}$ is uniformly expanded by $\wedge^k DX_t$.*

In particular, $E \oplus F$ is a singular hyperbolic splitting, where F is volume expanding for DX_t if, and only if, $\widetilde{E} \oplus \widetilde{F}$ is partially hyperbolic splitting for $\wedge^k DX_t$ such that $\widetilde{F}$ is uniformly expanded by $\wedge^k DX_t$.

Proof We consider the action of the cocycle $DX_t(x)$ on k-vector that is the k-exterior power $\wedge^k DX_t$ of the cocycle acting on $\wedge^k T_\Gamma M$.

Denote by $\| \cdot \|$ the standard norm on k-vectors induced by the Riemannian norm of M; see, e.g. [7]. We write $m = \dim M$.

Suppose that $T_\Gamma M$ admits a splitting $E_\Gamma \oplus F_\Gamma$ with $\dim E_\Gamma = m - k$ and $\dim F_\Gamma = k$.

We note that if $E \oplus F$ is a DX_t-invariant splitting of $T_\Gamma M$, with $\{e_1, \ldots, e_l\}$ a family of basis for E and $\{f_1, \ldots, f_k\}$ a family of basis for F, then $\widetilde{F} = \wedge^k F$ generated by $\{f_{i_1} \wedge \cdots \wedge f_{i_k}\}_{1 \leq i_1 < \cdots < i_k \leq k}$ is naturally $\wedge^k DX_t$-invariant by construction. Then, the dimension of $\widetilde{F}$ is one with basis given by the vector $f_1 \wedge \cdots \wedge f_k$.

Assume that F_Γ is volume expanding by DX_t. We must show that there exist C and $\lambda > 0$ such that $| \wedge^k DX_t|_P | \geq Ce^{\lambda t}$, for all $t > 0$, where P is spanned by $f_1 \wedge \cdots \wedge f_k$.

Note that

$$\| \wedge^k DX_t|_P \| = \| \wedge^k DX_t(f_1 \wedge \cdots \wedge f_k)\| = \|DX_t(f_1) \wedge \cdots \wedge DX_t(f_k)\|.$$

But $f_1, \cdots, f_k$ is a basis for F, by hypothesis there exist constants C and $\lambda > 0$ such that $| \det(DX_t|_F)| \geq C.e^{\lambda t}$ for all $t > 0$. So,

$$| \det(DX_t|_F)| = \mathrm{vol}(DX_t(f_1), \cdots, DX_t(f_k)) = \|DX_t(f_1) \wedge \cdots \wedge DX_t(f_k)\|.$$

The reciprocal statement is straightforward.

Given a basis $\{f_1, \cdots, f_k\}$ of F, we have that

$$|\det(DX_t|_F)| =$$
$$\text{vol}(DX_t(f_1), \cdots, DX_t(f_k)) = ||DX_t(f_1) \wedge \cdots \wedge DX_t(f_k)|| =$$
$$|| \wedge^k DX_t(f_1 \wedge \cdots \wedge f_k)|| = || \wedge^k DX_t|_P||$$

where P is spanned by $f_1 \wedge \cdots \wedge f_k$.

However, by hypothesis, there exist C and $\lambda > 0$ such that $|| \wedge^k DX_t|_P|| \geq Ce^{\lambda t}$, for all $t > 0$.

Corollary 3.5 *Assume that E is uniformly contracted by DX_t. $E \oplus F$ is a singular-hyperbolic splitting for DX_t if, and only if, $\widetilde{E} \oplus \widetilde{F}$ is partially hyperbolic splitting for $\wedge^k DX_t$ such that $\widetilde{F}$ is uniformly expanded by $\wedge^k DX_t$.*

Let M Riemannian manifold m-dimensional with $\langle \cdot, \cdot \rangle$ inner product in $T_\Gamma M$, and $\langle \cdot, \cdot \rangle_*$ the inner product in $\wedge^k T_\Gamma M$ induced by $\langle \cdot, \cdot \rangle$ where $\wedge^k T_\Gamma M = \bigcup_{x \in \Gamma} \wedge^k T_x M$. So for $x \in \Gamma$, we have that $\langle \cdot, \cdot \rangle$ is acting on $T_x M$, and $\langle \cdot, \cdot \rangle_*$ is acting on $\wedge^k T_x M$.

Lemma 3.6 *Let M be a riemannian m-dimensional manifold. Then, for all $[\cdot, \cdot]_*$ inner product in $\wedge^{(m-1)} T_\Gamma M$ there exists a inner product $[\cdot, \cdot]$ on $T_\Gamma M$ such that $[\cdot, \cdot]_*$ is induced by $[\cdot, \cdot]$.*

Proof Let M be a riemannian manifold m-dimensional with $\langle \cdot, \cdot \rangle$ an inner product in $T_\Gamma M$, and $\langle \cdot, \cdot \rangle_*$ the inner product in $\wedge^{(m-1)} T_\Gamma M$ induced by $\langle \cdot, \cdot \rangle$.

Take $[\cdot, \cdot]_{**}$ an arbitrary inner product in $\wedge^{(m-1)} T_\Gamma M$. Using that $[\cdot, \cdot]_{**}$ and $\langle \cdot, \cdot \rangle_*$ are inner products in $\wedge^{(m-1)} T_\Gamma M$ there exists $J : \wedge^{(m-1)} T_\Gamma M \to \wedge^{(m-1)} T_\Gamma M$ isomorphism linear such that $[u, v]_{**} = \langle J(u), J(v) \rangle_*$.

Define $\varphi : GL(T_\Gamma M) \to GL(\wedge^{(m-1)} T_\Gamma M)$ given by $A \mapsto \wedge^{(m-1)} A$.

Note that φ is an injective linear homomorphism, and due to the dimensions of the spaces, φ is a linear isomorphism.

Hence, there exists $A \in GL(T_\Gamma M)$ such that $\wedge^{(m-1)} A = J$.

Consider $[x, y] := \langle A(x), A(y) \rangle$ for $x, y \in T_\Gamma M$, then $[u, v]_* = \det([u_i, v_j])_{(m-1) \times (m-1)}$, where $u = u_1 \wedge \cdots \wedge u_{(m-1)}$ and $v = v_1 \wedge \cdots \wedge v_{(m-1)}$.

We have that

$$[u, v]_* = \det(\langle A(u_i), A(v_j) \rangle)_{(m-1) \times (m-1)}.$$

On the other hand,

$$[u, v]_{**} = \langle \wedge^{(m-1)} A(u), \wedge^{(m-1)} A(v) \rangle_* = \det(\langle A(u_i), A(v_j) \rangle)_{(m-1) \times (m-1)}.$$

Therefore, $[\cdot, \cdot]_* = [\cdot, \cdot]_{**}$, and we are done.

4 Proofs of Main Results

We are now able to prove our main results.

4.1 Proof of Theorem A

Proof Consider M is a m-manifold and Γ is a compact X_t-invariant subset having a singular-hyperbolic splitting $T_\Gamma M = E_\Gamma \oplus F_\Gamma$. By Theorem 2.8 we have a $\wedge^{(m-1)} DX_t$-invariant partial hyperbolic splitting $\wedge^{(m-1)} T_\Gamma M = \widetilde{E} \oplus \widetilde{F}$ with dim $\widetilde{F} = 1$ and $\widetilde{F}$ uniformly expanded. Following the proof of Theorem 2.8, if we write e for a unit vector in E_x and $\{u_1, u_2, \cdots, u_{m-1}\}$ an orthonormal base for F_x, $x \in \Gamma$, then $\widetilde{E}_x$ is a $(m-1)$-dimensional vector space spanned by set $\{e \wedge u_{i_1} \wedge u_{i_2} \wedge \cdots \wedge u_{i_{m-2}}\}$ with $i_1, \cdots, i_{m-2} \in \{1, \cdots, m-1\}$.

From Theorem 2.6 and the existence of adapted metrics (see e.g. [12]), there exists a field $\mathfrak{J}$ of quadratic forms so that X is $\mathfrak{J}$-non-negative, DX_t is strictly $\mathfrak{J}$-separated on a neighborhood U of Γ, E_Γ is a negative subbundle, F_Γ is a positive subbundle and these subspaces are almost orthogonal. In other words, there exists a function $\delta :$ $\Gamma \to \mathbb{R}$ such that $\tilde{\mathfrak{J}}_x - \delta(x)\mathfrak{J}_x > 0$, $x \in \Gamma$ and we can locally write $\mathfrak{J}(v) = \langle J(v), v \rangle$ where $J = \text{diag}\{-1, 1, \cdots, 1\}$ with respect to the basis $\{e, u_1, \cdots, u_{m-1}\}$ and $\langle \cdot, \cdot \rangle$ is the adapted inner product; see [3].

By Lemma 3.2, $\wedge^{(m-1)} A = \det(A) \cdot (A^{-1})^*$ with respect to the adapted inner product which trivializes $\mathfrak{J}$, for any linear transformation $A : T_x M \to T_y M$. Hence $\wedge^{m-1} DX_t(x) = \det(DX_t(x)) \cdot (DX_{-t} \circ X_t)^*$ and a straightforward calculation shows that the infinitesimal generator $D^{(m-1)}(x)$ of $\wedge^{(m-1)} DX_t$ equals $\text{tr}(DX(x) \cdot Id - DX(x)^*$.

Therefore, using the identification between $\wedge^{(m-1)} T_x M$ and $T_x M$ through the adapted inner product and Proposition 2.10

$$\hat{\mathfrak{J}}_x = \partial_t(-\mathfrak{J})(\wedge^{(m-1)} DX_t \cdot v)\,|_{t=0} = \langle -(J \cdot D^{(m-1)}(x) + D^{(m-1)}(x)^* \cdot J)v, v \rangle$$

$$= \langle [(\mathfrak{J} \cdot DX(x) + DX(x)^* \cdot \mathfrak{J}) - 2\,\text{tr}(DX(x))\mathfrak{J}]v, v \rangle$$

$$= (\tilde{\mathfrak{J}} - 2\,\text{tr}(DX(x))\mathfrak{J})(v). \tag{18}$$

To obtain strict $(-\mathfrak{J})$-separation of $\wedge^{(m-1)} DX_t$ we search a function $\delta_{(m-1)} : \Gamma \to \mathbb{R}$ so that

$$(\tilde{\mathfrak{J}} - 2\,\text{tr}(DX)\mathfrak{J}) - \delta_{(m-1)}(-\mathfrak{J}) > 0 \quad \text{or} \quad \tilde{\mathfrak{J}} - (2\,\text{tr}(DX) - \delta_{(m-1)})\mathfrak{J} > 0.$$

Hence it is enough to make $\delta_{(m-1)} = 2\,\text{tr}(DX) - \delta$. This shows that in our setting $\wedge^{(m-1)} DX_t$ is always strictly $(-\mathfrak{J})$-separated.

Finally, according to Proposition 2.13, to obtain the partial hyperbolic splitting of $\wedge^{(m-1)} DX_t$ which ensures singular-hyperbolicity, it is sufficient that either $\hat{\mathfrak{J}}_x$ is positive definite or $\widetilde{\Delta}_a^b(x) = \int_a^b \delta_{(m-1)}(X_s(x))\,ds$ satisfies item (1) of Proposition 2.13, for all $x \in \Gamma$. This amounts precisely to the sufficient condition in the statement of Theorem A and we are done. □

Finally, we present the proof of Theorem B.

4.2 Proof of Theorem B

Proof Let singular-hyperbolic set Γ for vector field with a partially hyperbolic splitting $T_\Gamma M = E \oplus F$ where E is uniformly contracted and F is volume expanding.

Suppose that $T_\Gamma M$ admits a splitting $E_\Gamma \oplus F_\Gamma$ with $\dim E_\Gamma = 1$ and $\dim F_\Gamma = k = m - 1$.

We note that if $E \oplus F$ is a DX_t-invariant splitting of $T_\Gamma M$, with $\{e_1\}$ a basis for E and $\{f_1, \ldots, f_k\}$ a family of basis for F, then $\widetilde{F} = \wedge^k F$ generated by $\{f_{i_1} \wedge \cdots \wedge f_{i_k}\}_{1 \le i_1 < \cdots < i_k \le k}$ is naturally $\wedge^k DX_t$-invariant by construction. Then, the dimension of $\widetilde{F}$ is one with basis given by the vector $f_1 \wedge \cdots \wedge f_k$.

By Corollary 3.5, we have a partially hyperbolic splitting $\widetilde{E} \oplus \widetilde{F}$ for $\wedge^k DX_t$ such that $\widetilde{F}$ is uniformly expanded by $\wedge^k DX_t$. Hence, from [12, Theorem 1], there exists an adapted inner product $\langle \cdot, \cdot \rangle_*$ for $\wedge^k DX_t$. There exists $\lambda > 0$ satisfying for all $x \in \Gamma$ and $t > 0$ such that $\| \wedge^k DX_t \mid_{\widetilde{F}_x} \| \ge e^{\lambda t}$ for all $t > 0$.

By Lemma 3.6, $\langle \cdot, \cdot \rangle_*$ is induced by an inner product $\langle \cdot, \cdot \rangle$ in $T_\Gamma M$. So, we have a partially hyperbolic splitting $\widetilde{E} \oplus \widetilde{F}$ for $\wedge^k DX_t$ such that $\widetilde{F}$ is uniformly expanded by $\wedge^k DX_t$. By Theorem 2.8, we have that $E \oplus F$ is a dominated splitting for DX_t. From Theorem 2.5, there exists C^1 field of quadratic $\mathcal{J}$ such that DX_t is strictly $\mathcal{J}$-separated.

But DX_t is strictly $\mathcal{J}$-separated, this ensures, in particular, by Lemma 3.1, that the norm
$$|w| = \sqrt{\mathcal{J}(w_E)^2 + \mathcal{J}(w_F)^2}$$ is adapted to the dominated splitting $E \oplus F$ for the cocycle DX_t, where $w = w_E + w_F \in E_x \oplus F_x$, $x \in \Gamma$. This means that there exists $\mu > 0$ such that $|DX_t \mid_{E_x}| \cdot |DX_{-t} \mid_{F_{X_t(x)}}| \le e^{-\mu t}$ for all $t > 0$.

Moreover, from the definition of the inner product and $\wedge$, it follows that
$$|\det(DX_t \mid_{F_x})| = \|(\wedge^k DX_t)(u_1 \wedge \cdots \wedge u_k)\| = \|(\wedge^k DX_t) \mid_{\widetilde{F}}\| \ge e^{\lambda t}$$ for all $t > 0$, so $|\cdot|$ is adapted to the volume expanding along F.

To conclude, we are left to show that E admits a constant $\omega > 0$ such that $|DX_t \mid_E| \le e^{-\omega t}$ for all $t > 0$.

But since E is uniformly contracted, we know that $X(x) \in F_x$ for all $x \in \Gamma$.

Lemma 4.1 *Let Γ be a compact invariant set for a flow X of a C^1 vector field X on M. Given a continuous splitting $T_\Gamma M = E \oplus F$ such that E is uniformly contracted, then $X(x) \in F_x$ for all $x \in \Gamma$.*

Proof See [1, Lemma 5.1] and [3, Lemma 3.3].

Define the norm $|\cdot|_* = \xi |\cdot|$ where ξ is a small constant such that $\sup\{|X(z)| : z \in \Gamma\} \le 1$. We note that the choice of the positive constant ξ does not change any of the previous relations involving $|\cdot|$.

On the one hand, on each non-singular point x of Γ we obtain for each $w \in E_x$

$$e^{-\mu t} \geq \frac{|DX_t \cdot w|}{|DX_t \cdot X(x)|} = \frac{|DX_t \cdot w|}{|X(X_t(x))|} \geq \frac{|DX_t \cdot w|}{\sup\{|X(z)| : z \in \Gamma\}} \geq |DX_t \cdot w|.$$

On the other hand, for $\sigma \in \Gamma$ such that $X(\sigma) = 0$, we fix $t > 0$ and, since Γ is a non-trivial invariant set, we can find a sequence $x_n \to \sigma$ of regular points of Γ. The continuity of the derivative cocycle ensures $|DX_t|_{E_\sigma}| = \lim_{n \to \infty} |DX_t|_{E_{x_n}}| \leq e^{-\lambda t}$. Since $t > 0$ was arbitrarily chosen, we see that $|\cdot|$ is adapted for the contraction along E_σ.

This shows that $\lambda = \mu$ and completes the proof.

Acknowledgements L.S. dedicates this article to the memory of Prof. Welington de Melo who taught her the profundity and beauty of hyperbolic dynamics. The authors are deeply grateful to the anonymous referee for the kind analysis of this paper and the excellent suggestions which helped us to improve our work. L.S. is partially supported by Fapesb-JCB0053/2013, PRODOC-UFBA/2014 and CNPq 2017 postdoctoral schoolarship at Universidade Federal do Rio de Janeiro, V.C. is supported by CAPES.

References

1. V. Araujo, A. Arbieto, L. Salgado, Dominated splittings for flows with singularities. Nonlinearity **26**(8), 2391 (2013)
2. V. Araújo, M. J. Pacifico, *Three-dimensional flows*, volume 53 of *Ergebnisse der Mathematik und ihrer Grenzgebiete. 3. Folge. A Series of Modern Surveys in Mathematics [Results in Mathematics and Related Areas. 3rd Series. A Series of Modern Surveys in Mathematics]*. Springer, Heidelberg (2010). With a foreword by Marcelo Viana
3. V. Araujo, L. Salgado, Infinitesimal Lyapunov functions for singular flows. Math. Z. **275**(3–4), 863–897 (2013)
4. V. Araujo, L. Salgado, Dominated splitting for exterior powers and singular hyperbolicity. J. Differ. Equ. **259**, 3874–3893 (2015)
5. A. Arbieto, Sectional lyapunov exponents. Proc. Am. Math. Soc. **138**, 3171–3178 (2010)
6. A. Arbieto, L. Salgado, On critical orbits and sectional hyperbolicity of the nonwandering set for flows. J. Differ. Equ. **250**, 2927–2939 (2010)
7. L. Arnold, *Random Dynamical Systems* (Springer, Heidelberg, 1998)
8. C. Bonatti, L.J. Díaz, M. Viana, *Dynamics Beyond Uniform Hyperbolicity: A Global Geometric and Probabilistic Perspective (Encyclopedia of Mathematical Sciences)*, vol. 102. Mathematical Physics, III. (Springer, Heidelberg, 2005)
9. R. Bowen, *Equilibrium States and the Ergodic Theory of Anosov Diffeomorphisms*, vol. 470. Lecture Notes in Mathematics. (Springer, Heidelberg, 1975)
10. R. Bowen, D. Ruelle, The ergodic theory of Axiom A flows. Invent. Math. **29**, 181–202 (1975)
11. E.L. Lima, *Cálculo Tensorial*. Publicações Matemáticas, IMPA (1964)
12. N. Gourmelon, Adapted metrics for dominated splittings. Ergod. Theory Dynam. Syst. **27**(6), 1839–1849 (2007)
13. M. Hirsch, C. Pugh, M. Shub, *Invariant Manifolds*. Lectures Notes in Mathematics, vol. 583. (Springer, Heidelberg, 1977)
14. A. Katok, Infinitesimal Lyapunov functions, invariant cone families and stochastic properties of smooth dynamical systems. Ergod Theory Dynam. Syst. **14**(4), 757–785 (1994). With the collaboration of Keith Burns

15. A. Katok, B. Hasselblatt, *Introduction to the Modern Theory of Dynamical Systems (Encyclopedia Mathematics and its Applications*, vol. 54. (Cambridge University Press, Cambridge, 1995)

16. J. Lewowicz, Lyapunov functions and topological stability. J. Differ. Equ. **38**(2), 192–209 (1980)

17. R. Metzger, C. Morales, Sectional-hyperbolic systems. Ergod. Theory Dynam. Syst. **28**, 1587–1597 (2008)

18. C.A. Morales, M.J. Pacifico, E.R. Pujals, Singular hyperbolic systems. Proc. Am. Math. Soc. **127**(11), 3393–3401 (1999)

19. C.A. Morales, M.J. Pacifico, E.R. Pujals, Robust transitive singular sets for 3-flows are partially hyperbolic attractors or repellers. Ann. Math. **160**(2):375–432 (2004)

20. S. Newhouse, Cone-fields, domination, and hyperbolicity, in *Modern Dynamical Systems and Applications*, pp. 419–432. (Cambridge University Press, Cambridge, 2004)

21. C. Robinson. *Dynamical Systems: Stability, Symbolic Dynamics, and Chaos*, 2nd edn. Studies in Advanced Mathematics. (CRC Press, Boca Raton, 1999)

22. C. Robinson, *An Introduction to Dynamical Systems: Continuous and Discrete* (Pearson Prentice Hall, Upper Saddle River, NJ, 2004)

23. L. Salgado. *Sobre hiperbolicidade fraca para fluxos singulares*. PhD thesis, UFRJ, Rio de Janeiro, 2012

24. M. Shub, *Global Stability of Dynamical Systems*. (Springer, New York, 1987)

25. W. Tucker. The Lorenz attractor exists. C. R. Acad. Sci. Paris **328**(12), 1197–1202 (1999)

26. M. Viana, What's new on Lorenz strange attractor. Math. Intell. **22**(3), 6–19 (2000)

27. S. Winitzki, *Linear Algebra: Via Exterior Products*, 1.2 edn. (Lulu.com, Raleigh, 2012)

28. M. Wojtkowski, Invariant families of cones and Lyapunov exponents. Ergod. Theory Dynam. Syst. **5**(1), 145–161 (1985)

29. M.P. Wojtkowski, Monotonicity, J-algebra of Potapov and Lyapunov exponents. in *Smooth Ergodic Theory and its Applications (Seattle, WA, 1999)*, vol. 69. *Proc. Sympos. Pure Math.*. (American Mathematical Society, Providence, 2001), pp. 499–521

On Conformal Measures and Harmonic Functions for Group Extensions

Manuel Stadlbauer

Abstract We prove a Perron-Frobenius-Ruelle theorem for group extensions of topological Markov chains based on a construction of σ-finite conformal measures and give applications to the construction of harmonic functions.

Keywords Group extension · Conformal measures · Harmonic functions

1 Introduction

The Perron-Frobenius-Ruelle theorem is a statement about the maximal eigenvalue of an operator L who preserves the cone of positive functions. Namely, it provides existence of a function f in this cone and ν in its dual, such that $Lf = \rho f$ and $L^*\nu = \rho\nu$, with ρ referring to the spectral radius of L. The first result of this type was obtained by Perron in [21] as a byproduct of his analysis of periodic continued fractions. He proved that, for a strictly positive $n \times n$-matrix A, the maximal eigenvalue ρ is simple. Moreover, his proof reveals that there exist strictly positive vectors x, $y \in \mathbb{R}^n$ such that $x^t A = \rho x^t$, $Ay = \rho y$ and that $\rho^{-n} A^n$ converges to $y \cdot x^t$. Even though there were many important contributions to the theory of positive operators in the following decades, e.g. by Doeblin-Fortet [10] or Birkhoff [4], whose methods are today standard tools in proving exponentially fast convergence of the iterates (see, e.g., [3, 17]), it was only at the end of the 60's when Ruelle obtained an analog of Perron's theorem for the one-dimensional Ising model with long range interactions from mathematical physics ([25, Theorem 3]). In the context of dynamical systems, as observed by Bowen, the result of Ruelle has the following formulation in terms of shift spaces. For a fixed $k \in \mathbb{N}$, let

$$\Sigma := \{(x_i : i \in \mathbb{N}) : x_i \in \{1, \ldots, k\} \; \forall i \in \mathbb{N}\},$$
$$\theta : \Sigma \to \Sigma \; (x_1, x_2, \ldots) \mapsto (x_2, x_3, \ldots)$$

M. Stadlbauer (✉)
Universidade Federal do Rio de Janeiro-UFRJ, Rio de Janeiro, Brazil
e-mail: manuel@im.ufrj.br

© Springer Nature Switzerland AG 2019
M. J. Pacifico and P. Guarino (eds.), *New Trends in One-Dimensional Dynamics*, Springer Proceedings in Mathematics & Statistics 285,
https://doi.org/10.1007/978-3-030-16833-9_15

and suppose that $\varphi : \Sigma \to (0, \infty)$ is a log-Hölder continuous function with respect to the shift metric (for details, see below). The associated operator, the Ruelle operator, is then defined by

$$L_\varphi(f)(x) := \sum_{\theta(y)=x} \varphi(y) f(y).$$

Ruelle's theorem states that there exist a strictly positive Hölder continuous function h and a probability measure μ such that $L_\varphi(f) = \rho f$, $L_\varphi^*(\mu) = \rho\mu$ and $\lim \rho^{-n} L_\varphi^n f = \int f d\mu \cdot h$. Furthermore, $\rho^{-n} L_\varphi^n f \to \int f d\mu \cdot h$ converges exponentially fast, which implies, among many other things, that h is unique and that the measure μ is exact (and, in particular, ergodic).

The aim of this note is to establish an analogue of the result for dynamical systems of the form

$$T : \Sigma \times G \to \Sigma \times G, (x, g) \mapsto (\theta x, g\psi(x)),$$

where G is a discrete group G, Σ a shift space with the b.i.p.-property as defined below and $\psi : \Sigma \to G$ a locally constant function. This kind of dynamical system is called group extension, or, as they first were considered by Rokhlin in [24], Rokhlin transformation. Even though one might be tempted to think of ψ as a cocycle, the probably most fruitful approach is to consider T as a kind of random walk on G. That is, by fixing a potential function φ which only depends on the first coordinate, $\varphi(x)$ stands for the transition probability to go from (x, g) to $(\theta x, g\psi(x))$, which is also reflected by the fact that the Ruelle operator $\mathcal{L}_\varphi$ associated to T has many similarities to the Markov operator of a random walk.

In this setting, it is possible to obtain the following operator theorem which is the main result (Theorem 5.1) of this note. Under a technical condition (which is satisfied if, e.g., Σ is compact), it is shown that there exists a Lipschitz continuous map $\mu \to \nu_\mu$ from the space of probability measures on $\Sigma \times G$ to the space of σ-finite, conformal measures, that is ν_μ is σ-finite and $\mathcal{L}_\varphi^*(\nu_\mu) = \rho\nu_\mu$. This map is constructed using the method by Denker-Urbanski in [8]. By adapting ideas of Patterson [20] and Sullivan [31] from hyperbolic geometry, one obtains by variation of μ a family of strictly positive, ρ-harmonic functions. In here, we refer to $h : \Sigma \times G \to \mathbb{R}$ as *harmonic* or ρ-*harmonic* if $\mathcal{L}_\varphi(h) = \rho h$. In particular, Theorem 5.1 gives rise to families of σ-finite, conformal measures $\{\nu_\mu\}$, ρ-harmonic, positive functions $\{h\}$, and T-invariant measures $\{h d\nu_\mu\}$. Furthermore, as the conformal measures are pairwise equivalent, the function $\mathbb{K}(\mu, z) := (d\nu_\mu/d\nu)(z)$ for a fixed conformal reference measure ν is defined and, as shown in Theorem 5.1, its logarithm is locally Lipschitz continuous with respect to the first coordinate and T-invariant with respect to the second.

It is important to point out that these families might not be one-dimensional. For example, the classical and general result of Zimmer in [34] (see [14] for a version for group extensions) states that ergodicity implies amenability. Hence, as G not necessarily is amenable, ν_μ might not be ergodic and the standard argument for uniqueness of conformal measures no longer is applicable. However, for the setting in here, a sharp criterium of classical flavour holds (Proposition 5.3). That is, ν_μ is conservative and ergodic if and only if

$$\sum_{n=1}^{\infty} \rho^{-n} \sum_{T^n(x,id)=(x,id)} \prod_{k=0}^{n-1} \varphi(\theta^k(x)) = \infty.$$

Hence, it is of interest to analyse the families of conformal measures and harmonic functions if T is non-ergodic. In order to have explicit examples at hand, we use the fact that a random walk with independent increments can be identified with a group extension. For T associated with the random walk on $\mathbb{Z}^d$ or the free group $\mathbb{F}_d$, the d-dimensional central limit theorem and the local limit theorem by Gerl and Woess in [12], respectively, allow to explicitly determine $\mathbb{K}$. For these specific examples, it turns out that the family of conformal measures is one-dimensional for $\mathbb{Z}^d$ and non-trivial for $\mathbb{F}_d$.

For a further analysis of the general setting, these conformal measures are employed to construct a positive map from the space of functions $\mathcal{C}$ whose logarithm is uniformly continuous to the space of harmonic functions $\mathcal{H}$ satisfying a certain local Lipschitz condition (Theorem 6.1). Then, in order to at least roughly determine the behaviour of harmonic functions and ν at infinity, further ideas from probability and ergodic theory are employed. Namely, for a given pair (h, ν) of a positive harmonic function and a conformal measure, $h d\nu$ is invariant and therefore, the natural extension of $(T, h d\nu)$ is well-defined. Therefore, through Martingale convergence, it is possible to show (Corollary 6.3) for G non-amenable and under a symmetry condition that

$$\nu_\mu\left(\left\{(x, \psi(\omega) \cdots \psi(\theta^{n-1}(\omega))) : x \in \Sigma\right\}\right) = o(\rho^n),$$

for a.e. $\omega \in \Sigma$ with respect to the equilibrium measure of $(\Sigma, \theta, \varphi)$. These results also have a canonical application to the dimension theory of graph directed Markov systems, which is outlined in Theorem 7.1.

Remark. After submitting the article, the author was made aware of the results inspired by Martin boundaries by Shwartz in [29]. In there, a complete description of harmonic functions and conformal measures on locally compact shift spaces was obtained.

2 Topological Markov Chains

We begin with defining the basic object of our analysis, that is topological Markov chains and their group extensions. For a countable alphabet $\mathcal{W}$ and a matrix $(a_{ij} : i, j \in \mathcal{W})$ with $a_{ij} \in \{0, 1\}$ for all $i, j \in \mathcal{W}$ and no rows and columns equal to 0, let the pair (Σ, θ) denote the associated one-sided topological Markov chain given by

$$\Sigma := \big\{(w_k : k = 1, 2, \ldots) \ : \ w_k \in \mathcal{W}, a_{w_k w_{k+1}} = 1 \ \forall i = 0, 1, \ldots \big\},$$
$$\theta : \Sigma \to \Sigma, \theta : (w_k : k = 1, 2, \ldots) \mapsto (w_k : k = 2, 3, \ldots).$$

A finite sequence $w = (w_1 w_2 \ldots w_n)$ with $n \in \mathbb{N}$, $w_k \in \mathcal{W}$ for $k = 1, 2, \ldots, n$ and $a_{w_k w_{k+1}} = 1$ for $k = 1, 2, \ldots, n - 1$ is referred to as *admissible* or as *word of length* n, the set of words of length n will be denoted by $\mathcal{W}^n$ and the set

$$[w] := \{(v_k) \in \Sigma \ : \ w_k = v_k \ \forall k = 1, 2, \ldots, n\}$$

is referred to as a *cylinder of length* n. Furthermore, $|w|$ denotes the length of a word and $\mathcal{W}^\infty = \bigcup_{n=1}^\infty \mathcal{W}^n$ the set of all admissible words. Since $\theta^n : [w] \to \theta^n([w])$ is a homeomorphism, observe that the inverse $\tau_w : \theta^n([w]) \to [w]$ is well defined.

As it is well known, Σ is a Polish space with respect to the topology generated by cylinders and Σ is compact with respect to this topology if and only if $\mathcal{W}$ is a finite set. Moreover, the topology generated by cylinders is compatible with the metric defined by, for $r \in (0, 1)$ and $(w_k), (v_k) \in \Sigma$,

$$d_r((w_k), (v_k)) := r^{\min(i : w_i \neq v_i) - 1}.$$

Observe that with respect to this definition, the r^n-neighbourhood of $(w_k) \in \Sigma$ is given by the cylinder $[w_1 w_2 \ldots w_n]$ of length n. Also recall that Σ is *topologically transitive* if for all $a, b \in \mathcal{W}$, there exists $n \in \mathbb{N}$ such that $\theta^n([a]) \cap [b] \neq \emptyset$ and is *topologically mixing* if for all $a, b \in \mathcal{W}$, there exists $N \in \mathbb{N}$ such that $\theta^n([a]) \cap [b] \neq \emptyset$ for all $n \geq N$. Moreover, a topological Markov chain is said to have *big images* and *big preimages* if there exists a finite set $\mathcal{I}_{\mathrm{bip}} \subset \mathcal{W}$ such that for all $v \in \mathcal{W}$, there exists $\beta_1, \beta_2 \in \mathcal{I}_{\mathrm{bip}}$ such that $(v\beta_1) \in \mathcal{W}^2$ and $(\beta_2 v) \in \mathcal{W}^2$. Finally, we say that a topological Markov chain satisfies the *big images and preimages (b.i.p.) property* if the chain is topologically mixing and has big images and preimages (see [27]). Note that the b.i.p. property coincides with the notion of finite irreducibility for topological mixing topological Markov chains as introduced by Mauldin and Urbanski [18].

Potentials. A further basic object for our analysis is a fixed, strictly positive function $\varphi : \Sigma \to \mathbb{R}$ which is referred to as a *potential*. This function might be seen as weight on the preimages of a point and in many applications, φ is defined as the conformal derivative of an underlying iterated function system. For $n \in \mathbb{N}$ and $w \in \mathcal{W}^n$, set $\Phi_n := \prod_{k=0}^{n-1} \varphi \circ \theta^k$ and $\Phi_w := \Phi_n \circ \tau_w$. We refer to φ as a potential of *(locally) bounded variation* if

$$\sup \left\{ \frac{\Phi_n(x)}{\Phi_n(y)} \ : \ n \in \mathbb{N}, \ w \in \mathcal{W}^n, \ x, y \in [w] \right\} < \infty.$$

From now on, for positive sequences $(a_n), (b_n)$ we will write $a_n \ll b_n$ if there exists $C > 0$ with $a_n \leq C b_n$ for all $n \in \mathbb{N}$, and $a_n \asymp b_n$ if $a_n \ll b_n \ll a_n$. For example, the above could be rewritten by $\Phi_{|w|}(x) \asymp \Phi_{|w|}(y)$ for all $w \in \mathcal{W}^\infty$ and $x, y \in [w]$. A further, stronger assumption on the variation is related to local Hölder continuity.

Recall that the n-th variation of a function $f : \Sigma \to \mathbb{R}$ is defined by

$$V_n(f) = \sup\{|f(x) - f(y)| : x_i = y_i,\ i = 0, 1, 2, \ldots, n - 1\}.$$

The function f is referred to as a *locally Hölder continuous function* if there exists $0 < r < 1$ and $C \geq 1$ such that $V_n(f) \ll r^n$ for all $n \geq 1$. Moreover, we refer to a locally Hölder continuous function with $\| f \|_\infty < \infty$ as a *Hölder continuous function*. We now recall a well-known estimate. For $n \leq m, x, y \in [w]$ for some $w \in \mathcal{W}^m$, and a locally Hölder continuous function f,

$$\left| \sum_{k=0}^{n-1} f \circ \theta^k(x) - f \circ \theta^k(y) \right| \ll \frac{1}{1 - r} r^{m-n}. \tag{1}$$

In particular, if $\log \varphi = f$ is locally Hölder continuous, then φ is a potential of bounded variation. Moreover, as $r^{m-n} = d(\theta^n(x), \theta^n(y))$, there exists $C_\varphi \geq 1$ such that

$$|\Phi_w(x)/\Phi_w(y) - 1| \leq C_\varphi d(x, y) \quad \text{and} \quad \Phi_w(x)/\Phi_w(y) \leq C_\varphi$$

for all $w \in \mathcal{W}^\infty$ and $x, y \in [w]$.

Conformal measures. In here, due to the fact that the constructions canonically lead to σ-finite measures, we will make use of a slightly more general definition of conformality by allowing infinite measures. We refer to a σ-finite Borel measure μ as a φ-*conformal* measure if

$$\mu(\theta(A)) = \int_A \frac{1}{\varphi} d\mu$$

for all Borel sets A such that $\theta|_A$ is injective. For $w = (w_1 \ldots w_n) \in \mathcal{W}^n$ and a potential of bounded variation, it then immediately follows that

$$\mu([w]) \asymp \Phi_n(x)\, \mu(\theta([w_n])) \tag{2}$$

for all $x \in [w]$. Note that this estimate implies that $P_G(\theta, \varphi) = 0$ is a necessary condition for the existence of a conformal measure with respect to a potential of bounded variation. Moreover, if $\mu(\theta([w])) \asymp 1$ (e.g., if μ is finite and θ has the big image property), we obtain that

$$\mu([w]) \asymp \Phi_n(x) \tag{3}$$

for all $n \in \mathbb{N}$, $w \in \mathcal{W}^n$ and $x \in [w]$. Also note that a probability measure satisfying (3) is referred to as a φ-*Gibbs measure*.

Ruelle's operator, b.i.p. and Gibbs-Markov maps. Ruelle's operator is defined, for $f : \Sigma \to \mathbb{R}$ in a suitable function space to be specified later, by

$$L_\varphi(f) = \sum_{v \in \mathcal{W}} \mathbf{1}_{\theta([v])} \cdot \varphi \circ \tau_v \cdot f \circ \tau_v.$$

Furthermore, there is an associated action on the space of σ-finite Borel measures defined through $\int f \, dL_\varphi^*(\nu) := \int L_\varphi(f) d\nu$, for each continuous $f : \Sigma \to [0, \infty)$. We then have that ν is a φ/ρ conformal measure if and only if $L_\varphi^*(\nu) = \rho\nu$. If, in addition, there is a measurable function $h : \Sigma \to [0, \infty)$ with $L_\varphi(h) = \rho h$, then $d\mu := h \, d\nu$ defines an invariant, σ-finite measure, that is $\mu = \mu \circ \theta^{-1}$. Moreover, for $\varphi' := \varphi h / (\rho h \circ \theta)$, we have $L_{\varphi'}(1) = 1$.

An important consequence of the b.i.p. property is a Perron-Frobenius-Ruelle theorem in case of an infinite alphabet $\mathcal{W}^1$ (see [18, 27]). That is, if (Σ, θ) has the b.i.p. property, $\log \varphi$ is Hölder continuous and $\|L_\varphi(1)\|_\infty < \infty$, then there exists a Gibbs measure μ and a Hölder continuous, strictly positive eigenfunction h of L_φ, which is uniformly bounded from above and below. Moreover, in this situation, (Σ, θ, μ) has the *Gibbs-Markov property*, that is μ is a Borel probability measure, for all $w \in \mathcal{W}^1$, μ and $\mu \circ \tau_w$ are equivalent, $\inf \{\mu(\theta([w])) : w \in \mathcal{W}^1\} > 0$ and there exists $0 < r < 1$ such that, for all $m, n \in \mathbb{N}$, $v \in \mathcal{W}^m$, $w \in \mathcal{W}^n$ with $(vw) \in \mathcal{W}^{m+n}$,

$$\sup_{x,y \in [w]} \left| \log \frac{d\mu \circ \tau_v}{d\mu}(x) - \log \frac{d\mu \circ \tau_v}{d\mu}(y) \right| \ll r^n. \tag{4}$$

As it is well known, the action on the space of bounded continuous functions of the transfer operator with respect to μ coincides with $L_{\varphi/\rho}$ and, with h referring to the function given by the Perron-Frobenius-Ruelle, $h \, d\mu$ is an invariant probability measure with exponential decay of correlations and associated transfer operator given by $L_{(\varphi h)/(\rho h \circ \theta)}$ (see [3, 27]).

Furthermore, several arguments in here are based on an inequality in the flavour of Doeblin-Fortet or Lasota-Yorke for arbitrary topological Markov chains (Σ, θ) and potentials φ such that $\log \varphi$ is locally Hölder continuous. For $f : \Sigma \to \mathbb{R}$, define

$$D(f) : \Sigma \to [0, \infty), \quad (x_1, x_2 \ldots) \mapsto \sup_{y, \tilde{y} \in [x_1]} \frac{|f(y) - f(\tilde{y})|}{d_r(y, \tilde{y})}.$$

That is, $D(f)(x)$ is the local Hölder coefficient of the function f restricted to $[a]$, with $x \in [a]$. Now assume that $L_\varphi^n(f)$ is well-defined. Then, for x, y in the same cylinder,

$$\left| L_\varphi^n(f)(x) - L_\varphi^n(f)(y) \right|$$

$$= \left| \sum_{v \in \mathcal{W}^n} \left(1 - \frac{\Phi_v(y)}{\Phi_v(x)} \right) \Phi_v(x) f \circ \tau_v(x) + \Phi_v(y) (f \circ \tau_v(x) - f \circ \tau_v(y)) \right|$$

$$\leq C_\varphi L_\varphi^n(|f|)(x) d_r(x, y) + r^n L_\varphi^n(D(f))(y) d_r(x, y)$$

$$\leq C_\varphi d_r(x, y) L_\varphi^n \left(|f| + r^n D(f) \right)(x) \tag{5}$$

If, in addition, for all $a \in \mathcal{W}^1$, either $f(x) = 0$ for all $x \in [a]$ or $|f(x)/f(y) - 1| \leq C_a d_r(x, y)$ for all $x, y \in [a]$, set $LD(f)(x) := 0$ in the first case and $LD(f)(x) := \sup\{|f(x)/f(y) - 1|/d_r(x, y) : x, y \in [a]\}$ in the second case. By the same arguments,

$$
\left| L_\varphi^n(f)(x) - L_\varphi^n(f)(y) \right|
$$

$$
\leq C_\varphi \, d_r(x, y) \, L_\varphi^n \left(|f| \right)(x) + C_\varphi \sum_{v \in \mathcal{W}^n} \Phi_v(x) \, |f \circ \tau_v(x)| \left| \frac{f \circ \tau_v(y)}{f \circ \tau_v(x)} - 1 \right|
$$

$$
\leq C_\varphi \, d_r(x, y) \, L_\varphi^n \left(|f| \left(1 + r^n LD(f) \right) \right)(x) \tag{6}
$$

3 Group Extensions of Topological Markov Chains

Fix a countable group G and a map $\psi : \Sigma \to G$ such that ψ is constant on $[w]$ for all $w \in \mathcal{W}^1$. Then, for $X := \Sigma \times G$ equipped with the product topology of Σ and the discrete topology on G, the *group extension or G-extension* (X, T) of (Σ, θ) is defined by

$$
T : X \to X, (x, g) \mapsto (\theta x, g\psi(x)).
$$

Note that (X, T) is a topological Markov chain with respect to the alphabet $\mathcal{W}^1 \times G$ and the following transition rule: $((a, g), (b, h))$ is admissible if and only if $(ab) \in \mathcal{W}^2$ and $g\psi(a) = h$, where $\psi(a) := \psi(x)$, for some $x \in [a]$. Furthermore, set $X_g := \Sigma \times \{g\}$ and

$$
\psi_n(x) := \psi(x)\psi(\theta x) \cdots \psi(\theta^{n-1} x)
$$

for $n \in \mathbb{N}$ and $x \in \Sigma$. Observe that $\psi_n : \Sigma \to G$ is constant on cylinders of length n which implies that $\psi_k(w) := \psi_k(x)$, for some $x \in [w]$, $k \leq n$ and $w \in \mathcal{W}^n$, is well defined. If $k = n$, we will write $\psi_w := \psi_n(w)$. It is then easy to see that the finite words of (X, T) can be identified with $\mathcal{W}^\infty \times G$ by

$$
((w_0, \ldots, w_n), g) \equiv ((w_0, g), (w_1, g\psi_1(w)), \ldots, (w_n, g\psi_n(w))).
$$

Also observe that topologically transitivity of (X, T) implies that $\{\psi(a) : a \in \mathcal{W}^1\}$ is a generating set for G as a semigroup.

Throughout, we now fix a topological mixing topological Markov chain (Σ, θ), and a topological transitive G-extension (X, T). Furthermore, we fix a (positive) potential $\varphi : \Sigma \to \mathbb{R}$ with $P_G(\theta, \varphi) = 0$. Note that φ lifts to a potential φ^* on X by setting $\varphi^*(x, g) := \varphi(x)$. For ease of notation, we will not distinguish between φ^* and φ. Moreover, for $v \in \mathcal{W}^\infty$, the inverse branch given by $[v, \cdot]$ will be as well denoted by τ_v, that is $\tau_v(x, g) := (\tau_v(x), g\psi(v)^{-1})$. In order to distinguish between the Ruelle operator of θ and T, these objects for the group extension will be written in calligraphic letters. That is, for $a \in \mathcal{W}, \xi \in [a] \times \{id\}, (\eta, g) \in X$, and $n \in \mathbb{N}$,

$$\mathcal{L}(f)(\xi, g) := \sum_{v \in \mathcal{W}} \varphi(\tau_v(\xi)) f \circ \tau_v(\xi, g).$$

Remark 3.1 In the context of topological transitivity, it is natural to ask whether (X, T) is ergodic with respect to the product of the Gibbs measure on Σ and the counting measure. For example, a classical result of Zimmer in [34] (see also [14]) states that ergodicity of (X, T) implies that G is amenable, that is, there exists a sequence (K_n) of finite subsets of G with $\bigcup_n K_n = G$ such that

$$\lim_{n \to \infty} |g K_n \triangle K_n| / |K_n| = 0 \quad \forall g \in G,$$

where $\triangle$ refers to the symmetric difference and $|\cdot|$ to the cardinality of a set. Moreover, it was shown in [30] for this class of extensions that $P_G(T) = P_G(\theta)$ implies that G is amenable. Hence, if G is a non-amenable group, then $P_G(T) < P_G(\theta)$ and (X, T) is not ergodic. In particular, by bounded distortion, T has to be totally dissipative. For a further criterion for ergodicity, we also refer to Corollary 5.3 below. Also note that the classical result of Varopoulos on recurrent groups motivates the conjecture that a group extension only can be ergodic if G is a finite extension of the trivial group, $\mathbb{Z}$ or $\mathbb{Z}^2$.

Symmetric extensions. In several interesting applications, group extensions are satisfying a certain notion of symmetry. In here, we will use a pathwise notion (as in [30]) in contrast to the more general notion in [13]. Namely, we say that (Σ, θ, ψ) is *symmetric* if there exists $\mathcal{W}^1 \to \mathcal{W}^1$, $w \mapsto w^\dagger$ with the following properties.

1. For $w \in \mathcal{W}^1$, $(w^\dagger)^\dagger = w$.
2. For $v, w \in \mathcal{W}^1$, the word (vw) is admissible if and only if $(w^\dagger v^\dagger)$ is admissible.
3. $\psi(v^\dagger) = \psi(v)^{-1}$ for all $v \in \mathcal{W}^1$.

Moreover, we refer to $(\Sigma, \theta, \psi, \varphi)$ as a *symmetric group extension* if (Σ, θ, ψ) is symmetric and, with $\dagger : \mathcal{W}^\infty \to \mathcal{W}^\infty$ defined by $(w_1 \ldots w_n)^\dagger := (w_n^\dagger \ldots w_1^\dagger)$,

$$\sup_{n \in \mathbb{N}} \sup_{x \in [w], y \in [w^\dagger]} \frac{\Phi_n(x)}{\Phi_n(y)} < \infty.$$

4   Conformal σ-Finite Measures

As a first step towards a Ruelle theorem for group extensions, we now adapt ideas from [8, 20] in order to obtain invariant measures for the dual of the Ruelle operator. In contrast to [8, 20], the method in here gives rise to conformal σ-finite measures, which seems to be advantageous as group extensions in many cases are totally dissipative dynamical systems and therefore might not admit finite invariant measures. We now fix $\xi \in \Sigma$ and, for $n \in \mathbb{N}$, set

$$\mathcal{Z}^n(\xi) = \sum_{\theta^n(x)=\xi, \psi_n(x)=id} \Phi_n(x) = \mathcal{L}_\varphi^n(\mathbf{1}_{X_{id}})(\xi, id).$$

Since the construction relies on the divergence of a power series at its radius of convergence, recall that, for a sequence of positive real numbers (a_n), the radius of convergence of $\sum_n a_n x^n$ is equal to $1/\rho$ where, by Hadamard's formula,

$$\rho := \limsup_{n \to \infty} \sqrt[n]{a_n}.$$

We now ensure divergence at the radius of convergence by pointwise multiplication with a slowly diverging sequence as given by the following result. For the proof, we refer to [8].

Lemma 4.1 *For a positive sequence (a_n) with $\rho < \infty$, there exists a nondecreasing sequence $(b_n : n \in \mathbb{N})$ with $b_n \geq 1$ for all $n \in \mathbb{N}$ such that $\lim_{n\to\infty} b_n/b_{n+1} = 1$ and for all $s \geq 0$,*

$$\sum_{n=1}^\infty b_n a_n s^{-n} \begin{cases} = \infty & s \leq \rho \\ < \infty & s > \rho. \end{cases}$$

Moreover, there exists a non-increasing sequence $(\lambda(n) : n \in \mathbb{N})$ with $\lambda(n) \geq 1$ and $\lambda(n) \to 1$ such that $b_n = \prod_{k=1}^n \lambda(k)$.

Now suppose that $\rho = \limsup \sqrt[n]{\mathcal{Z}^n(\xi)} < \infty$. Then, for (b_n) given by Lemma 4.1 applied to $a_n = \mathcal{Z}^n(\xi)$, we have that

$$\mathcal{P}(s) := \sum_{n \in \mathbb{N}} s^{-n} b_n \mathcal{Z}^n(\xi).$$

diverges as $s \searrow \rho$. Furthermore, for $\rho < s < \infty$, set

$$m_s := \frac{1}{\mathcal{P}(s)} \sum_{n \in \mathbb{N}} s^{-n} b_n \sum_{T^n(z)=(\xi,id)} \Phi_n(x) \delta_z, \tag{7}$$

where δ_z refers to the Dirac measure supported in z. Note that, by construction, $m_s(X_{id}) = 1$ for all $s > \rho$. In order to construct a σ-finite, conformal measure, we consider an accumulation point ν of $\{m_s\}$ in the weak* topology, i.e. convergence of $\int f dm_s$ to $\int f d\nu$ for every bounded and continuous function f. For ease of notation, we now identify Σ with X_{id} and, for $B \subset \Sigma$ with $T^k|_{B \times \{id\}}$ invertible and $T^k(B \times \{id\}) \subset X_{id}$, the restriction $T^k|_{B \times \{id\}}$ with $\theta^k|_B$.

Lemma 4.2 *Assume that, for $s_l \searrow \rho$, there exists a probability measure m on Σ which is the weak*-limit of $(m_{s_l} : l \in \mathbb{N})$. Then, for each pair (B, k) with $B \in \mathcal{B}(\Sigma)$, $k \in \mathbb{N}$ such that $T^k|_{B \times \{id\}}$ is invertible and $T^k(B \times \{id\}) \subset X_{id}$,*

$$m(\theta^k(B)) = \int_B \rho^k/\Phi_k dm. \tag{8}$$

Proof Suppose that B is a cylinder, that is $B = [w]$ for some $w \in \mathcal{W}^m$ and $m > k$. Since T^k is injective on $B \times \{id\}$, we have, for $s > \rho$, that

$$m_s(\theta^k(B)) = \frac{1}{P(s)} \sum_{n\in\mathbb{N}} \sum_{x\in\theta^n(B)\cap E_n} \frac{b_n\Phi_n(x)}{s^n} = \frac{1}{P(s)} \sum_{n\in\mathbb{N}} \sum_{x\in B\cap\theta^{-k}(E_n)} \frac{b_n\Phi_n(\theta^k x)}{s^n}$$

$$= \frac{1}{P(s)} \sum_{n\in\mathbb{N}} \sum_{x\in B\cap E_{n+k}} \frac{b_{n+k}\Phi_{n+k}(x)}{s^{n+k}} \frac{b_n s^k}{b_{n+k}\Phi_k(x)}$$

In particular this gives

$$\left| m_s(\theta^k(B)) - \int_B \frac{s^k}{\Phi_k(x)} dm_s \right| \le \frac{1}{P(s)} \sum_{n\in\mathbb{N}} \left| \frac{b_n}{b_{n+k}} - 1 \right| \sum_{x\in B\cap E_{n+k}} \frac{b_{n+k}\Phi_{n+k}(x)}{s^{n+k}}$$

$$+ \frac{1}{P(s)} \sum_{n=1}^{k} \sum_{x\in B\cap E_n} b_n\Phi_{k-n}(\theta^n(x))s^{k-n}.$$

By Lemma 4.1, it follows that $\lim P(s_l) = \infty$, and hence the second term of the right hand side tends to zero as $l \to \infty$. Since $\lim_{n\to\infty} b_n/b_{n+k} = 1$, we then obtain that the first summand also tends to zero. Moreover, by applying the Portmanteau theorem to the open and closed set $[w]$, it follows that (8) holds for $[w]$. As $\mathcal{B}(\Sigma)$ is generated by cylinders, the lemma follows.   □

As it seems to be impossible to show the existence of a weak*-accumulation point of (m_s) in full generality, the following condition is introduced.

Definition 4.3 We say that the group extension $(\Sigma, \theta, \varphi)$ satisfies property (C) if there exists (b_n) as in Lemma 4.1 and (s_k) with $s_k \searrow \rho$ such that (m_{s_k}) converges weakly* to some probability measure on X_{id} as $k \to \infty$.

In order to obtain criteria for property (C), recall that Prohorov's theorem states that a sequence (m_{s_k}) has a weak*-accumulation point if and only if for each $\epsilon > 0$ there exists a compact set K and $k_0 \in \mathbb{N}$ such that $m_{s_k}(K) \ge 1 - \epsilon$ for all $k \ge k_0$, or in other words, if (m_{s_k}) is tight. In particular, if Σ is a subshift of finite type, then the property is always satisfied. By lifting the limit from Σ to X as in [7, 9] we arrive at a conformal, not necessarily finite measure for T.

Theorem 4.4 *Assume that (Σ, θ) satisfies the b.i.p. property, $\log \varphi$ is Hölder continuous, $\|L_\varphi(1)\|_\infty < \infty$ and that (X, T) be a topologically transitive group extension with property (C). Then there exists a σ-finite, nonatomic, (ρ/φ)-conformal measure ν with $\nu(X_g) < \infty$, for each $g \in G$. Furthermore, there exists a sequence (s_k) with $s_k \searrow \rho$, such that, for each non-negative, continuous function $f : X \to \mathbb{R}$,*

$$\int f d\nu = \lim_{k \to \infty} \frac{1}{\mathcal{P}(s_k)} \sum_{n \in \mathbb{N}} b_n s_k^{-n} (\mathcal{L}_\varphi^n f)(\xi, id). \tag{9}$$

Before giving the proof, recall that the conditions on $(\Sigma, \theta, \varphi)$ are equivalent to the existence of a probability measure μ such that (Σ, θ, μ) is a Gibbs-Markov map with the b.i.p. property. Hence, the above theorem holds in verbatim for topologically transitive group extensions of Gibbs-Markov maps, with μ playing the rôle of a reference measure.

Proof By property (C), there exists $s_k \searrow \rho$ and m such that m is the weak*-limit of $(m_{s_k} : k \in \mathbb{N})$. Using Eq. (8) in Lemma 4.2, we extend m to a measure ν on $\mathcal{B}(X)$ as follows. For $b \in \mathcal{W}^1$ and $g \in G$, there exists by transitivity $j \in \mathbb{N}$ and $u \in \mathcal{W}^{j+1}$ with $T^j([u, id]) = [b, g]$. The restriction of ν on $[b, g]$ is now defined by

$$\int_{[b,g]} f(x) d\nu(x, h) := \int_{[u]} f \circ \theta^j \rho^j / \Phi_j dm,$$

for each bounded and continuous function $f : \Sigma \to \mathbb{R}$. In particular, if f is supported on $[b]$, then by the same arguments as in the proof of Lemma 4.2,

$$\int_{X_g} f(x) d\nu(x, h) = \int_{[u]} f \circ \theta^j \rho^j / \Phi_j dm$$

$$= \lim_{k \to \infty} \frac{1}{\mathcal{P}(s_k)} \sum_{n \in \mathbb{N}} b_n \rho^j s_k^{-n} \sum_{x \in E_n \cap [u]} f \circ \theta^j(x) (\Phi_j(x))^{-1} \cdot \Phi_n(x)$$

$$= \lim_{k \to \infty} \frac{1}{\mathcal{P}(s_k)} \sum_{n > j} b_{n-j} s_k^{j-n} \sum_{(y,g) \in T^{j-n}(\{(\xi, id)\}) \cap [b,g]} \Phi_{n-j}(y) f(y).$$

This proves Eq. (9). Finally, using the construction of ν from m and the big preimages property, it easily can be seen that $\nu(X_g) < \infty$ for each $g \in G$. $\qquad \square$

We now collect several immediate consequences from conformality and the b.i.p.-property in the base.

Proposition 4.5 *For the measure ν given by Theorem 4.4, the following holds.*

1. *If $\lim_n \mathcal{Z}^n(\xi)\rho^{-n} = 0$, then $\nu(X) = \infty$.*
2. *If $L_\varphi(1) = 1$, then $d\nu \circ T^{-1} = \rho^{-1} d\nu$.*
3. *For $w \in \mathcal{W}^n$, $x \in [w]$ and $g \in G$, we have $\rho^n \nu([w, g]) \asymp \Phi_n(x)\nu(X_{g\psi_n(x)})$.*
4. *If the extension is symmetric, then*

$$\nu(X_g) \asymp \nu(X_{g^{-1}}), \quad \nu([w^\dagger, \psi_w^{-1} g^{-1} \psi_w]) \asymp \nu([w, g]).$$

Proof The first assertion follows from (9) applied to $f = 1$. In order to prove part 2, note that $L_\varphi(1) = 1$ implies that $\mathcal{L}_\varphi(1) = 1$. Hence, for $f \in L^1(\nu)$, we have

$$\frac{1}{\mathcal{P}(s)} \sum_{n \in \mathbb{N}} b_n s^{-n} (\mathcal{L}_\varphi^n f \circ T)(\xi, id) = \frac{1}{\mathcal{P}(s)} \sum_{n \in \mathbb{N}} b_n s^{-n} (\mathcal{L}_\varphi^{n-1} f)(\xi, id)$$

$$= \frac{s^{-1}}{\mathcal{P}(s)} \sum_{n \in \mathbb{N}} \frac{b_n}{b_{n-1}} b_{n-1} s^{-n+1} (\mathcal{L}_\varphi^{n-1} f)(\xi, id).$$

Since $\mathcal{P}(s) \nearrow \infty$ as $s \to \rho$ and $\lim_n b_n/b_{n-1} = 1$ as $n \to \infty$, we obtain that $\int f \circ T d\nu = \rho^{-1} \int f d\nu$. Part 3 is a consequence of conformality and the b.i.p. property. Namely, by (2),

$$\rho^n \nu([w, g]) \asymp \Phi_n(x) \nu \left(\theta^n([w]) \times \{g\psi_n(x)\} \right) \leq \Phi_n(x) \nu \left(X_{g\psi_n(x)} \right).$$

Furthermore, by the big images property, there exists $a \in \mathcal{I}_{\text{bip}}$ such that $[a] \subset \theta^n([w])$. Hence, it remains to show that $\nu([a, h]) \asymp \nu(X_h)$ for all $h \in G$. By the big preimages property, for each $y \in \Sigma$, there exists $b \in \mathcal{I}_{\text{bip}}$ such that $y \in \theta([b])$. Hence, by transitivity of T, there exists a finite word w such that awb is admissible and $\psi_{awb} = id$. Hence, $\nu([a, h]) \geq \nu([awb, h]) \asymp \nu(\theta([b]) \times \{h\})$ with respect to a constant only depending on awb, which implies that

$$|\mathcal{I}_{\text{bip}}| \nu([a, h]) \gg \sum_{b \in \mathcal{I}_{\text{bip}}} \nu(\theta([b]) \times \{h\}) \geq \nu(X_h).$$

The proof of the remaining assertion relies on a similar argument. For each $w \in \mathcal{W}^\infty$ with $\psi_w = g$ and $\xi \in \theta^{|w|}([w])$, there exists by transitivity a finite word u such that such that wu is admissible, $\xi \in \theta^{|w|+|u|}([w^\dagger u])$ and $\psi_u = id$. As $\mathcal{I}_{\text{bip}}$ is finite, u can be chosen from a finite set. Hence, by the definition of ν and the symmetry of φ, we have $\nu(X_{g^{-1}}) \ll \nu(X_g)$ which implies that $\nu(X_{g^{-1}}) \asymp \nu(X_g)$. The second assertion follows from this and part 3. $\qquad \square$

5   The Ruelle-Perron-Frobenius Theorem for Group Extensions

In order to prove the existence of eigenfunctions for the Ruelle operator, we make use of a well-known idea from hyperbolic geometry (see [20, 31]): As the reference point for the construction in Theorem 4.4 was chosen arbitrarily, there exists a family $\{\nu_\zeta : \zeta \in X\}$ of conformal measures. It is then relatively easy to show that $\zeta \mapsto d\nu_\zeta/d\nu$ defines an eigenfunction, provided that $\{\nu_\zeta\}$ is a family of pairwise equivalent measures. In here, this approach is partially generalized by constructing a conformal measure ν_μ for a given probability measure μ on X. In order to do so, recall that the Vaserstein distance W of two probability measures $\mu, \tilde{\mu}$ is a metric compatible with the weak convergence and is equal to, by Kantorovich's duality,

$$W(\mu, \tilde{\mu}) = \sup \left\{ \int f d(\mu - \tilde{\mu}) : D(f) \le 1 \right\},$$

where $D(f) := \sup\{|f(\zeta) - f(\tilde{\zeta})|/d(\zeta, \tilde{\zeta}) : \zeta, \tilde{\zeta} \in X\}$ denotes the Lipschitz coefficient with respect to the metric defined by $d((x, g), (y, g)) = d_r(x, y)$ and $d((x, g), (x, h)) = 1$ for $g \ne h$. In the following theorem, ν refers to the σ-finite, conformal measure on X given by Theorem 4.4 with respect to some fixed base point in X_{id}.

Theorem 5.1 *Let (X, T) be a topologically transitive group extension with property (C) of a Gibbs-Markov map with the b.i.p. property. Then there exists a sequence (s_k) with $s_k \searrow \rho$ such that for each $\mu \in \mathcal{M}(X)$,*

$$\nu_\mu := \lim_{k \to \infty} \frac{1}{\mathcal{P}(s_k)} \sum_{n \in \mathbb{N}} b_n s_k^{-n} (\mathcal{L}_\varphi^n)^*(\mu) \tag{10}$$

exists. Furthermore, $\{\nu_\mu : \mu \in \mathcal{M}(X)\}$ is a family of pairwise equivalent measures and for the Radon-Nikodym derivative $\mathbb{K} : \mathcal{P}(X) \times X \to \mathbb{R}$, $(\mu, z) \mapsto (d\nu_\mu/d\nu)(z)$, we have the following.

1. *There exists $D > 0$ such that for ν-a.e. $z \in X$, $g \in G$ and probability measures μ_1, μ_2 supported on X_g,*

$$|\log \mathbb{K}(\mu_1, z) - \log \mathbb{K}(\mu_2, z)| \le DW(\mu_1, \mu_2).$$

2. *For all $\mu \in \mathcal{M}(X)$, we have $\mathbb{K}(\mathcal{L}_\varphi^*(\mu), z) = \rho \mathbb{K}(\mu, z)$ for ν-a.e. z.*
3. *For each $\mu \in \mathcal{M}(X)$, the map $\mathbb{K}(\mu, \cdot)$ is T-invariant, that is $\mathbb{K}(\mu, z) = \mathbb{K}(\mu, T(z))$ for ν-a.e. $z \in X$. In particular, if T is ergodic with respect to ν, then ν_μ is a multiple of ν, $\mathbb{K}(\mu, z)$ is constant with respect to z and $\{\nu_\mu : \mu \in \mathcal{M}(X)\}$ is one-dimensional.*

Remark 5.2 Before giving the proof, we discuss a relation to Ruelle's operator theorem. Namely, by considering the restriction $X \to \mathcal{M}_\sigma(X)$, $x \mapsto \nu_x := \nu_{\delta_x}$, part (ii) of the above implies that $h_z : x \mapsto \mathbb{K}(\delta_x, z)$ satisfies $\mathcal{L}_\varphi(h_z) = \rho h_z$. Hence, the above gives rise to the construction of a family of σ-finite, conformal measures $\{\nu_\mu : \mu \in \mathcal{M}(X)\}$ and a family of eigenfunctions $\{h_z : z \in X\}$. If ν is ergodic, these families are one dimensional, that is, they are subsets of $\{t\nu : t > 0\}$ and $\{(x \to t\nu_x(X_{id}) : t > 0\}$, respectively.

Proof We begin with the construction of ν_μ for the case that μ is a Dirac measure δ_ζ. So assume that $\zeta \in E(\xi) := \bigcup_{n \in \mathbb{N}} T^{-n}(\{(\xi, id)\})$, for $\xi \in \Sigma$ and define, for a non-negative, continuous function $f : X \to \mathbb{R}$ and $s > \rho$,

$$m_\zeta^s(f) := \frac{1}{\mathcal{P}(s)} \sum_{n \in \mathbb{N}} b_n s^{-n} (\mathcal{L}_\varphi^n f)(\zeta),$$

where $\mathcal{P}(s)$ is given by (7). It follows from property (C) that m_ζ^s restricted to functions on X_{id} defines a tight family of measures, and hence that, for a suitable subsequence $(s_{k_j} : j \in \mathbb{N})$ of (s_k) given by property (C),

$$m_\zeta(f) := \lim_{j \to \infty} \frac{1}{\mathcal{P}(s_{k_j})} \sum_{n \in \mathbb{N}} b_n s_{k_j}^{-n} (\mathcal{L}_\varphi^n f)(\zeta) \tag{11}$$

exists for each non-negative, continuous function $f : X \to \mathbb{R}$. In particular, m_ζ defines a measure. Since $E(\xi)$ is countable it is moreover possible to choose the subsequence (s_{k_j}) such that the limit in (11) exists for all $\zeta \in E(\xi)$ and f non-negative and continuous. Moreover, as $\lim_n b_n/b_{n+k} = 1$ for each $k \in \mathbb{N}$, it follows that

$$m_\zeta(f) = \rho^{-k} \sum_{v \in \mathcal{W}^k : \zeta \in \theta^k([v]) \times \{g\}} \Phi_k(\tau_v(\zeta)) m_{\tau_v(\zeta)}(f) = \rho^{-k} \mathcal{L}_\varphi^k(m_\cdot(f)(\zeta)). \tag{12}$$

Hence, for ζ with $T^n(\zeta) = (\xi, id)$, it follows from (12) that, for each Borel set A, $m_{\xi, id}(A) = \nu(A) \geq \rho^{-n} \Phi_n(\zeta) m_\zeta(A)$. On the other hand, it follows from transitivity that there exist $v \in \mathcal{W}^m$ and $m \in \mathbb{N}$ such that $\tau_v(\zeta)$ and (ξ, id) are in the same cylinder. Hence, by combining the above argument with bounded distortion, we obtain that

$$\rho^m \Phi_m(\xi)^{-1} m_\zeta(A) \gg \nu(A) \geq \rho^{-n} \Phi_n(\zeta) m_\zeta(A).$$

In particular, the measures are equivalent and the Radon-Nikodym derivative $\mathbb{K}(\zeta, \cdot) := dm_\zeta/d\nu$ exists and is a.s. strictly positive.

We now prove that $m_{\zeta_1}(A) \asymp m_{\zeta_2}(A)$ whenever the second coordinates coincide, that is $\zeta_1, \zeta_2 \in E(\xi) \cap X_g$ for some $g \in G$. In order to do so, assume that $\zeta_2 \in [a, g]$ for some $a \in \mathcal{I}_{\mathrm{bip}}$. By the b.i.p.-property, there exist $b \in \mathcal{I}_{\mathrm{bip}}$ and $h \in G$ such that $\zeta_1 \in T([b, h])$ and by transitivity a finite word w such that awb is admissible with $\psi_{awb} = id$. As above, it follows that $\Phi_{|awb|}(x) m_{\zeta_2}(A) \ll m_{\zeta_1}(A)$ for any $x \in [awb]$. Hence, as $\mathcal{I}_{\mathrm{bip}}$ is finite, $m_{\zeta_2}(A) \ll m_{\zeta_1}(A)$ with respect to a constant which does not depend on ζ_1 and $a \in \mathcal{I}_{\mathrm{bip}}$.

In order to prove the opposite direction, for each $b \in \mathcal{I}_{\mathrm{bip}}$ choose $x_b \in [b]$. Also note that, for each $v \in \mathcal{W}^1$ with $\zeta_1 \in \theta([v]) \times \{g\}$, there exists $b(v) \in \mathcal{I}_{\mathrm{bip}}$ such that $vb(v)$ is admissible. As $\varphi(\tau_v(\zeta_1)) \asymp \varphi(\tau_v((x_{b(v)}, g)))$, we have by the above that

$$m_{\zeta_1}(A) = \rho^{-1} \sum_{v : \zeta_1 \in \theta([v]) \times \{g\}} \varphi(\tau_v(\zeta_1)) \, m_{\tau_v(\zeta_1)}(A)$$

$$\asymp \rho^{-1} \sum_{v : \zeta_1 \in \theta([v]) \times \{g\}} \varphi(\tau_v((x_{b(v)}, g))) \, m_{\tau_v((x_{b(v)}, g))}(A)$$

$$\leq \sum_{b \in \mathcal{I}_{\mathrm{bip}}} m_{(x_b, g)}(A) \ll |\mathcal{I}_{\mathrm{bip}}| \, m_{\zeta_2}(A).$$

Hence, $m_{\zeta_1}(A) \asymp m_{\zeta_2}(A)$ which implies that

$$\sup\left(\left\{\frac{\mathbb{K}((x_1, g), z)}{\mathbb{K}((x_2, g), z)} \ : \ (x_1, g), (x_2, g) \in E(\xi), g \in G, z \in X\right\}\right) < \infty. \quad (13)$$

In order to extend $\mathbb{K}(\cdot, \cdot)$ to a globally defined function, we now show that $\zeta \mapsto \mathbb{K}(\zeta, z)$ is log Hölder. For $k, n \in \mathbb{N}$, $\zeta_1, \zeta_2 \in [a, g] \cap E(\xi)$, $a \in \mathcal{W}^1$, $b \in \mathcal{W}^k$ with $k \leq n$ and $h \in G$, we obtain by (5) that

$$|\mathcal{L}^n_\varphi(\mathbf{1}_{[b,h]})(\zeta_1) - \mathcal{L}^n_\varphi(\mathbf{1}_{[b,h]})(\zeta_2)| \leq C_\varphi d(\zeta_1, \zeta_2)\mathcal{L}^n_\varphi(\mathbf{1}_{[b,h]})(\zeta_1), \quad (14)$$

with C_φ only depending on the Hölder constant of φ. Hence,

$$|m^s_{\zeta_1}([b, h]) - m^s_{\zeta_2}([b, h])| \leq C_\varphi m^s_{\zeta_1}([b, h])d(\zeta_1, \zeta_2)$$

and, by taking the limit,

$$|m_{\zeta_1}([b, h]) - m_{\zeta_2}([b, h])| \leq C_\varphi m_{\zeta_1}([b, h])d(\zeta_1, \zeta_2).$$

Since cylinder sets are generating the Borel algebra and are stable under intersections it follows by taking the limit as $[b, h] \to z \in X$ that $|(dm_{\zeta_1}/dm_{\zeta_2})(z) - 1| \ll d(\zeta_1, \zeta_2)$ for ν-a.e. $z \in X$. Furthermore, as $\zeta_1, \zeta_2 \in X_g$, it follows from (13) that $dm_{\zeta_1}/dm_{\zeta_2} \asymp 1$. Hence, $|\log(dm_{\zeta_1}/dm_{\zeta_2})(z)| \ll d(\zeta_1, \zeta_2)$, which proves that the function $\zeta \mapsto \log \mathbb{K}(\zeta, z)$ is Lipschitz continuous on $E(\xi) \cap [a, g]$ with respect to a Lipschitz coefficient which is independent from z and $[a, g]$. By a further application of (13), there is a uniform bound for $|\log(dm_{\zeta_1}/dm_{\zeta_2})(z)|$ which is independent from z and g. As $E(\xi)$ is dense by transitivity, there exists a unique locally Lipschitz continuous extension of $\zeta \mapsto \log \mathbb{K}(\zeta, z))$ to X. By taking the exponential of this extension, we obtain a globally defined function which, for ease of notation, will also be denoted by $\mathbb{K}(\cdot, \cdot)$. As the function has the same regularity as the one defined on $E(\xi)$, we have shown that there exists $D > 0$ such that, for all $g \in G$, $\zeta_1, \zeta_2 \in X_g$ and ν-a.e. $z \in X$,

$$|\log \mathbb{K}(\zeta_1, z) - \log \mathbb{K}(\zeta_2, z)| \leq Dd(\zeta_1, \zeta_2).$$

In order to obtain the representation (10), note that the construction of m_ζ through (11) extends to all $\zeta \in X$ by the estimate (14) and the fact that $E(\xi)$ is dense in X. The next step is to verify that (11) extends to an arbitrary Borel probability measure μ on X. In analogy to the above, define

$$M_\mu^s(f) := \frac{1}{\mathcal{P}(s)} \sum_{n\in\mathbb{N}} b_n s^{-n} \int f d(\mathcal{L}_\varphi^n)^*(\mu)$$

$$= \frac{1}{\mathcal{P}(s)} \sum_{n\in\mathbb{N}} b_n s^{-n} \int \mathcal{L}_\varphi^n(f) d\mu = \int m_\zeta^s(f) d\mu(\zeta),$$

where the last equality follows from monotone convergence. By a further application of monotone convergence, it follows that $\lim_k M_\mu^{s_k}(f) = \int m_\zeta(f) d\mu$ which proves that (11) defines a measure and that $\nu_\mu(f) := \int m_\zeta(f) d\mu$. Moreover, as

$$\int f d\nu_\mu = \int m_\zeta(f) d\mu = \iint f(z) \mathbb{K}(\zeta, z) d\nu(z) d\mu(\zeta), \tag{15}$$

it follows that $d\nu_\mu/d\nu = \int \mathbb{K}(\zeta, z) d\mu(\zeta)$, which will also be denoted by $\mathbb{K}(\mu, z)$, by a slight abuse of notation. This finishes the proof of the existence of ν_μ. Part 1 of the theorem then follows from the definition of W through Kantorovich's duality.

In order to prove part 2, note that (10) implies that, for $\zeta \in X$ and each positive and continuous function f, that

$$\int f(z) \mathbb{K}(\zeta, z) d\nu(z) = \int f d\nu_\zeta = \lim_{k\to\infty} \frac{1}{\mathcal{P}(s_k)} \sum_{n\in\mathbb{N}} b_n s_k^{-n} (\mathcal{L}_\varphi^n(f))(\zeta)$$

$$= \sum_{v\in W} \varphi \circ \tau_v(\zeta) \lim_{k\to\infty} \frac{1}{\mathcal{P}(s_k)} \sum_{n\in\mathbb{N}} b_n s_k^{-n} (\mathcal{L}_\varphi^{n-1}(f))(\tau_v(\zeta))$$

$$= \sum_{v\in W} \varphi \circ \tau_v(\zeta) \rho^{-1} \int f d\nu_{\tau_v(\zeta)} = \sum_{v\in W} \varphi \circ \tau_v(\zeta) \rho^{-1} \int f \mathbb{K}(\tau_v(\zeta), \cdot) d\nu$$

$$= \rho^{-1} \int f(z) \mathcal{L}_\varphi(\mathbb{K}(\cdot, z))(\zeta) d\nu(z) \tag{16}$$

where the last identity follows from monotone convergence. Hence, by (15),

$$\rho \int f \mathbb{K}(\mu, \cdot) d\nu = \int f(z) \int \mathcal{L}_\varphi(\mathbb{K}(\cdot, z))(\zeta) d\mu(\zeta) d\nu(z)$$

$$= \int f(z) \int \mathbb{K}(\cdot, z) d\mathcal{L}_\varphi^*(\mu) d\nu(z) = \int f \mathbb{K}(\mathcal{L}_\varphi^*(\mu), \cdot) d\nu.$$

As f is arbitrary, $\rho \mathbb{K}(\mu, z) = \mathbb{K}(\mathcal{L}_\varphi^*(\mu), z)$ almost surely, which is part 2 of the theorem.

For the proof of part 3, note that X is a Besicovitch space and that each ν_μ is conformal. Therefore, we have for ν-a.e. $((w_i), g) \in X$, that

$$\mathbb{K}(\mu, ((w_i), g)) = \lim_{n\to\infty} \frac{\nu_\mu([(w_1 \dots w_n), g])}{\nu([(w_1 \dots w_n), g])} = \lim_{n\to\infty} \frac{\rho^{-1} \int_{[(w_2\dots w_n), g\psi_{w_1}]} \varphi d\nu_\mu}{\rho^{-1} \int_{[(w_2\dots w_n), g\psi_{w_1}]} \varphi d\nu}.$$

It hence follows from continuity of φ that $\mathbb{K}$ is T-invariant in the second coordinate. The second statement is a standard application of the ergodic theorem. $\qquad\square$

We now give a brief characterization of the measures given by above theorem in case of an ergodic extension (as e.g. in Example 1 below for $d = 1, 2$). For ease of exposition, we assume that the base transformation is a Gibbs-Markov map with respect to the invariant probability μ on Σ. In this situation, the product measure μ_G of μ and the counting measure on G clearly is $1/\varphi$-conformal and T-invariant, i.e. $\mu_G = \mu_G \circ T^{-1}$. However, note that μ_G in many cases is totally dissipative, e.g. if G is non-amenable [14, 34].

If T is conservative with respect to μ_G, then T also is ergodic and $\sum_n \mathcal{Z}_w^n(\xi) = \infty$ (see [1, 3] and the proof of Proposition 5.3 below). In particular, $\rho = 1$ and G is amenable by a result in [30]. Since μ_G and ν_ζ are both $1/\varphi$-conformal, as observed by Sullivan, $d\nu/d\mu_G$ exists, is T-invariant and hence constant. This then implies that the measures ν_ξ are all multiples of the product measure μ_G. If T is conservative with respect to ν and $\rho \leq 1$, then the same arguments show that the measures ν_ξ are again all multiples of ν. In this situation, a result by Jaerisch [14] shows that the invariant measure $h(x, z)d\nu(x)$ is unique and is the product of another measure on Σ and counting measure on G.

As a corollary of the existence of $\rho^{-1}\mathcal{L}_\varphi$-invariant functions as shown in Remark 5.2, one obtains the following criterion of classical flavor for ergodicity.

Proposition 5.3 *The map T is either conservative or totally dissipative with respect to ν. If T is conservative, then T is ergodic. Furthermore, T is conservative and ergodic if and only if*

$$\sum_n \rho^{-n} \mathcal{Z}^n(\xi) = \infty.$$

Proof Observe that T is a transitive topological Markov chain and that it follows from

$$(d\nu/d\nu \circ T)(x, g) = \rho\phi(x)$$

that $d\nu/d\nu \circ T$ is a potential of bounded variation. Hence, (T, ν) is a Markov fibered system with the bounded distortion property as in [3]. In particular, (T, ν) either is totally dissipative or conservative and if (T, ν) is conservative, then it is ergodic. Note that $\rho^{-1}\mathcal{L}_\varphi$ acts as the transfer operator on $L^1(\nu)$. It hence follows from the definition of the transfer operator that, for all W measurable and $n \in \mathbb{N}$,

$$\int \mathbf{1}_W \rho^{-n} \mathcal{L}_\varphi^n(x, g) d\nu(x, g) = \int \mathbf{1}_W \circ T^n \mathbf{1}_{X_{id}} d\nu = \nu\left(T^{-n}(W) \cap X_{id}\right).$$

Now assume that $\sum_n \rho^{-n} \mathcal{Z}^n(\xi) = \infty$. It follows from bounded variation and transitivity that the sum diverges for all $\xi \in \Sigma$. For $W := \{z \in X_{id} : T^n(z) \notin X_{id} \forall n \geq 1\}$, we hence have that $\nu(W) = 0$. Hence, the first return map

$$T_{X_{id}} : X_{id} \to X_{id}, \ (x, id) \mapsto T^{n_x}(x, g), \ n_x := \min\{n \geq 1 : T^n(x, id) \in X_{id}\}$$

is well defined. By substituting ν with an equivalent, invariant measure given by the above theorem, an application of Poincaré's recurrence theorem gives that $T_{X_{id}}$ is conservative. It is then easy to see that T also is conservative, and hence ergodic. The remaining assertion is a consequence of the standard result in ergodic theory, that T is ergodic and conservative if and only if $\sum_n \rho^{-n} \mathcal{L}_\varphi^n(f)$ diverges for all $f \geq 0$, $\int f \, d\nu > 0$ (see [1, Proposition 1.3.2]). $\square$

6 Harmonic Functions

By applying Theorem 5.1 to Dirac measures, it is possible to construct a map $\Theta : \mathcal{C} \to \mathcal{H}$ from a subspace of the continuous functions to a subspace of ρ-harmonic functions. In here, we refer to $f : X \to \mathbb{R}$ as ρ-*harmonic* if $\mathcal{L}_\varphi(f) = \rho f$. In order to define $\mathcal{C}$, fix a reference point $\xi_0 \in X_{id}$ and set $\nu_{\mathbf{0}} := \nu_{\xi_0}$. The space $\mathcal{C}$ is now defined by

$$\mathcal{C} := \left\{ f : X \to \mathbb{R} : \nu_{\mathbf{0}}(|f|) < \infty, \lim_{n \to \infty} C_n(f) = 0 \right\}, \text{ where}$$

$$C_n(f) := \inf\left(\{C : |f(z_1) - f(z_2)| \leq C|f(z_1)| \forall z_1, z_2 \in [w, g], w \in \mathcal{W}^n, g \in G\}\right).$$

The space might be alternatively characterised as the space of log-uniformly continuous functions with an integrability condition. Namely, if $C_n(f) < \infty$, then for $[w, g], w \in \mathcal{W}^n$ and $g \in G$ either $f|_{[w,g]} = 0$ or $f(z) \neq 0$ for $z \in [w, g]$. In particular, with $0/0 = 1$, it follows that

$$|f(z_1)/f(z_2) - 1| < C_n(f), \quad \forall z_1, z_2 \in [w, g].$$

Hence, if $C_n(f) < 1$, then $f(z_1)/f(z_2) > 0$, that is the sign of f is constant on $[w, g]$. These arguments show that $f \in \mathcal{C}$ if and only if $\log f_+$ and $\log f_-$ are uniformly continuous, with $f_\pm$ referring to the strictly positive and negative parts of f and $\{f \neq 0\}$ is a union of cylinders of length n, for some n depending on f.

In order to define $\mathcal{H}$, recall that d_r refers to the shift metric on X, with $r \in (0, 1)$ adapted to the Hölder continuity of $\log \varphi$. In order to be able to not only consider positive ρ-harmonic functions, the following coefficients for the local regularity of a function $f : X \to \mathbb{R}$ are useful.

$$D_x(f) := \sup \{|f(x) - f(z)|/d_r(x, z) : d_r(x, z) < 1\}$$

$$LD(f) := \sup \left\{ \left|\frac{f(z_1)}{f(z_2)} - 1\right| / d_r(z_1, z_2) : d_r(z_1, z_2) < 1 \right\}$$

The space $\mathcal{H}$ is now defined through a control of the local Lipschitz constant $D_x(f)$ as follows.

$$\mathcal{H}^+ := \left\{ (f : X \to [0, \infty)) : \mathcal{L}_\varphi(f) = \rho f, \ LD(f) < \infty \right\},$$
$$\mathcal{H} := \left\{ (f : X \to \mathbb{R}) : \mathcal{L}_\varphi(f) = \rho f, \ \exists h \in \mathcal{H}^+ \text{ s.t. } D_x(f) \le h(x) \forall x \in X \right\}.$$

The map Θ is then defined by, for $f \in \mathcal{C}$,

$$\Theta(f)(z) := \nu_z(f) = \int \mathbb{K}(\delta_z, y) f(y) d\nu_0(y).$$

Based on a slightly more involved version of the argument used in the proof of log-Hölder continuity of $\mathbb{K}$ in Theorem 5.1 we are now in position to prove that Θ is well defined and that LD is always bounded by

$$C_\varphi := \sup \left\{ \left| \frac{\Phi_n \circ \tau_v(z_1)}{\Phi_n \circ \tau_v(z_2)} - 1 \right| \Big/ d_r(z_1, z_2) : d_r(z_1, z_2) < 1 \right\} < \infty.$$

Theorem 6.1 *The map* $\Theta : \mathcal{C} \to \mathcal{H}$ *is well defined. If* $f \in \mathcal{H}$ *and* $f \ge 0$, *then* $LD(f) \le C_\varphi$ *and, in particular,* $f \in \mathcal{H}^+$.

Proof Suppose that $f \in \mathcal{C}$. By applying the arguments in (16) to f shows that $\mathcal{L}_\varphi(\Theta(f)) = \rho\Theta(f)$. Hence, it remains to obtain a bound on $D_x(f)$. For ease of notation, set $f_v := f \circ \tau_v$, for $v \in \mathcal{W}^n$ and $n \in \mathbb{N}$. Suppose that $z_1, z_2 \in [w, g]$ with $w \in \mathcal{W}^k$, $g \in G$ and that n is sufficiently large such that for all $v \in \mathcal{W}^n$, either $f_v(z_1) = f_v(z_2) = 0$ or $f_v(z_1), f_v(z_2) \ne 0$. Setting $0/0 := 1$ and $A_n := \sup_{v \in \mathcal{W}^n} \left| \frac{f_v(z_1)}{f_v(z_2)} - 1 \right|$, we obtain by a similar argument as in (5) that

$$\left| \mathcal{L}_\varphi^n(f)(z_1) - \mathcal{L}_\varphi^n(f)(z_2) \right|$$
$$\le \sum_{v \in \mathcal{W}^n} \left| (\Phi_{n,v}(z_1) - \Phi_{n,v}(z_2)) f_v(z_1) \right| + \sum_{v \in \mathcal{W}^n} \left| \Phi_{n,v}(z_2) (f_v(z_1) - f_v(z_2)) \right|$$
$$\le \sum_{v \in \mathcal{W}^n} \left| \left(1 - \frac{\Phi_{n,v}(z_2)}{\Phi_{n,v}(z_1)} \right) \Phi_{n,v}(z_1) f_v(z_1) \right| + \sum_{v \in \mathcal{W}^n} \left| \Phi_{n,v}(z_2) f_v(z_2) \left(\frac{f_v(z_1)}{f_v(z_2)} - 1 \right) \right|$$
$$\le C_\varphi d_r(z_1, z_2) \cdot \mathcal{L}_\varphi^n(|f|)(z_1) + A_n \cdot \mathcal{L}_\varphi^n(|f|)(z_2).$$

Since $\lim_{n \to \infty} A_n = 0$, we have

$$|\Theta(f)(z_1) - \Theta(f)(z_2)| = |\nu_{z_1}(f) - \nu_{z_2}(f)|$$
$$\le C_\varphi d_r(z_1, z_2) \nu_{z_1}(|f|) = C_\varphi d_r(z_1, z_2) \Theta(|f|)(z_1).$$

Hence, $D_{z_1}(f) \le C_\varphi \Theta(|f|)(z_1)$. By dividing with $\Theta(|f|)$ and substituting f with $|f|$, the same argument shows that $LD(\Theta(|f|)) \le C_\varphi$. In particular, $\Theta(|f|) \in \mathcal{H}^+$. Now assume that $f \in \mathcal{H}$ and $f \ge 0$. Then there exists $\hat{h} \ge 0$ with $D_z(h) \le \hat{h}$ and $\mathcal{L}_\varphi(\hat{h}) = \rho\hat{h}$. By similar arguments,

$$|h(z_1) - h(z_2)| = \rho^{-n} \left| \mathcal{L}_\varphi^n(h)(z_1) - \mathcal{L}_\varphi^n(h)(z_2) \right|$$
$$\leq \rho^{-n} \left(C_\varphi d_r(z_1, z_2) \mathcal{L}_\varphi^n(h)(z_1) + r^n d_r(z_1, z_2) \mathcal{L}_\varphi^n(D.(h))(z_2) \right)$$
$$\leq C_\varphi d_r(z_1, z_2) \left(h(z_1) + r^n \hat{h}(z_2) \right).$$

Since n is arbitrary and $r \in (0, 1)$, $LD(h) < C_\varphi$. □

The classical Martin boundary of a random walk on a group is a quotient of the space of paths, where two paths (g_k), (h_k) in G are identified if $\lim_k K(\cdot, g_k) = \lim_k K(\cdot, h_k)$, where K refers to the Martin kernel (see, e.g., [33]). In the context of group extensions, the natural candidate for a path in G is given by $(\psi_k(x))$, for some $x \in \Sigma$, whereas the function $(z, g) \mapsto \nu_z(X_g)/\nu_0(X_g)$ might serve as the analogue of the Martin kernel.

Here, the situation is different. Assume that $(x, g) = ((w_k), g) \in X$. Using the conformality of ν in Proposition 4.5, we have by Theorem 5.1 that, for $f_n := \mathbf{1}_{[w_1 \ldots w_n, g]}/\nu_0([w_1 \ldots w_n, g])$,

$$\frac{\nu_z(X_{g\psi_n(x)})}{\nu_0(X_{g\psi_n(x)})} \asymp \frac{\nu_z([w_1 \ldots w_n, g])}{\nu_0([w_1 \ldots w_n, g])} = \Theta(f_n)(z) \xrightarrow{n \to \infty} \mathbb{K}(z, (x, g)).$$

6.1 Natural Extensions and Immediate Implications

In order to obtain information on the asymptotic behavior of elements of $\mathcal{H}$, we now employ ideas from the theory of Markov processes, which are similar but somehow dual to the ones for Markov maps. Namely, in order to obtain a stochastic process associated with (X, T), we consider the process with transition probability $(dm \circ \tau_v/dm)(x)$ for transitions from x to $\tau_v(x)$, where m is an T-invariant measure. Hence, the appropriate object are the left-infinite sequences with respect to an invariant measure $\widehat{m}$ constructed from m. That is, the stochastic process is the left half of the natural extension of (X, T, m) whose construction in case of an underlying shift space we recall now. Set

$$Y := \left\{ ((w_i, g_i) : i \in \mathbb{Z}) : w_i \in \mathcal{W}^1, g_i \in G, a_{w_i w_{i+1}} = 1, g_{i+1} = g_i \psi(w_i) \right\},$$
$$S : Y \to Y, \; ((w_i, g_i)) \to ((w_i', g_i')), \text{ with } w_i' = w_{i+1}, \; g_i' = g_{i+1} \; \forall i \in \mathbb{Z}.$$

In other words, S is the left shift on the two sided shift space Y. The cylinder sets of Y are given by, for $(w_0 w_1 \cdots w_n) \in \mathcal{W}^{n+1}$, $h_i \in G$ and $k \in \mathbb{Z}$,

$$[((w_0, g_0) \cdots (w_n, g_n))]_k$$
$$:= \left\{ ((v_i, h_i)) \in Y : (v_{k+j}, h_{k+j}) = (w_j, g_j), \text{ for } j = 0, 1, \ldots, n \right\}.$$

If m is T-invariant, then $\widehat{m}([((w_0, g_0) \cdots (w_n, g_n))]_k) := m([(w_0 w_1 \cdots w_n), g_0])$ defines a measure $\widehat{m}$ on Y. As it easily can be seen, we then have, for

$$\pi : Y \to X, ((w_i, g_i)) \to ((w_i : i \geq 0), g_0),$$

that $\pi \circ S = T \circ \pi$, $\widehat{m} = m \circ \pi^{-1}$, S is invertible, $\widehat{m}$ is S-invariant and $(Y, S, \widehat{m})$ is minimal in the sense that the σ-algebra $\mathcal{F}$ generated by the cylinder sets of Y is generated by $\{S^n(\pi^{-1}(\mathcal{B})) : n \in \mathbb{Z}\}$, with $\mathcal{B}$ referring to the σ-algebra generated by the cylinder sets of X. In particular, $(Y, S, \mathcal{F}, \widehat{m})$ is the natural extension of $(X, T, \mathcal{B}, m)$ (see, e.g., [6]).

Observe that there are several canonical choices for the invariant measure m. Either μ is θ-invariant and m is the product μ_G of μ and the counting measure on G, or $dm = h d\nu_0$, for some $h \in \mathcal{H}^+$. However, in both cases, it is possible to identify martingales with respect to the filtration $(\mathcal{F}_n : n \in \mathbb{N})$, where $\mathcal{F}_n := S^n \circ \pi^{-1}(\mathcal{B})$. We begin with the analysis of (Y, S) with respect to $\widehat{\mu}_G$.

Proposition 6.2 *Suppose that μ is θ-invariant and that $h \in \mathcal{H}^+$. Then, for $\widehat{\mu}_G$-a.e. $z \in Y$,*

$$h_\infty(z) := \lim_{n \to \infty} \rho^{-n} h \circ \pi \circ S^{-n}(z)$$

exists. If $\rho < 1$, then $h_\infty = 0$, and if $\rho = 1$, then $h_\infty = h_\infty \circ S$ and $h_\infty < \infty$ a.s.

Proof Set $W_n := \rho^{-n} h \circ \pi \circ S^{-n}$. Since $\widehat{\mu}_G$ is S-invariant, $\int f \circ T g d\mu_G = \int f \mathcal{L}(g) d\mu_G$ and $\mathcal{L}_\varphi(h) = \rho h$, we have for all $A \in \mathcal{B}$ that

$$\int_{S^n(\pi^{-1}(A))} \mathbb{E}(W_{n+1}|\mathcal{F}_n) d\widehat{\mu}_G$$

$$= \rho^{-n-1} \int \mathbf{1}_A \circ \pi \circ S^{-n} \, h \circ \pi \circ S^{-n-1} d\widehat{\mu}_G$$

$$= \rho^{-n-1} \int \mathbf{1}_A \circ T \, h d\mu_G = \rho^{-n} \int \mathbf{1}_A h d\mu_G$$

$$= \rho^{-n} \int \mathbf{1}_A \circ \pi \circ S^{-n} h \circ \pi \circ S^{-n} d\widehat{\mu}_G = \int_{S^n(\pi^{-1}(A))} W_n d\widehat{\mu}_G.$$

Hence, $\mathbb{E}(W_{n+1}|\mathcal{F}_n) = W_n$ and $(W_n, \mathcal{F}_n)$ is a positive martingale. In particular, $h_\infty := \lim_n W_n$ by Doob's convergence theorem. As it easily can be verified, we have $h_\infty = \rho h_\infty \circ S$. Furthermore, by Fatou's Lemma and the martingale property, $\int_{\pi^{-1}A} h_\infty d\widehat{\mu} \leq \int_A h d\mu$ for all measurable sets $A \subset X$, which implies that $h_\infty < \infty$ a.s. $\square$

By applying the proposition to $\Theta(\mathbf{1}_{X_{id}})$, we obtain the decay of ν along μ-a.s. path as $n \to -\infty$. If the extension is symmetric, the result also transfers to paths with $n \to \infty$.

Corollary 6.3 *If $\rho < 1$ and μ_G is invariant, then, for $\widehat{\mu}_G$-a.e. $((w_i), g) \in Y$,*

$$\lim_{n\to\infty} \nu_{\mathbf{0}}(X_{g\psi_{w_{-n}}\cdots\psi_{w_{-1}}})/\rho^n = 0, \quad \lim_{n\to\infty} \frac{\nu_{\mathbf{0}}([w_{-n}\dots w_{-1}, g])}{\mu([w_{-n}\dots w_{-1}])} = 0.$$

Moreover, if the group extension is symmetric, then for μ-a.e. $x \in \Sigma$ and $g \in G$,

$$\lim_{n\to\infty} \nu_{\mathbf{0}}(X_{\psi_n(x)})/\rho^n = 0, \quad \lim_{n\to\infty} \frac{\nu_{\mathbf{0}}([w_1\dots w_n, \psi_n(x)g\psi_n(x)^{-1}])}{\mu([w_1\dots w_n])} = 0.$$

Proof The first two assertions follow from $\nu_{(x,g)}(X_{id}) \asymp \nu_{\mathbf{0}}(X_{g^{-1}})$ and (iii) of Proposition 4.5, whereas the last two assertions are a consequence of the fact that $Y \to Y$, $((w_i), g) \mapsto ((w^\dagger_{-i}), g)$ is a non-singular automorphism.   □

By considering the natural extension of the invariant version of $\nu_{\mathbf{0}}$, we obtain a further convergence. That is, as the measure $dm_h := h d\nu_{\mathbf{0}}$ is T-invariant, there exists a unique extension to an invariant, σ-finite, S-invariant measure $\widehat{m}_h$ on Y. The analogue of Proposition 6.2 is as follows.

Proposition 6.4 *Suppose that $f, h \in \mathcal{H}$, $h > 0$ such that $\| f/h\|_\infty < \infty$. Then, for $\widehat{m}_h$-a.e. $z \in Y$,*

$$\Xi_h(f)(z) := \lim_{n\to\infty} \frac{f \circ \pi \circ S^{-n}(z)}{h \circ \pi \circ S^{-n}(z)}$$

exists, $\Xi_h(f) \circ S = \Xi_h(f)$ and $\mathbb{E}_{\widehat{m}_h}(\Xi_h(f)|\mathcal{F}_0) = f \circ \pi/h \circ \pi$. Moreover, for the signed invariant measure $\widehat{m}_f$, we have $d\widehat{m}_f/d\widehat{m}_h = \Xi_h(f)$.

Proof The proof that $(f \circ \pi \circ S^{-n}/h \circ \pi \circ S^{-n}|\mathcal{F}_n)$ is a bounded martingale and is the same as above and therefore omitted. Hence, $\Xi_h(f)$ is well defined and by bounded convergence, we have for $A \in \mathcal{B}$ and $k \in \mathbb{N}$ that

$$\int_{S^k\pi^{-1}(A)} \Xi_h(f) d\widehat{m}_h$$

$$= \lim_{n\to\infty} \int \mathbf{1}_A \circ \pi \circ S^{-k} \frac{f \circ \pi \circ S^{-n}}{h \circ \pi \circ S^{-n}} d\widehat{m}_h$$

$$= \lim_{n\to\infty} \int \mathbf{1}_A \circ T^{n-k} f d\nu_{\mathbf{0}} = \int \mathbf{1}_A f d\nu_{\mathbf{0}} = m_f(A) = \widehat{m}_f(S^k\pi^{-1}(A)).$$

Since $\mathcal{F}$ is generated by $\{\mathcal{F}_n : n \in \mathbb{N}\}$, we have $\Xi_h(f) d\widehat{m}_h = d\widehat{m}_f$. The remaining assertion in the conditional expectation is a consequence of the above for $k = 0$. □

7 Applications and Examples

The construction of conformal measures has the following application to conformal graph directed Markov systems. In order to have a zero of the pressure function, we have to assume that there exists $h > 0$ such that

$$\limsup_{n \to \infty} \sqrt[n]{\mathcal{L}_{\varphi^h}^n(\mathbf{1}_{X_{id}})(\xi, id)} = 1, \tag{17}$$

$$\|L_{\varphi^h}(\mathbf{1})\|_\infty < \infty \tag{18}$$

are satisfied. It follows from standard arguments that the expression on the left hand side of (17), seen as a function of h, is continuous and strictly decreasing to 0 on its domain of definition. Hence, if there exists h' such that the left hand side of (17) is finite and greater than or equal to 1 and (18) holds, then there exists a zero of the pressure function. In the context of graph directed Markov systems, this property is known as strong regularity (see [19]). Furthermore, if $|\mathcal{W}^1| < \infty$, then this is true for $h' = 0$, and in particular there always exists a zero of the pressure function in this case.

Now let δ be given by (17) and set $\rho_\delta := \exp(P_G(\theta, \varphi^\delta)) \geq 1$. It then follows from the Ruelle-Perron-Frobenius theorem for systems with the b.i.p. property (see, e.g., [27]) that there exists a $\rho_\delta/\varphi^\delta$-conformal probability measure μ_δ and a Hölder continuous function h_δ with $L_{\varphi^\delta}(h_\delta) = \rho_\delta h_\delta$ such that θ has the Gibbs-Markov property with respect to the invariant measure given by $h_\delta d\mu_\delta$. As an application of Theorem 4.4 and Proposition 4.5 we obtain that there exists a σ-finite measure ν on X which is $1/\varphi^\delta$-conformal, and which satisfies, for $w \in \mathcal{W}^n$ and $x \in [w]$,

$$\nu([w, g]) \asymp \Phi_n^\delta(x)\nu(X_{g\psi_n(x)}). \tag{19}$$

Theorem 7.1 *Assume that the group extension is symmetric and that property (C), (17) and (18) are satisfied. Then, for μ_δ-a.e. $(w_k) \in \Sigma$,*

$$\lim_{n \to \infty} \frac{\log(\nu([w_1 \cdots w_n, id]))}{\log \Phi_n(x)} = \delta + \frac{P_G(\theta, \varphi^\delta)}{\int (\log \varphi)h_\delta d\mu_\delta}.$$

Moreover, the group G is amenable if and only if the above limit is equal to δ. If G is non-amenable, then, for μ_δ-a.e. $(w_k) \in \Sigma$,

$$\lim_{n \to \infty} \rho_\delta^n \frac{\nu([w_1 \cdots w_n, id]))}{(\Phi_n(x))^\delta} = 0.$$

Before giving the proof, we sketch a straight forward application to conformal dynamical systems. Namely, if Σ is given by a conformal iterated function system, the inverse branch τ_w corresponds to a conformal map and $\Phi_{|w|} \circ \tau_w$ to its conformal derivative. In this situation, the above limit can be identified with the ν-dimension $\dim_\nu$ of the support of μ_δ. Hence, with $H(h_\delta d\mu_\delta)$ referring to the entropy of $h_\delta d\mu_\delta$, it follows from the variational principle that

$$\dim_\nu(\mathrm{supp}(\mu_\delta)) = \delta + \frac{P_G(\theta, \varphi^\delta)}{\int (\log \varphi)h_\delta d\mu_\delta} = 2\delta + \frac{H(h_\delta d\mu_\delta)}{\int (\log \varphi)h_\delta d\mu_\delta}.$$

Moreover, note that in many regular situations, δ is equal to the Hausdorff dimension $\dim(K)$ of the attractor K of the iterated function system. In this situation, the amenability of G is equivalent to $\dim_\nu(\text{supp}(\mu_\delta)) = \dim(K)$.

Proof of Theorem 7.1 By symmetry and Proposition 6.2, $\lim_n (\log \nu(X_{\psi_n(x)}))/n = \log \rho^\delta$. Hence, by (19),

$$\lim_{n \to \infty} \frac{\log(\nu([w_n, id]))}{\log \Phi_n(x)} = \delta + \lim_{n \to \infty} \frac{\log \nu(X_{\psi_n(x)})}{\log \Phi_n(x)} = \delta + \frac{P_G(\theta, \varphi^\delta)}{\lim_{n \to \infty}(\log \Phi_n(x))/n}.$$

The above limit exists by application of the ergodic theorem. The amenability criterion is an immediate corollary of Kesten's criterion for group extensions in [30], where it is shown that $P_G(\theta, \varphi^\delta) = 0$ if and only if G is amenable. For the remaining assertion, note that $\rho_\delta < 1$ by non-amenability. The assertion then follows from Corollary 6.3. □

In order to have concrete examples of the σ-finite measure at hand, we give two examples from probability theory, where known local limit theorems give rise to explicit expressions.

Example 1 The first example is Polya's random walk on $\mathbb{Z}^d$. Choose $(p_i \in (0, 1) : i \in \{\pm 1, \ldots, \pm d\})$ with $\sum_{i=1}^d (p_i + p_{-i}) = 1$ and consider the random walk on $\mathbb{Z}^d$ with transition probabilities $P(\pm e_i) = p_{\pm i}$, where e_i refers to the i-th element of the canonical basis of $\mathbb{Z}^d$.

This random walk has an equivalent description through the following group extension. Let Σ be the full shift with $2d$ symbols $\{-d, \ldots, -1, 1, \ldots, d\}$ and φ the locally constant function defined by $\varphi|_{[\pm i]} := p_{\pm i}$. Note that $\sum_{i=1}^d (p_i + p_{-i}) = 1$ implies that $L_\varphi(1) = 1$. Moreover, it is well known that the measure defined by $\mu([i_1 \ldots i_n]) := p_{i_1} \cdots p_{i_n}$ is θ-invariant, ergodic and $1/\varphi$-conformal. The associated group extension is defined through

$$\psi : \Sigma \to \mathbb{Z}^d, \ (i_1 i_2 \cdots) \mapsto \begin{cases} e_{i_1} & : i_1 > 0 \\ -e_{-i_1} & : i_1 < 0 \end{cases}.$$

As Σ is the full shift and φ is constant on cylinders, it follows from the construction that $\nu_{(x,g)} = \nu_{(y,g)}$ for all $x, y \in \Sigma$ and $g \in G$. Therefore, we only will write ν_g for $\nu_{(x,g)}$. In order to apply known local limit theorems from probability theory, observe that

$$\mathcal{L}_\varphi^n(\mathbf{1}_{X_{id}})(x, g) = \sum_{w \in \mathcal{W}^n: \ \psi_n(w) = g} \phi_n(\tau_w(x)) = P(X_n = g),$$

where $X_n = h$ refers to the random walk at time n started in the identity with distribution (p_i) and P to the probability of the associated Markov process. By the local limit theorem for Polya's random walk ([32, Theorem 13.12]), we have that, for $(k_1, \ldots, k_d) \in \mathbb{Z}^d$ and $n \in \mathbb{N}$ such that $n - (k_1 + \cdots + k_d)$ is even,

$$P(X_n = (k_1, \ldots, k_d)) \sim Cn^{-d/2} \left(2 \sum_{i=1}^{d} \sqrt{p_i p_{-i}}\right)^n \prod_{i=1}^{d} \left(\sqrt{p_i/p_{-i}}\right)^{k_i}.$$

Hence, $\rho = 2 \sum_{i=1}^{d} \sqrt{p_i p_{-i}}$ and, with $\lambda_i := \sqrt{p_i/p_{-i}}$,

$$\mathcal{L}_\varphi^n(\mathbf{1}_{X_{(k_1,\ldots,k_d)}})(x, id) \sim Cn^{-d/2} \rho^n \prod_{i=1}^{d} \lambda_i^{-k_i}.$$

Recall that a random walk is called symmetric if $p_i = p_{-i}$ for all $i = 1, \ldots, d$. The estimate then implies that $\rho = 1$ if and only if the random walk is symmetric. Furthermore, by Proposition 5.3, the term $n^{-d/2}$ implies that the group extension is ergodic and conservative with respect to ν if and only if $d = 1$ or $d = 2$. It is remarkable that this conclusion is independent of symmetry. In order to determine ν_{id} explicitly, note that the local limit theorem implies that

$$\nu_{id}(X_{(k_1,\ldots,k_d)}) = \lim_{k \to \infty} \frac{\sum_{n \in \mathbb{N}} b_n s_k^{-n}(\mathcal{L}_\varphi^n \mathbf{1}_{X_{(k_1,\ldots,k_d)}})(x, id)}{\sum_{n \in \mathbb{N}} b_n s_k^{-n}(\mathcal{L}_\varphi^n \mathbf{1}_{X_{id}})(x, id)} = \prod_{i=1}^{d} \lambda_i^{-k_i}.$$

Using conformality then gives that, for a cylinder $[(i_1, \ldots, i_n), z]$ in $\Sigma \times \mathbb{Z}^d$,

$$\nu_{id}([(i_1 \ldots i_n), z]) = \rho^{-n} p_{i_1} \cdots p_{i_n} \nu_{id}(X_{z+\psi_n(i_1 \ldots i_n)})$$

$$= \rho^{-n} p_{i_1} \cdots p_{i_n} \nu_{id}(X_z) \prod_{k=1}^{n} \lambda_{i_k}^{-1} = \rho^{-n} \nu_{id}(X_z) \prod_{k=1}^{n} \sqrt{p_{i_k} p_{-i_k}}$$

$$= \frac{1}{2^n} \nu_{id}(X_z) \prod_{k=1}^{n} \frac{\sqrt{p_{i_k} p_{-i_k}}}{\sum_{i=1}^{d} \sqrt{p_i p_{-i}}} \tag{20}$$

In particular, the last term in (20) reveals the local symmetry

$$\nu_{id}([(i_1 \ldots i_k \ldots i_n), z]) = \nu_{id}([(i_1 \ldots - i_k \ldots i_n), z]), \quad (k \in 1, \ldots, n),$$

whereas globally, the measure is multiplicative with respect to the last component, that is

$$\nu_{id}([(i_1 \ldots i_n), z_1 + z_2]) = \nu_{id}([(i_1 \ldots i_n), z_1]) \nu_{id}([(i_1 \ldots i_n), z_2]).$$

Furthermore, (20) implies that the the function $\mathbb{K}$ from Theorem 5.1 is given by

$$\mathbb{K}(\delta_{(x,g)}, (y, h)) = \frac{d\nu_g}{d\nu_{id}}(y, h) = \nu(X_g).$$

These considerations might be summarized as follows. If φ is symmetric, then $\rho = 1$ and $\nu(X_g) = 1$ for all $g \in \mathbb{Z}^d$. If φ is not symmetric, then $\rho < 1$ and $\{\nu_{id}(X_g) : g \in \mathbb{Z}^d\}$ neither is bounded from below nor from above. Moreover, the function h defined by $h(x, g) := \nu_g(X_{id})$ is an $\mathcal{L}_\varphi$-proper function by Remark 5.2. Therefore, $dm := h d\nu$ is T-invariant. However, as it easily can be verified, $m(X_g) = 1$ for all $g \in \mathbb{Z}^d$ and, in particular, m is the measure associated to the symmetric random walk with transition probabilities $P(\pm e_i) = \sqrt{p_i p_{-i}}/(2 \sum_k \sqrt{p_k p_{-k}})$.

Example 2 In this example, we replace the group $\mathbb{Z}^d$ with the free group $\mathbb{F}_d$ with d generators $g_1, \ldots, g_d$. As above, the transition probabilities are given by $P(g_{\pm i}) = p_{\pm i}$, where $g_{-i} := g_i^{-1}$. The construction of the associated group extension then has to be adapted only by changing ψ to

$$\psi : \Sigma \to \mathbb{F}_d, \ (i_1 i_2 \cdots) \mapsto g_{i_1}.$$

As above, we now apply a local limit theorem. The result of Gerl and Woess in [12] is applicable in full generality, however, for ease of exposition, we restrict ourselves to the special case where $q := \sqrt{p_i p_{-i}}$ does not depend on i. Then, by (5.3) and (5.4) in [12], we have that $\rho = 2q\sqrt{2d-1}$ and that

$$\lim_{n \to \infty} \frac{P(X_n = g_{i_1} \cdots g_{i_k})}{P(X_n = id)} = \left(1 + \tfrac{d-1}{d}k\right)(2d-1)^{-k/2} \prod_{i=1}^{k} \lambda_{i_k}, \qquad (21)$$

for n and k even and $g_{i_1} \cdots g_{i_k}$ in reduced form, that is $i_l \neq -i_{l+1}$, for $l = 1, \ldots, n-1$. Also note that there is a misprint in Eq. (5.4) in [12]. In there, one has to replace $d/(d-1)$ in the first factor by its inverse as in (21). As above, the right hand side in (21) is equal to $\nu_{id}(X_{g_{i_1} \cdots g_{i_k}})$. Using the identities for q and ρ and setting $C_k := 1 + k(d-1)/d$, this gives that

$$\nu_{id}(X_{g_{i_1} \cdots g_{i_k}}) = C_k(2/\rho)^k \prod_{i=1}^{k} q\lambda_{-i_k} = C_k(2/\rho)^k \prod_{i=1}^{k} p_{-i_k}.$$

Since the identity requires that $g = g_{i_1} \cdots g_{i_k}$ is in reduced form, we have to introduce the following operations on finite words in order to obtain a formula for arbitrary cylinders. For $w = (i_1 \ldots i_n) \in \mathcal{W}^n$, there exists a unique $k \leq n$ and a word $(j_1 \ldots j_k) \in \mathcal{W}^k$ such that $\psi_n(w) = g_{j_1} \cdots g_{j_k}$ is in reduced form. We will refer to $\mathfrak{r}(w) := (j_1 \ldots j_k)$ as the *active part* of w, whereas the word which is obtained by deleting the entries of $\mathfrak{r}(w)$ from w is referred to as the *inactive part* $\mathfrak{i}(w) \in \mathcal{W}^{n-k}$ of w. Note that $\psi_k(\mathfrak{r}(w)) = \psi_n(w)$ and $\psi_{n-k}(\mathfrak{i}(w)) = id$. Moreover, for a given word $v = (i_1 \ldots i_n) \in \mathcal{W}^n$, we will refer to $\kappa(v) := (-i_n, \ldots, -i_2, -i_1)$ as the *inverse word* of v. For ease of notation, we also will make use of the Bernoulli measure on Σ defined through $\mu([i_1 \ldots i_n]) = p_{i_1} \cdots p_{i_n}$.

As it will be shown below, the measure of a cylinder $[w, g]$, for $w \in \mathcal{W}^n$ and $g \in G$ and the function $\mathbb{K}$ given by Theorem 5.1 depend on possible cancelations of the concatenation of the path to $g \in G$ and w. So, let $v_g \in \mathcal{W}^m$ be given by $\psi_m(v_g) = g$ and $\mathrm{i}(v_g) = \emptyset$, that is v_g is given by the reduced form of g. With $k := |\mathrm{r}(v_g w)|$, the conformality of ν_{id} implies that

$$\nu_{id}([w, g]) = \rho^{-n} \mu([w]) \nu_{id}(X_{g\psi_n(w)}) = \rho^{-n} \mu([w]) C_k (2/\rho)^k \mu([\kappa(\mathrm{r}(v_g w))])$$

in case that $\mathrm{r}(v_g w) \neq \emptyset$. If $\mathrm{r}(v_g w) \neq \emptyset$, then $\nu_{id}([w, g]) = \rho^{-n} \mu([w])$ by the same arguments. The identity now allows to determine the function $\mathbb{K}$ explicitly. That is, for $g_1, g_2 \in G$, $x \in \Sigma$ and $(w_{(n)})$ with $w_{(n)} \in \mathcal{W}^n$ and $x = \lim_n [w_{(n)}]$, we have ν_{id} a.s., that

$$\mathbb{K}(g_1, (x, g_2)) = \lim_{n \to \infty} \frac{\nu_{g_1}([w_{(n)}, g_2])}{\nu_{id}([w_{(n)}, g_2])} = \lim_{n \to \infty} \frac{\nu_{id}([w_{(n)}, g_1^{-1} g_2])}{\nu_{id}([w_{(n)}, g_2])}$$

$$= \lim_{n \to \infty} \frac{C_{|\mathrm{r}(v_{g_1^{-1} g_2} w_{(n)})|}}{C_{|\mathrm{r}(v_{g_2} w)|}} \left(\frac{2}{\rho}\right)^{|\mathrm{r}(v_{g_1^{-1} g_2} w_{(n)})| - |\mathrm{r}(v_{g_2} w_{(n)})|} \frac{\mu([\kappa(\mathrm{r}(v_{g_1^{-1} g_2} w_{(n)}))])}{\mu([\kappa(\mathrm{r}(v_{g_2} w_{(n)}))])}.$$

Observe that total dissipativity implies that $(\psi_n(x) : n \in \mathbb{N}) = (\psi_n(w_{(n)}) : n \in \mathbb{N})$ will almost surely only return finitely many times to a finite subset of G. Hence, the first term in the product converges to 1 whereas the second and third eventually are constant. By setting $k_{g_1, g_2}(x) := \lim_{n \to \infty} |g_1^{-1} g_2 \psi_n(x)| - |g_2 \psi_n(x)|$, analyzing the cancelations in $v_{g_1^{-1}} v_{g_2} w_{(n)}$ and $v_{g_2} w_{(n)}$ and using that $q^2 = p_i p_{-i}$, it follows that

$$\mathbb{K}(g_1, (x, g_2)) = (2/\rho)^{k_{g_1, g_2}(x)} \lim_{n \to \infty} \frac{\mu([\kappa(\mathrm{r}(v_{g_1^{-1} g_2} w_{(n)}))])}{\mu([\kappa(\mathrm{r}(v_{g_2} w_{(n)}))])}$$

$$= (2/\rho)^{k_{g_1, g_2}(x)} \cdot \frac{\mu([v_{g_1}])}{q^{|g_1| - k_{g_1, g_2}(x)}} = (2d - 1)^{-\frac{k_{g_1, g_2}(x)}{2}} \sqrt{\mu([v_{g_1}])/\mu([\kappa v_{g_1}])}.$$

The regularity of h can now be analyzed through k_{g_1, g_2}. In order to do so, for each open subset U of X and $g = g_1 \in G$, observe that $k_{g, g_2}(U)$ is equal to $\{-|g|, 2 - |g|, \ldots, |g| - 2, |g|\}$. This implies that

$$\sup_{z, \tilde{z} \in U} \frac{\mathbb{K}(g, z)}{\mathbb{K}(g, \tilde{z})} = (2d - 1)^{|g|} \sqrt{\frac{\mu([v_g])}{\mu([\kappa v_g])}}.$$

In particular, the fluctuations of $h(g, \cdot)$ only depend on $|g|$ and the quantity $\mu([v_g])/\mu([\kappa v_g])$, which measures the asymmetry of the random walk. If the random walk is symmetric, that is $p_i = p_{-i} = 1/2d$ for all $i = 1, \ldots, d$, then this simplifies to

$$\rho = \sqrt{2d - 1}/d, \quad \nu_{id}(X_g) = C_{|g|}(2d - 1)^{-\frac{|g|}{2}}, \quad \mathbb{K}(g_1, (x, g_2)) = (2d - 1)^{-\frac{k_{g_1, g_2}(x)}{2}}.$$

Acknowledgements The author acknowledges support by CNPq through PQ 310883/2015-6 and Projeto Universal 426814/2016-9.

References

1. J. Aaronson, *An Introduction to Infinite Ergodic Theory, Mathematical Surveys and Monographs*, vol. 50 (American Mathematical Society, Providence, RI, 1997)
2. J. Aaronson, M. Denker, The Poincaré series of $\mathbb{C}\backslash\mathbb{Z}$. Ergod. Theory Dynam. Syst. **19**(1), 1–20 (1999)
3. J. Aaronson, M. Denker, M. Urbański, Ergodic theory for Markov fibred systems and parabolic rational maps. Trans. Am. Math. Soc. **337**(2), 495–548 (1993)
4. G. Birkhoff, Extensions of Jentzsch's theorem. Trans. Am. Math. Soc. **85**, 219–227 (1957)
5. R. Brooks, The bottom of the spectrum of a Riemannian covering. J. Reine Angew. Math. **357**, 101–114 (1985)
6. I.P. Cornfeld, S.V. Fomin, Y.G. Sinaĭ, *Ergodic theory, Grundlehren der Mathematischen Wissenschaften*, vol. 245 (Springer, New York, 1982)
7. M. Denker, Y. Kifer, M. Stadlbauer, Thermodynamic formalism for random countable Markov shifts. Discrete Contin. Dyn. Syst. **22**(1–2), 131–164 (2008)
8. M. Denker, M. Urbański, On the existence of conformal measures. Trans. Am. Math. Soc. **328**(2), 563–587 (1991)
9. M. Denker, M. Yuri, A note on the construction of nonsingular Gibbs measures. Colloq. Math. **84/85**(2), 377–383 (2000). Dedicated to the memory of Anzelm Iwanik
10. W. Doeblin, R. Fortet, Sur des chaînes à liaisons complètes. Bull. Soc. Math. France **65**, 132–148 (1937)
11. G. Frobenius, Über Matrizen aus nicht negativen Elementen. Berl. Ber. **1912**, 456–477 (1912)
12. P. Gerl, W. Woess, Local limits and harmonic functions for nonisotropic random walks on free groups. Probab. Theory Relat. Fields **71**(3), 341–355 (1986)
13. J. Jaerisch, Fractal models for normal subgroups of Schottky groups. Trans. Am. Math. Soc. **366**(10), 5453–5485 (2014)
14. J. Jaerisch, Recurrence and pressure for group extensions. Ergod. Theory Dynam. Syst. (2014)
15. H. Kesten, Full Banach mean values on countable groups. Math. Scand. **7**, 146–156 (1959)
16. F. Ledrappier, O. Sarig, Invariant measures for the horocycle flow on periodic hyperbolic surfaces. Isr. J. Math. **160**(1), 281–315 (2007)
17. C. Liverani, Decay of correlations. Ann. Math. **142**(2), 239–301 (1995)
18. R.D. Mauldin, M. Urbański, Gibbs states on the symbolic space over an infinite alphabet. Isr. J. Math. **125**, 93–130 (2001)
19. R.D. Mauldin, M. Urbański, *Graph Directed Markov Systems: Geometry and Dynamics of Limit Sets*, vol. 148, Cambridge Tracts in Mathematics (Cambridge University Press, Cambridge, 2003)
20. S.J. Patterson, The limit set of a Fuchsian group. Acta Math. **136**(3–4), 241–273 (1976)
21. O. Perron, Zur Theorie der Matrices. Math. Ann. **64**(2), 248–263 (1907)
22. G. Pólya, Über eine Aufgabe der Wahrscheinlichkeitsrechnung betreffend die Irrfahrt im Straßennetz. Math. Ann. **84**(1–2), 149–160 (1921)
23. M. Rees, Checking ergodicity of some geodesic flows with infinite Gibbs measure. Ergod. Theory Dynam. Syst. **1**(1), 107–133 (1981)
24. V.A. Rohlin, On the fundamental ideas of measure theory. Am. Math. Soc. Transl. **1952**(71), 55 (1952)
25. D. Ruelle, Statistical mechanics of a one-dimensional lattice gas. Comm. Math. Phys. **9**, 267–278 (1968)
26. O.M. Sarig, Thermodynamic formalism for countable Markov shifts. Ergod. Theory Dyn. Syst. **19**(6), 1565–1593 (1999)
27. O.M. Sarig, Existence of Gibbs measures for countable Markov shifts. Proc. Am. Math. Soc. **131**(6), 1751–1758 (2003)
28. O.M. Sarig, Invariant Radon measures for horocycle flows on Abelian covers. Invent. Math. **157**(3), 519–551 (2004)
29. O. Shwartz. Thermodynamic formalism for transient potential functions. Comm. Math. Phys. **366**, 737–779 (2019)

30. M. Stadlbauer, An extension of Kesten's criterion for amenability to topological Markov chains. Adv. Math. **235**, 450–468 (2013)
31. D. Sullivan, The density at infinity of a discrete group of hyperbolic motions. Inst. Hautes Études Sci. Publ. Math. **50**, 171–202 (1979)
32. W. Woess, *Random walks on infinite graphs and groups*, vol. 138, Cambridge Tracts in Mathematics (Cambridge University Press, Cambridge, 2000)
33. W. Woess, *Denumerable Markov chains*, EMS Textbooks in Mathematics (European Mathematical Society (EMS), Zürich, 2009)
34. R.J. Zimmer, Amenable ergodic group actions and an application to Poisson boundaries of random walks. J. Funct. Anal. **27**(3), 350–372 (1978)

The Boundaries of Golden-Mean Siegel Disks in the Complex Quadratic Hénon Family Are Not Smooth

Michael Yampolsky and Jonguk Yang

Abstract As was recently shown by the first author and others in Gaidashev et al. (Renormalization and Siegel disks for complex Henon maps, [12]), golden-mean Siegel disks of sufficiently dissipative complex quadratic Hénon maps are bounded by topological circles. In this paper we investigate the geometric properties of such curves, and demonstrate that they cannot be C^1-smooth.

Keywords Henon map · Renormalization · Siegel disk · Complex dynamics

1 Introduction

Up to a biholomorphic conjugacy, a complex quadratic Hénon map can be written as

$$H_{c,a}(x, y) = (x^2 + c + ay, ax) \quad \text{for} \quad a \neq 0;$$

this form is unique modulo the change of coordinates $(x, y) \mapsto (x, -y)$, which conjugates $H_{c,a}$ with $H_{c,-a}$. In this paper we will always assume that the Hénon map is dissipative, $|a| < 1$. Note that for $a = 0$, the map $H_{c,a}$ degenerates to

$$(x, y) \mapsto (f_c(x), 0),$$

where $f_c(x) = x^2 + c$ is a one-dimensional quadratic polynomial. Thus for a fixed small value of a_0, the one parameter family H_{c,a_0} can be seen as a small perturbation of the quadratic family.

M. Yampolsky (✉)
Mathematics Department, University of Toronto, Toronto, ON, Canada
e-mail: yampol@math.toronto.edu

J. Yang
Mathematics Department, Stony Brook University, Stony Brook, NY, USA
e-mail: jonguk.yang@stonybrook.edu

M. J. Pacifico and P. Guarino (eds.), *New Trends in One-Dimensional Dynamics*, Springer Proceedings in Mathematics & Statistics 285,
https://doi.org/10.1007/978-3-030-16833-9_16

As usual, we let $K^{\pm}$ be the sets of points that do not escape to infinity under forward, respectively backward iterations of the Hénon map. Their topological boundaries are $J^{\pm} = \partial K^{\pm}$. Let $K = K^+ \cap K^-$ and $J = J^- \cap J^+$. The sets $J^{\pm}$, $K^{\pm}$ are unbounded, connected sets in $\mathbb{C}^2$ (see [3]). The sets J and K are compact (see [13]). In analogy to one-dimensional dynamics, the set J is called the Julia set of the Hénon map.

Note that a Hénon map $H_{c,a}$ is determined by the multipliers μ and ν at a fixed point uniquely up to changing the sign of a. In particular,

$$\mu\nu = -a^2,$$

the parameter c is a function of a^2 and μ:

$$c = (1 - a^2)\left(\frac{\mu}{2} - \frac{a^2}{2\mu}\right) - \left(\frac{\mu}{2} - \frac{a^2}{2\mu}\right)^2.$$

Hence, we sometimes write $H_{\mu,\nu}$ instead of $H_{c,a}$, when convenient.

When $\nu = 0$, the Hénon map degenerates to

$$H_{\mu,0}(x, y) = (P_\mu(x), 0), \text{ where } P_\mu(x) = x^2 + \mu/2 - \mu^2/4. \tag{1}$$

We say that a dissipative Hénon map $H_{c,a}$ has a *semi-Siegel fixed point* (or simply that $H_{c,a}$ is semi-Siegel) if the eigenvalues of the linear part of $H_{c,a}$ at that fixed point are $\mu = e^{2\pi i\theta}$, with $\theta \in (0, 1)\backslash\mathbb{Q}$ and ν, with $|\nu| < 1$, and $H_{c,a}$ is locally biholomorphically conjugate to the linear map

$$L(x, y) = (\mu x, \nu y).$$

The classic theorem of Siegel states, in particular, that $H_{\mu,\nu}$ is semi-Siegel whenever θ is Diophantine, that is $q_{n+1} < cq_n^d$, where p_n/q_n are the continued fraction convergents of θ. The existence of a linearization is a local result, however, in this case there exists a linearizing biholomorphism $\phi : \mathbb{D} \times \mathbb{C} \to \mathbb{C}^2$ sending $(0, 0)$ to the semi-Siegel fixed point,

$$H_{\mu,\nu} \circ \phi = \phi \circ L,$$

such that the image $\phi(\mathbb{D} \times \mathbb{C})$ is *maximal* (see [15]). We call $\phi(\mathbb{D} \times \mathbb{C})$ the *Siegel cylinder*; it is a connected component of the interior of K^+ and its boundary coincides with J^+ (see [4]). We let

$$\Delta = \phi(\mathbb{D} \times \{0\}),$$

and by analogy with the one-dimensional case call it the *Siegel disk* of the Hénon map. Clearly, the Siegel cylinder is equal to the stable manifold $W^s(\Delta)$, and $\Delta \subset K$ (which is always bounded). Moreover, $\partial\Delta \subset J$, the Julia set of the Hénon map.

Remark 1.1 Let q be the semi-Siegel fixed point of the Hénon map. Then $\Delta \subset W^c(q)$, the center manifold of q (see e.g. [18] for a definition of W^c). The center manifold is not unique in general, but all center manifolds $W^c(q)$ must coincide on the Siegel disk (it is unknown if they may extend beyond its boundary). This phenomenon is nicely illustrated in [16], Fig. 5.

In a recent paper [12] it was shown that:

Theorem 1.2 ([12]) *There exists $\epsilon > 0$ such that the following holds. Let $\theta_* = (\sqrt{5} - 1)/2$ be the inverse golden mean, $\mu_* = e^{2\pi i \theta_*}$, and let $|\nu| < \epsilon$. Then the boundary of the Siegel disk of $H_{\mu_*, \nu}$ is a homeomorphic image of the circle.*
   Furthermore, the linearizing map

$$\phi : \mathbb{D} \times \{0\} \to \Delta \qquad (2)$$

extends continuously and injectively to the boundary. However, the restriction

$$\phi : S^1 \times \{0\} \to \partial\Delta$$

is not C^1-smooth.

This is the first result of its kind on the structure of the boundaries of Siegel disks of complex Hénon maps. It is based on a renormalization theory for two-dimensional dissipative Hénon-like maps, developed in [10]. While renormalization technique is new in the study of two-dimensional Siegel disks and Siegel Julia sets, it has a history of one-dimensional applications (see e.g. [14, 21]). Below, we will briefly review the relevant renormalization results.

Theorem 1.2 raises a natural question whether the boundary $\partial\Delta$ can ever lie on a smooth curve. Classical results (see [20]) imply that the smoothness of $\partial\Delta$ must be less than $C^{1+\epsilon}$ – otherwise, ϕ would have a $C^{1+\epsilon}$ extension to the boundary, contradicting Theorem 1.2. However, we can ask, whether $\partial\Delta$ can be a C^1-smooth curve. In the present note we answer this in the negative:

Main Theorem *Let $\epsilon > 0$ be as in Theorem 1.2 and $|\nu| < \epsilon$. Then the boundary of the Siegel disk of $H_{\mu_*, \nu}$ is not C^1-smooth.*

We note that the question of smoothness of Siegel disk boundaries for polynomial maps of $\mathbb{C}$ has a rich history. For quadratic polynomials

$$Q_\theta(z) = z^2 + e^{2\pi i \theta} z, \ \theta \in \mathbb{R},$$

with Brjuno rotation numbers θ, in particular, it is known that the boundary of the Siegel disk D_θ at 0 can be smooth for some rotation numbers (see [1] and references therein). In contrast, for $\theta = \theta_*$ (and more generally, for rotation numbers of bounded type), it is known that ∂D_θ is not smooth, is a quasicircle, and contains the critical point 0 [7]. The proof of the latter theorem, due to A. Douady, E. Ghys, M. Herman, and M. Shishikura, is based on the technique of quasiconformal surgery, which is very specifically one-dimensional, and does not generalize to maps of $\mathbb{C}^2$.

Let us also mention that Bedford and Kim [2] have recently shown that the set J^+ cannot be smooth. Note, however, that the Julia set of a quadratic polynomial Q_θ is never smooth [17], however, as mentioned above, it may possess a Siegel disk with a smooth boundary.

2 Review of Renormalization Theory for Siegel Disks

In this section we give a brief summary of the relevant statements on renormalization of Siegel disks; we refer the reader to [10] for the details.

2.1 One-Dimensional Renormalization: Almost-Commuting Pairs

For a domain $Z \subset \mathbb{C}$, we denote $\mathcal{H}(Z)$ the Banach space of bounded analytic functions $f : Z \to \mathbb{C}$ equipped with the norm

$$\|f\| = \sup_{x \in Z} |f(x)|. \tag{3}$$

Denote $\mathcal{H}(Z, W)$ the Banach space of bounded pairs of analytic functions $\zeta = (\eta, \xi)$ from domains $Z \subset \mathbb{C}$ and $W \subset \mathbb{C}$ respectively to $\mathbb{C}$ equipped with the norm

$$\|\zeta\| = \frac{1}{2} (\|\eta\| + \|\xi\|) . \tag{4}$$

Henceforth, we assume that the domains Z and W contain 0.

For a pair $\zeta = (\eta, \xi)$, define the *rescaling map* as

$$\Lambda(\zeta) := (s_\zeta^{-1} \circ \eta \circ s_\zeta, s_\zeta^{-1} \circ \xi \circ s_\zeta), \tag{5}$$

where

$$s_\zeta(x) := \lambda_\zeta x \quad \text{and} \quad \lambda_\zeta := \xi(0).$$

Definition 2.1 We say that $\zeta = (\eta, \xi) \in \mathcal{H}(Z, W)$ is a *critical pair* if η and ξ have a simple unique critical point at 0. The space of critical pairs is denoted by $\mathcal{C}(Z, W)$.

Definition 2.2 We say that $\zeta = (\eta, \xi) \in \mathcal{C}(Z, W)$ is a *commuting pair* if

$$\eta \circ \xi = \xi \circ \eta.$$

Definition 2.3 We say that $\zeta = (\eta, \xi) \in C(Z, W)$ is an *almost commuting pair* (cf. [5, 19]) if

$$\frac{d^i (\eta \circ \xi - \xi \circ \eta)}{dx^i}(0) = 0 \quad \text{for} \quad i = 0, 2,$$

and

$$\xi(0) = 1.$$

The space of almost commuting pairs is denoted by $B(Z, W)$.

Note that if $\zeta = (\eta, \xi) \in C(Z, W)$, then the first order commuting relation is automatically satisfied:

$$\frac{d(\eta \circ \xi - \xi \circ \eta)}{dx}(0) = \eta'(\xi(0))\xi'(0) - \xi'(\eta(0))\eta'(0) = 0.$$

Proposition 2.4 (cf. [10]) *The spaces $C(Z, W)$ and $B(Z, W)$ have the structure of an immersed Banach submanifold of $\mathcal{H}(Z, W)$ of codimension 2 and 5 respectively.*

Denote

$$c(x) := \bar{x}.$$

Definition 2.5 Let $\zeta = (\eta, \xi) \in B(Z, W)$. The *pre-renormalization* of ζ is defined as:

$$p\mathcal{R}((\eta, \xi)) := (\eta \circ \xi, \eta).$$

The *renormalization* of ζ is defined as:

$$\mathcal{R}((\eta, \xi)) := \Lambda(c \circ \eta \circ \xi \circ c, c \circ \eta \circ c).$$

It is easy to see that

Proposition 2.6 *The renormalization of an (almost) commuting pair is an (almost) commuting pair (on different domains).*

The following is shown in [10]:

Theorem 2.7 *There exist topological disks $\hat{Z} \supseteq Z$ and $\hat{W} \supseteq W$, and an almost commuting pair $\zeta_* = (\eta_*, \xi_*) \in B(Z, W)$ such that the following hold:*

(1) There exists a neighbourhood $\mathcal{N}(\zeta_*)$ of ζ_* in the submanifold $B(Z, W)$ such that

$$\mathcal{R} : \mathcal{N}(\zeta_*) \to B(\hat{Z}, \hat{W})$$

is an anti-analytic operator.
(2) The pair ζ_* is the unique fixed point of $\mathcal{R}$ in $\mathcal{N}(\zeta_*)$.

(3) *The differential $D\mathcal{R}^2|_{\zeta_*}$ is a compact linear operator. It has a single, simple eigenvalue with modulus greater than 1. The rest of its spectrum lies inside the open unit disk $\mathbb{D}$ (and hence is compactly contained in $\mathbb{D}$ by the spectral theory of compact operators).*

2.2   Renormalization of Two-Dimensional Maps

For a domain $\Omega \subset \mathbb{C}^2$, we denote $\mathcal{H}(\Omega)$ the Banach space of bounded analytic functions $F : \Omega \to \mathbb{C}^2$ equipped with the norm

$$\|F\| = \sup_{(x,y)\in\Omega} \|F(x, y)\|. \tag{6}$$

Define

$$\|F\|_y := \sup_{(x,y)\in\Omega} \|\partial_y F(x, y)\|. \tag{7}$$

Moreover, for

$$F = \begin{bmatrix} f_1 \\ f_2 \end{bmatrix},$$

define

$$\|F\|_{\text{diag}} := \sup_{(x,y)\in\Omega} \|f_1(x, y) - f_2(x, y)\|. \tag{8}$$

Denote $\mathcal{H}(\Omega, \Gamma)$ the Banach space of bounded pairs of analytic functions $\Sigma = (A, B)$ from domains $\Omega \subset \mathbb{C}^2$ and $\Gamma \subset \mathbb{C}^2$ respectively to $\mathbb{C}^2$ equipped with the norm

$$\|\Sigma\| = \frac{1}{2} \left(\|A\| + \|B\| \right). \tag{9}$$

Define

$$\|\Sigma\|_y := \frac{1}{2} \left(\|A\|_y + \|B\|_y \right). \tag{10}$$

Moreover,

$$\|\Sigma\|_{\text{diag}} := \frac{1}{2} \left(\|A\|_{\text{diag}} + \|B\|_{\text{diag}} \right). \tag{11}$$

Henceforth, we assume that

$$\Omega = Z \times Z \quad \text{and} \quad \Gamma = W \times W,$$

where Z and W are subdomains of $\mathbb{C}$ containing 0.

For a pair $\Sigma = (A, B) \in \mathcal{H}(\Omega, \Gamma)$, define the *rescaling map* as

$$\Lambda(\Sigma) := (s_\Sigma^{-1} \circ A \circ s_\Sigma, s_\Sigma^{-1} \circ B \circ s_\Sigma), \tag{12}$$

where

$$s_\Sigma(x, y) := (\lambda_\Sigma x, \lambda_\Sigma y) \quad \text{and} \quad \lambda_\Sigma := p_1 B(0).$$

Let $U = Z$ or W, and consider $F : U \times U \to \mathbb{C}^2$ given by

$$F(x, y) := \begin{bmatrix} f_1(x, y) \\ f_2(x, y) \end{bmatrix}.$$

Define $p_1 F : U \to \mathbb{C}$ and $p_2 F : U \to \mathbb{C}$ as

$$p_1 F(x) := f_1(x, 0) \quad \text{and} \quad p_2 F(x) := f_2(x, 0).$$

The operator p_1 is a projection map from the set of two-dimensional maps to the set of one-dimensional maps.

Let $f : U \to \mathbb{C}$ be a one-dimensional map. Define $\iota(f) : U \times U \to \mathbb{C}^2$ by

$$\iota(f)(x, y) = \begin{bmatrix} f(x) \\ f(x) \end{bmatrix}.$$

Note that we have

$$\|\iota(f)\|_y = \|\iota(f)\|_{\text{diag}} = 0,$$

and

$$p_1 \circ \iota = \text{Id}.$$

The operator ι is an embedding from the set of one-dimensional maps to the set of two-dimensional maps.

Let $\zeta = (\eta, \xi)$ and $\Sigma = (A, B)$ be a one-dimensional pair and a two-dimensional pair respectively. Define

$$\iota(\zeta) := (\iota(\eta), \iota(\xi)) \quad \text{and} \quad p_1(\Sigma) := (p_1 A, p_1 B).$$

Definition 2.8 For $0 < \kappa \leq \infty$, we say that $\Sigma = (A, B) \in \mathcal{H}(\Omega, \Gamma)$ is a *κ-critical pair* if $p_1 A$ and $p_1 B$ have a simple unique critical point which is contained in a κ-neighbourhood of 0. The space of κ-critical pairs in $\mathcal{H}(\Omega, \Gamma)$ is denoted by $C_2(\Omega, \Gamma, \kappa)$.

Definition 2.9 We say that $\Sigma = (A, B) \in C_2(\Omega, \Gamma, \kappa)$ is a *commuting pair* if

$$A \circ B = B \circ A.$$

Definition 2.10 We say that $\Sigma = (A, B) \in \mathcal{C}_2(\Omega, \Gamma, \kappa)$ is an *almost commuting pair* if

$$\frac{d^i p_1[A, B]}{dx^i}(0) := \frac{d^i p_1(A \circ B - B \circ A)}{dx^i}(0) = 0 \quad \text{for} \quad i = 0, 2,$$

and

$$p_1 B(0) = 1.$$

The space of almost commuting pairs in $\mathcal{C}_2(\Omega, \Gamma, \kappa)$ is denoted by $\mathcal{B}_2(\Omega, \Gamma, \kappa)$.

Proposition 2.11 (cf. [10]) *The space $\mathcal{B}_2(\Omega, \Gamma, \kappa)$ has the structure of an immersed Banach submanifold of $\mathcal{H}(\Omega, \Gamma)$ of codimension 3.*

For $0 < \epsilon, \delta \leq \infty$, let $\mathcal{H}(\Omega, \Gamma, \epsilon, \delta)$ be the open subset of $\mathcal{H}(\Omega, \Gamma)$ consisting of pairs $\Sigma = (A, B)$ such that the following holds:

(1) $\|\Sigma\|_y < \epsilon$, and
(2) $\|\Sigma\|_{\mathrm{diag}} < \delta$.

We denote

$$\mathcal{C}_2(\Omega, \Gamma, \epsilon, \delta, \kappa) := \mathcal{H}(\Omega, \Gamma, \epsilon, \delta) \cap \mathcal{C}_2(\Omega, \Gamma, \kappa), \tag{13}$$

and

$$\mathcal{B}_2(\Omega, \Gamma, \epsilon, \delta, \kappa) := \mathcal{H}(\Omega, \Gamma, \epsilon, \delta) \cap \mathcal{B}_2(\Omega, \Gamma, \kappa). \tag{14}$$

Define $\mathcal{C}_2(\Omega, \Gamma, 0, 0, 0)$ as the set of pairs ζ such that $\zeta \in \mathcal{C}_2(\Omega, \Gamma, \epsilon, \delta, \kappa)$ for all $\epsilon, \delta, \kappa > 0$. The definition of $\mathcal{B}_2(\Omega, \Gamma, 0, 0, 0)$ is similar. Note that

$$\iota(\mathcal{C}(Z, W)) = \mathcal{C}_2(\Omega, \Gamma, 0, 0, 0), \quad \text{and} \quad \iota(\mathcal{B}(Z, W)) = \mathcal{B}_2(\Omega, \Gamma, 0, 0, 0),$$

where $\mathcal{C}(Z, W)$ and $\mathcal{B}(Z, W)$ denotes the space of one-dimensional critical pairs and almost commuting pairs respectively.

Proposition 2.12 (cf. [10]) *If ϵ, δ, and κ are sufficiently small, then there exists an analytic map $\Pi_{\mathrm{ac}} : \mathcal{C}_2(\Omega, \Gamma, \epsilon, \delta, \kappa) \to \mathcal{B}_2(\Omega, \Gamma, \epsilon, \delta, \kappa)$ such that*

$$\Pi_{ac}|_{\mathcal{B}_2(\Omega, \Gamma, \epsilon, \delta, \kappa)} \equiv \mathrm{Id}. \tag{15}$$

We are now ready to define the 2D renormalization operator **R**. Our approach is to extend the action of the 1D operator $\mathcal{R}$ on $\iota(\mathcal{B}(Z, W)) = \mathcal{B}_2(\Omega, \Gamma, 0, 0, 0)$ to nearby 2D pairs $\Sigma = (A, B) \in \mathcal{B}_2(\Omega, \Gamma, \delta, \kappa)$. It turns out that in order to ensure the hyperbolicity of **R**, the definition of $\mathbf{R}(\Sigma)$ must incorporate non-linear changes of coordinates that are only well-defined away from the critical values of $p_1(\Sigma) = (p_1 A, p_1 B)$. This requires us to take four iterates of $\mathcal{R}$ before extending it to 2D pairs.

Notation 2.13 Let $\mathcal{I}$ be the space of all finite multi-indexes

$$\overline{\omega} = (a_1, \ldots, a_{2n}) \in (\{0\} \cup \mathbb{N})^{2n} \quad \text{for some } n \in \mathbb{N}.$$

For a pair $\zeta = (\eta, \xi)$ and a multi-index $\overline{\omega} = (a_1, \ldots, a_{2n}) \in \mathcal{I}$, denote

$$\zeta^{\overline{\omega}} = \xi^{a_{2n}} \circ \eta^{a_{2n-1}} \circ \ldots \circ \xi^{a_2} \circ \eta^{a_1}$$

Similarly, for a pair $\Sigma = (A, B)$, denote

$$\Sigma^{\overline{\omega}} = B^{a_{2n}} \circ A^{a_{2n-1}} \circ \ldots \circ B^{a_2} \circ A^{a_1}.$$

Let $\zeta = (\eta, \xi) \in \mathcal{B}(Z, W)$ be a four-times 1D renormalizable pair. Define multi-indexes $\overline{a}_1, \overline{b}_1, \overline{a}'_1,$ and $\overline{b}'_1$ by

$$(\zeta^{\overline{a}_1}, \zeta^{\overline{b}_1}) := p\mathcal{R}^4(\zeta) = (\eta \circ \xi \circ \eta^2 \circ \xi \circ \eta \circ \xi \circ \eta, \eta \circ \xi \circ \eta^2 \circ \xi) =: (\eta \circ \xi \circ \zeta^{\overline{a}'_1}, \eta \circ \xi \circ \zeta^{\overline{b}'_1}). \tag{16}$$

Let $\mathcal{D}(\Omega, \Gamma, 0)$ be the subset of $\mathcal{H}(\Omega, \Gamma)$ consisting of pairs $\Sigma = (A, B)$ such that the following holds:

(1) The functions $A : \Omega \to \mathbb{C}^2$ and $B : \Gamma \to \mathbb{C}^2$ are of the form

$$A(x, y) = \begin{bmatrix} \eta(x) \\ h(x) \end{bmatrix} \quad \text{and} \quad B(x, y) = \begin{bmatrix} \xi(x) \\ g(x) \end{bmatrix}.$$

(2) The pair $\zeta := (\eta, \xi)$ is contained in $\mathcal{B}(Z, W)$ and is four-times 1D renormalizable.

(3) The function g is conformal on $\zeta^{\overline{a}_1}(U) \cup \zeta^{\overline{b}_1}(U)$, where

$$U := \lambda_{p\mathcal{R}^4(\zeta)} Z \cup W.$$

Let $\mathcal{D}(\Omega, \Gamma, \epsilon) \subset \mathcal{H}(\Omega, \Gamma, \epsilon, \infty)$ be a neighbourhood of $\mathcal{D}(\Omega, \Gamma, 0)$ consisting of pairs $\Sigma = (A, B)$ such that for

$$\tilde{\Sigma} := (\Sigma^{\overline{a}_1}, \Sigma^{\overline{b}_1}), \tag{17}$$

the pair $\Lambda(\tilde{\Sigma})$ is a well-defined element of $\mathcal{H}(\Omega, \Gamma)$. Moreover, for

$$V := \lambda_{\tilde{\Sigma}} Z \cup W,$$

the following holds:

(1) $p_1 A$ is conformal on $(p_1 A)^{-1}(V)$,
(2) $p_1(A \circ B)$ is conformal on $(p_1(A \circ B))^{-1}(V)$, and
(3) $p_2 B$ is conformal on $p_1 \Sigma^{\overline{a}'_1}(V)$ and $p_1 \Sigma^{\overline{b}'_1}(V)$.

Consider the fixed point $\zeta_* = (\eta_*, \xi_*) \in \mathcal{B}(Z, W)$ of the 1D renormalization operator $\mathcal{R}$ given in Theorem 1.2. Fix $\epsilon > 0$, and let $\widehat{\mathcal{N}}(\iota(\zeta_*)) \subset \mathcal{D}(\Omega, \Gamma, \epsilon)$ be a neighbourhood of $\iota(\zeta_*)$ whose closure is contained in $\mathcal{D}(\Omega, \Gamma, \epsilon)$. Let

$$\Sigma = (A, B) = \left(\begin{bmatrix} a \\ h \end{bmatrix}, \begin{bmatrix} b \\ g \end{bmatrix} \right)$$

be a pair contained in $\widehat{\mathcal{N}}(\iota(\zeta_*))$. Denote

$$\eta_i(x) := p_i A(x) \quad \text{and} \quad \xi_i(x) := p_i B(x) \quad, \quad \text{for } i \in \{1, 2\},$$

and let

$$\zeta := (\eta_1, \xi_1).$$

Denote

$$a_y(x) := a(x, y),$$

and consider the following non-linear changes of coordinates:

$$H(x, y) := \begin{bmatrix} a_y(x) \\ y \end{bmatrix} \quad \text{and} \quad V(x, y) := \begin{bmatrix} x \\ \eta_1 \circ \xi_1 \circ \xi_2^{-1}(y) \end{bmatrix}. \tag{18}$$

Observe that

$$A \circ H^{-1}(x, y) = \begin{bmatrix} a_y \circ a_y^{-1}(x) \\ g(a_y^{-1}(x), y) \end{bmatrix} = \begin{bmatrix} x \\ g(a_y^{-1}(x), y) \end{bmatrix}.$$

Furthermore,

$$V \circ H \circ B = \begin{bmatrix} a_g \circ b \\ \eta_1 \circ \xi_1 \circ \xi_2^{-1} \circ g \end{bmatrix}.$$

Thus, we have

$$\|A \circ H^{-1}\|_y < O(\epsilon) \quad \text{and} \quad \|V \circ H \circ B - \iota(\eta_1 \circ \xi_1)\| < O(\epsilon).$$

Let

$$A_1 := V \circ H \circ A^{-1} \circ \Sigma^{\bar{a}_1} \circ A \circ H^{-1} \circ V^{-1},$$

and

$$B_1 := V \circ H \circ A^{-1} \circ \Sigma^{\bar{b}_1} \circ A \circ H^{-1} \circ V^{-1}.$$

Define the *pre-renormalization* of Σ as

$$p\mathbf{R}(\Sigma) = \Sigma_1 := (A_1, B_1). \tag{19}$$

By the definition of $\mathcal{D}(\Omega, \Gamma, \epsilon)$, the pair $\Lambda(p\mathbf{R})$ is a well-defined element of $\mathcal{H}(\Omega, \Gamma)$. From the above inequalities, it follows that

$$\|p\mathbf{R}(\Sigma) - \iota(p\mathcal{R}^4(\zeta))\| < O(\epsilon) \quad \text{and} \quad \|p\mathbf{R}(\Sigma)\|_y < O(\epsilon^2). \tag{20}$$

By the argument principle, if ϵ is sufficiently small, then the function $p_1 B_1 \circ A_1$ has a simple unique critical point c_a near 0. Set

$$T_a(x, y) := (x + c_a, y). \tag{21}$$

Likewise, the function $p_1 T_a^{-1} \circ A_1 \circ B_1 \circ T_a$ has a simple unique critical point c_b near 0. Set

$$T_b(x, y) := (x + c_b, y). \tag{22}$$

Note that if Σ is a commuting pair (i.e. $A \circ B = B \circ A$), then $T_b \equiv \text{Id}$.
    Define the *critical projection* of $p\mathbf{R}(\Sigma)$ as

$$\Pi_{\text{crit}} \circ p\mathbf{R}(\Sigma) = (A_2, B_2) := (T_b^{-1} \circ T_a^{-1} \circ A_1 \circ T_a, T_a^{-1} \circ B_1 \circ T_a \circ T_b). \tag{23}$$

Note that

$$0 = \left(p_1(B_2 \circ A_2)\right)'(0) = (p_1 A_2)'(0) + O(\epsilon^2),$$

and likewise

$$0 = \left(p_1(B_2 \circ A_2)\right)'(0) = (p_1 B_2)'(0) + O(\epsilon^2).$$

Hence,

$$(p_1 A_2)'(0) = O(\epsilon^2) \quad \text{and} \quad (p_1 B_2)'(0) = O(\epsilon^2). \tag{24}$$

It follows that there exists a uniform constant $C > 0$ such that the rescaled pair $\Lambda \circ \Pi_{\text{crit}} \circ p\mathbf{R}(\Sigma)$ is contained in $\mathcal{C}_2(\Omega, \Gamma, C\epsilon^2, C\epsilon, C\epsilon^2)$ (recall that this means $\Lambda \circ \Pi_{\text{crit}} \circ p\mathbf{R}(\Sigma)$ is a $C\epsilon^2$-critical pair with $C\epsilon^2$ dependence on y that is $C\epsilon$ away from the diagonal; see (13)).
    Finally, define the *2D renormalization* of Σ as

$$\mathbf{R}(\Sigma) := \Pi_{\text{ac}} \circ \Lambda \circ \Pi_{\text{crit}} \circ p\mathbf{R}(\Sigma), \tag{25}$$

where the projection map Π_{ac} is given in Proposition 2.12.

Proposition 2.14 *If $\Sigma = (A, B) \in \mathcal{D}(\Omega, \Gamma, \epsilon)$ is a commuting pair (i.e. $A \circ B = B \circ A$), then $\mathbf{R}(\Sigma)$ is conjugate to $(\Sigma^{\bar{a}_1}, \Sigma^{\bar{b}_1})$.*

Proof If Σ is a commuting pair, then Π_{crit} is equal to conjugation by the translation map

$$T_a(x, y) := (x + c_a, y).$$

Moreover, by Proposition 2.12, Π_{ac} is equal to the identity when restricted to almost commuting pairs. The claim follows. $\square$

Theorem 2.15 ([10]) *Let ζ_* be the fixed point of the 1D renormalization given in Theorem 1.2. For any sufficiently small $\epsilon > 0$, let $\widehat{N}(\iota(\zeta_*)) \subset \mathcal{D}(\Omega, \Gamma, \epsilon)$ be a neighbourhood of $\iota(\zeta_*)$ whose closure is contained in $\mathcal{D}(\Omega, \Gamma, \epsilon)$. Then there exists a uniform constant $C > 0$ depending on $\widehat{N}(\iota(\zeta_*))$ such that the 2D renormalization operator*

$$\mathbf{R} : \mathcal{D}(\Omega, \Gamma, \epsilon) \to \mathcal{H}(\Omega, \Gamma),$$

is a well-defined compact analytic operator satisfying the following properties:

(1) $\mathbf{R}|_{\widehat{N}(\iota(\zeta_*))} : \widehat{N}(\iota(\zeta_*)) \to \mathcal{B}_2(\Omega, \Gamma, C\epsilon^2, C\epsilon, C\epsilon^2)$.
(2) *If $\Sigma = (A, B) \in \widehat{N}(\iota(\zeta_*))$ and $\zeta := (p_1 A, p_1 B)$, then*

$$\|\mathbf{R}(\Sigma) - \iota(\mathcal{R}^4(\zeta))\| < C\epsilon.$$

Consequently, if $N(\zeta_*) \subset \mathcal{B}(Z, W)$ is a neighbourhood of ζ_* such that $\iota(N(\zeta_*)) \subset \widehat{N}(\iota(\zeta_*))$, then

$$\mathbf{R} \circ \iota|_{N(\zeta_*)} \equiv \iota \circ \mathcal{R}^4|_{N(\zeta_*)}.$$

(3) *The pair $\iota(\zeta_*)$ is the unique fixed point of $\mathbf{R}$ in $\widehat{N}(\iota(\zeta_*))$.*
(4) *The differential $D_{\iota(\zeta_*)}\mathbf{R}$ is a compact linear operator whose spectrum coincides with that of $D_{\zeta_*}\mathcal{R}^4$. More precisely, in the spectral decomposition of $D_{\iota(\zeta_*)}\mathbf{R}$, the complement to the tangent space $T_{\iota(\zeta_*)}(\iota(N(\zeta_*)))$ corresponds to the zero eigenvalue.*

We denote the stable manifold of the fixed point $\iota(\zeta_*)$ for the 2D renormalization operator $\mathbf{R}$ by $W^s(\iota(\zeta_*)) \subset \mathcal{D}(\Omega, \Gamma, \epsilon)$.

Let $H_{\mu_*, \nu}$ be the Hénon map with a semi-Siegel fixed point $\mathbf{q}$ of multipliers $\mu_* = e^{2\pi i \theta_*}$ and ν, where $\theta_* = (\sqrt{5} - 1)/2$ is the inverse golden mean rotation number, and $|\nu| < \epsilon$. We identify $H_{\mu_*, \nu}$ as a pair in $\mathcal{D}(\Omega, \Gamma, \epsilon)$ as follows:

$$\Sigma_{H_{\mu_*, \nu}} := \Lambda(H_{\mu_*, \nu}^2, H_{\mu_*, \nu}). \tag{26}$$

The following is shown in [12]:

Theorem 2.16 *The pair $\Sigma_{H_{\mu_*, \nu}}$ is contained in the stable manifold $W^s(\iota(\zeta_*)) \subset \mathcal{D}(\Omega, \Gamma, \epsilon)$ of the fixed point $\iota(\zeta_*)$ for the 2D renormalization operator $\mathbf{R}$.*

3   Proof of Main Theorem

3.1   Preliminaries

Let

$$\zeta_* = (\eta_*, \xi_*)$$

be the fixed point of the 1D renormalization operator $\mathcal{R}$ given in Theorem 1.2. By Theorem 2.15, the fixed point of the 2D renormalization operator

$$\mathbf{R} : \widehat{\mathcal{N}}(\iota(\zeta_*)) \to \mathcal{B}_2(\Omega, \Gamma, C\epsilon^2, C\epsilon).$$

is the diagonal embedding $\iota(\zeta_*)$ of ζ_*. Thus, we have

$$\iota(\zeta_*) = \mathbf{R}(\iota(\zeta_*)) = (s_*^{-1} \circ \iota(\zeta)^{\bar{a}_1} \circ s_*, \ s_*^{-1} \circ \iota(\zeta)^{\bar{b}_1} \circ s_*),$$

where

$$s_*(x, y) := (\lambda_* x, \lambda_* y) \quad , \quad |\lambda_*| < 1.$$

Let $\Sigma = (A, B)$ be a pair contained in the stable manifold $W^s(\iota(\zeta_*))$ of the fixed point $\iota(\zeta_*)$. Assume that Σ is commuting, so that

$$A \circ B = B \circ A.$$

Set

$$\Sigma_n = (A_n, B_n) = \left(\begin{bmatrix} a_n \\ h_n \end{bmatrix}, \begin{bmatrix} b_n \\ g_n \end{bmatrix} \right) := \mathbf{R}^n(\Sigma).$$

Let

$$\eta_n(x) := p_1 A_n(x) = a_n(x, 0) \quad \text{and} \quad \xi_n(x) := p_1 B_n(x) = b_n(x, 0).$$

By Theorem 2.15, we may express

$$A_n = \iota(\eta_n) + E_n \quad \text{and} \quad B_n = \iota(\xi_n) + F_n \tag{27}$$

where the error terms E_n and F_n satisfy

$$\|E_n\| < C\epsilon^{2^{n-1}} \quad \text{and} \quad \|F_n\| < C\epsilon^{2^{n-1}}. \tag{28}$$

Hence, the sequence of pairs $\{\Sigma_n\}_{n=0}^{\infty}$ converges to $\mathcal{B}_2(\Omega, \Gamma, 0, 0, 0)$ *super-exponentially*.

Let

$$H_n(x, y) := \begin{bmatrix} a_n(x, y) \\ y \end{bmatrix} \quad \text{and} \quad V_n(x, y) := \begin{bmatrix} x \\ \eta_n \circ \xi_n \circ (p_2 B_n)^{-1}(y) \end{bmatrix}$$

be the non-linear changes of coordinates given in (18), let

$$T_n(x, y) := (x + d_n, y),$$

be the translation map given in (21), and let

$$s_n(x, y) := (\lambda_n x, \lambda_n y), \quad |\lambda_n| < 1$$

be the scaling map so that if

$$\phi_{n+1} := H_n^{-1} \circ V_n^{-1} \circ T_n \circ s_n, \tag{29}$$

then by Proposition 2.14, we have

$$A_{n+1} = \phi_{n+1}^{-1} \circ A_n^{-1} \circ \Sigma_n^{\bar{a}_1} \circ A_n \circ \phi_{n+1}$$

and

$$B_{n+1} = \phi_{n+1}^{-1} \circ A_n^{-1} \circ \Sigma_n^{\bar{b}_1} \circ A_n \circ \phi_{n+1}.$$

For $k > n$, denote

$$\Phi_n^k := \phi_{n+1} \circ \phi_{n+1} \circ \ldots \circ \phi_{k-1} \circ \phi_k.$$

Notation 3.1 Consider the space $\mathcal{I}$ of finite multi-indexes defined in Notation 2.13. We endow $\mathcal{I}$ with a partial order relation $\prec$ defined as follows:

$$(a_1, a_1, \ldots, a_{2k}, b, c) \prec (a_0, a_1, \ldots, a_{2n}, a_{2n+1}, a_{2n+2})$$

if either $k < n$ and

(1) $b \leq a_{2k+1}$ and $c = 0$, or
(2) $b = a_{2k+1}$ and $c \leq a_{2k+2}$;

or $k = n$ and

(1) $b < a_{2n+1}$ and $c = 0$, or
(2) $b = a_{2n+1}$ and $c < a_{2n+2}$.

Notation 3.2 We denote by

$$p\Sigma_n = (pA_n, pB_n) \quad \text{for} \quad n \in \mathbb{N}$$

the sequence of pairs of iterates of $\Sigma = (A, B)$ defined as follows:

(1) let $p\Sigma_0 := \Sigma$, and
(2) for $n \geq 0$, let

$$p\Sigma_{n+1} = (pA_{n+1}, pB_{n+1}) := (pA_n^{-1} \circ p\Sigma_n^{\bar{a}_1} \circ pA_n, pA_n^{-1} \circ p\Sigma_n^{\bar{b}_1} \circ pA_n).$$

Define multi-indexes $\overline{\alpha}_n$ and $\overline{\beta}_n$ by

$$(\Sigma^{\overline{\alpha}_n}, \Sigma^{\overline{\beta}_n}) := p\Sigma_n.$$

It is easy to see that for $k > n$, we have

$$\Sigma^{\overline{\alpha}_k} = p\Sigma_n^{\overline{\alpha}_{k-n}} \quad \text{and} \quad \Sigma^{\overline{\beta}_k} = p\Sigma_n^{\overline{\beta}_{k-n}}. \tag{30}$$

Lemma 3.3 *Let $k > n$. Then*

$$\mathbf{R}^{k-n}(\Sigma_n) = \left((\Phi_n^k)^{-1} \circ \Sigma_n^{\overline{\alpha}_{k-n}} \circ \Phi_n^k, (\Phi_n^k)^{-1} \circ \Sigma_n^{\overline{\beta}_{k-n}} \circ \Phi_n^k \right).$$

Proof By replacing Σ with Σ_n if necessary, we can assume that $n = 0$. We need to show

$$\mathbf{R}^k(\Sigma) = (A_k, B_k) = \left((\Phi_0^k)^{-1} \circ \Sigma^{\overline{\alpha}_k} \circ \Phi_0^k, (\Phi_0^k)^{-1} \circ \Sigma^{\overline{\beta}_k} \circ \Phi_0^k \right)$$
$$= \left((\Phi_0^k)^{-1} \circ pA_k \circ \Phi_0^k, (\Phi_0^k)^{-1} \circ pB_k \circ \Phi_0^k \right).$$

Clearly, it is true for $k = 1$. Assume it is true for $k \in \mathbb{N}$. For

$$\mathbf{R}^{k+1}(\Sigma) = (A_{k+1}, B_{k+1}),$$

we have

$$A_{k+1} = \phi_{k+1}^{-1} \circ A_k^{-1} \circ \Sigma_k^{\overline{a}_1} \circ A_k \circ \phi_{k+1}$$

and

$$B_{n+1} = \phi_{k+1}^{-1} \circ A_k^{-1} \circ \Sigma_k^{\overline{b}_1} \circ A_k \circ \phi_{k+1}.$$

By the induction hypothesis, we see that

$$A_k^{-1} \circ \Sigma_k^{\overline{a}_1} \circ A_k = (\Phi_0^k)^{-1} \circ (pA_k)^{-1} \circ p\Sigma_k^{\overline{a}_1} \circ pA_k \circ \Phi_0^k = (\Phi_0^k)^{-1} \circ pA_{k+1} \circ \Phi_0^k$$

and

$$B_k^{-1} \circ \Sigma_k^{\overline{b}_1} \circ B_k = (\Phi_0^k)^{-1} \circ (pB_k)^{-1} \circ p\Sigma_k^{\overline{b}_1} \circ pB_k \circ \Phi_0^k = (\Phi_0^k)^{-1} \circ pB_{k+1} \circ \Phi_0^k.$$

The statement follows.   □

For $k > n$, denote

$$\Omega_n^k := \Phi_n^k(\Omega) \quad \text{and} \quad \Gamma_n^k := \Phi_n^k(\Gamma).$$

Define

$$U_n^k := \bigcup_{\overline{\omega} \prec \overline{\alpha}_{k-n}} \Sigma_n^{\overline{\omega}}(\Omega_n^k) \quad \text{and} \quad V_n^k := \bigcup_{\overline{\omega} \prec \overline{\beta}_{k-n}} \Sigma_n^{\overline{\omega}}(\Gamma_n^k).$$

It is not hard to see that $\{U_n^k \cup V_n^k\}_{k=n+1}^{\infty}$ form a nested sequence of open domains in $\mathbb{C}^2$:

$$U_n^{n+1} \cup V_n^{n+1} \Subset U_n^{n+2} \cup V_n^{n+2} \Subset \ldots .$$

Define the *renormalization arc* of Σ_n as

$$\gamma_n := \bigcap_{k=n+1}^{\infty} U_n^k \cup V_n^k. \tag{31}$$

Proposition 3.4 *For any $k > n$, denote*

$$p_n^k := \bigcup_{\overline{\omega} \prec \overline{\alpha}_{k-n}} \Sigma_n^{\overline{\omega}}(\Phi_n^k(\gamma_k \cap \Omega)) \quad and \quad q_n^k := \bigcup_{\overline{\omega} \prec \overline{\beta}_{k-n}} \Sigma_n^{\overline{\omega}}(\Phi_n^k(\gamma_k \cap \Gamma)).$$

Then

$$\gamma_n = p_n^k \cup q_n^k.$$

Proof By replacing Σ with Σ_n if necessary, we can assume that $n = 0$. Let $m > k$. By Lemma 3.3 and (30), for any multi-index $\overline{\omega} \prec \overline{\alpha}_{m-k}$, we have

$$\Phi_0^k \circ \Sigma_k^{\overline{\omega}}(\Omega_k^m) = p\Sigma_k^{\overline{\omega}} \circ \Phi_0^k(\Omega_k^m) = p\Sigma_k^{\overline{\omega}}(\Omega_0^m) = \Sigma^{\overline{\sigma}}(\Omega_0^m)$$

for some $\overline{\sigma} \prec \overline{\alpha}_m$. Similarly, for any multi-index $\overline{\kappa} \prec \overline{\beta}_{m-k}$, we have

$$\Phi_0^k \circ \Sigma_k^{\overline{\kappa}}(\Gamma_k^m) = p\Sigma_k^{\overline{\kappa}} \circ \Phi_0^k(\Gamma_k^m) = p\Sigma_k^{\overline{\kappa}}(\Gamma_0^m) = \Sigma^{\overline{\tau}}(\Gamma_0^m)$$

for some $\overline{\tau} \prec \overline{\beta}_m$. It follows that $\Phi_0^k(\gamma_k) \subset \gamma_0$.

Conversely, let $\overline{\sigma} \prec \overline{\alpha}_m$ and $\overline{\tau} \prec \overline{\beta}_m$. By (30), we see that

$$\Sigma^{\overline{\sigma}}(\Omega_0^m) = \Sigma^{\overline{\sigma}'} \circ p\Sigma_k^{\overline{\omega}}(\Omega_0^m) = \Sigma^{\overline{\sigma}'} \circ \Phi_0^k \circ \Sigma_k^{\overline{\omega}}(\Omega_k^m)$$

and

$$\Sigma^{\overline{\tau}}(\Gamma_0^m) = \Sigma^{\overline{\tau}'} \circ p\Sigma_k^{\overline{\kappa}}(\Gamma_0^m) = \Sigma^{\overline{\tau}'} \circ \Phi_0^k \circ \Sigma_k^{\overline{\kappa}}(\Gamma_k^m),$$

where

(1) $\overline{\omega} \prec \overline{\alpha}_{m-k}$;
(2) $\overline{\kappa} \prec \overline{\beta}_{m-k}$;
(3) $\overline{\sigma}' \prec \overline{\alpha}_k$ and $\Sigma_k^{\overline{\omega}}(\Omega_k^m) \subset \Omega$, or $\overline{\sigma}' \prec \overline{\beta}_k$ and $\Sigma_k^{\overline{\omega}}(\Omega_k^m) \subset \Gamma$; and
(4) $\overline{\tau}' \prec \overline{\alpha}_k$ and $\Sigma_k^{\overline{\kappa}}(\Gamma_k^m) \subset \Omega$, or $\overline{\tau}' \prec \overline{\beta}_k$ and $\Sigma_k^{\overline{\kappa}}(\Gamma_k^m) \subset \Gamma$.

The result follows. $\qquad\qquad\qquad\qquad\qquad\qquad\qquad\qquad\qquad\qquad\qquad\qquad\qquad$ $\square$

Let $\theta_* = (\sqrt{5} - 1)/2$ be the golden mean rotation number, and let

$$I_L := [-\theta_*, 0] \quad \text{and} \quad I_R := [0, 1].$$

Define $L : I_L \to \mathbb{R}$ and $R : I_R \to \mathbb{R}$ as

$$L(t) := t + 1 \quad \text{and} \quad R(t) := t - \theta_*.$$

The pair (R, L) represents rigid rotation of $\mathbb{R}/\mathbb{Z}$ by angle θ_*.

The following is a classical result about the renormalization of 1D pairs.

Proposition 3.5 *Suppose* $\|\Sigma\|_y = 0$. *Then for every* $n \geq 0$, *there exists a quasi-symmetric homeomorphism between* $I_L \cup I_R$ *and the renormalization arc* γ_n *that conjugates the action of* $\Sigma_n = (A_n, B_n)$ *and the action of* (R, L). *Moreover, the renormalization arc* γ_n *contains the unique critical point* $c_n = 0$ *of* η_n.

The following is shown in [12].

Theorem 3.6 *Let* $\Sigma = (A, B)$ *be a commuting pair contained in the stable manifold* $W^s(\iota(\zeta_*))$ *of the 2D renormalization fixed point* $\iota(\zeta_*)$. *Then for every* $n \geq 0$, *there exists a homeomorphism between* $I_L \cup I_R$ *and the renormalization arc* γ_n *that conjugates the action of* $\Sigma_n = (A_n, B_n)$ *and the action of* (R, L). *Moreover, this conjugacy cannot be* C^1 *smooth.*

Theorem 1.2 follows from the above statement and the following:

Theorem 3.7 ([12]) *Suppose*

$$\Sigma = \Sigma_{H_{\mu_*, \nu}},$$

where $\Sigma_{H_{\mu_*, \nu}}$ *is the renormalization of the Hénon map given in Theorem 2.16. Then the linear rescaling of the renormalization arc* $s_0(\gamma_0)$ *is contained in the boundary of the Siegel disc* Δ *of* $H_{\mu_*, \nu}$. *In fact, we have*

$$\partial \Delta = s_0(\gamma_0) \cup H_{\mu_*, \nu} \circ s_0(\gamma_0).$$

Henceforth, we consider the renormalization arc of Σ_n as a continuous curve $\gamma_n = \gamma_n(t)$ parameterized by $I_L \cup I_R$. The components of γ_n are denoted

$$\gamma_n(t) = \begin{bmatrix} \gamma_n^x(t) \\ \gamma_n^y(t) \end{bmatrix}.$$

Lastly, denote the renormalization arc of $\iota(\zeta_*)$ by

$$\gamma_*(t) = \begin{bmatrix} \gamma_*^x(t) \\ \gamma_*^y(t) \end{bmatrix}.$$

The following are consequences of Theorem 2.15.

Corollary 3.8 *As* $n \to \infty$, *we have the following convergences (each of which occurs at a geometric rate):*

(1) $\eta_n \to \eta_*$,

(2) $\lambda_n \to \lambda_*$ (hence $s_n \to s_*$),

(3) $\phi_n \to \psi_*$, where

$$\psi_*(x, y) = \begin{bmatrix} \eta_*^{-1}(\lambda_* x) \\ \eta_*^{-1}(\lambda_* y) \end{bmatrix}, \quad and$$

(4) $\gamma_n \to \gamma_*$ (hence $|\gamma_n^x(0)| \to 0$).

3.2 Normality of the Compositions of Scope Maps

Define

$$\psi_{n+1}(x, y) := \begin{bmatrix} \eta_n^{-1}(\lambda_n x) \\ \eta_n^{-1}(\lambda_n y) \end{bmatrix}.$$

For $k > n$, denote

$$\Psi_n^k := \psi_{n+1} \circ \psi_{n+2} \circ \ldots \circ \psi_{k-1} \circ \psi_k.$$

Let

$$\begin{bmatrix} \sigma_n^k & 0 \\ 0 & \sigma_n^k \end{bmatrix} := (D_{(0,0)} \Psi_n^k)^{-1}.$$

Proposition 3.9 *There exists a domain $U \subset \mathbb{C}^2$ that contains $\Omega \cup A_k(\Omega) \cup \Gamma \cup B_k(\Gamma)$ for all k sufficiently large such that the family $\{\sigma_n^k \Psi_n^k\}_{k=n}^{\infty}$ is normal on U.*

Proof By Corollary 3.8, there exists a domain $U \subset \mathbb{C}^2$ and a uniform constant $\lambda < 1$ such that for all k sufficiently large, the map ψ_k is well defined on U, and

$$\Omega \cup A_k(\Omega) \cup \Gamma \cup B_k(\Gamma) \Subset \lambda U.$$

Thus, by choosing a smaller domain U if necessary, we can assume that ψ_k and hence, Ψ_n^k extends to a strictly larger domain $V \ni U$. It follows from applying Koebe distortion theorem to the first and second coordinate that $\{\sigma_n^k \Psi_n^k\}_{k=n}^{\infty}$ is a normal family on U. □

Proposition 3.10 *There exists a uniform constant $M > 0$ such that*

$$\|\phi_n - \psi_n\| < M \epsilon^{2^{n-1}}.$$

Proof The result follows readily from (27) and (28). □

Proposition 3.11 *There exists a uniform constant $K > 0$ such that*

$$\sigma_n^k \|\Phi_n^k - \Psi_n^k\| < K \epsilon^{2^n}.$$

Proof By Proposition 3.10, we have

$$\phi_{k-1} = \psi_{k-1} + \tilde{E}_{k-1} \quad \text{and} \quad \phi_k = \psi_k + E_k,$$

where $||\tilde{E}_{k-1}|| < M\epsilon^{2^{k-2}}$ and $||E_k|| < M\epsilon^{2^{k-1}}$. Observe that

$$\begin{aligned}
\phi_{k-1} \circ \phi_k &= \phi_{k-1} \circ (\psi_k + E_k) \\
&= \phi_{k-1} \circ \psi_k + \bar{E}_k \\
&= (\psi_{k-1} + \tilde{E}_{k-1}) \circ \psi_k + \bar{E}_k \\
&= \psi_{k-1} \circ \psi_k + \tilde{E}_{k-1} \circ \psi_k + \bar{E}_k,
\end{aligned}$$

where $||\bar{E}_k|| < L\epsilon^{2^{k-1}}$ for some uniform constant $L > 0$ by Corollary 3.8. Let

$$E_{k-1} := \tilde{E}_{k-1} + \bar{E}_k \circ \psi_k^{-1}.$$

By Corollary 3.8, ψ_k^{-1} is uniformly bounded, and hence, we have

$$||E_{k-1}|| < M\epsilon^{2^{k-2}} + 2L\epsilon^{2^{k-1}} < 2M\epsilon^{2^{k-2}}.$$

Thus, we have

$$\phi_{k-1} \circ \phi_k = \psi_{k-1} \circ \psi_k + E_{k-1} \circ \psi_k.$$

Proceeding by induction, we obtain

$$\Phi_n^k = \Psi_n^k + E_{n+1} \circ \Psi_{n+1}^k,$$

where

$$||E_{n+1}|| < 2M\epsilon^{2^n}.$$

By definition, we have

$$\sigma_n^k D_{(0,0)} \Psi_n^k = \text{Id}.$$

Factor the scaling constant as

$$\sigma_n^k := \tilde{\sigma}_n^k \sigma_{n+1}^k,$$

so that

$$\tilde{\sigma}_n^k D_{\Psi_{n+1}^k(0,0)} \psi_{n+1} = \text{Id},$$

and

$$\sigma_{n+1}^k D_{(0,0)} \Psi_{n+1}^k = \text{Id}.$$

Let

$$C := \sup_{x \in Z} \eta_*'(x).$$

Observe that $\tilde{\sigma}_n^k$ is uniformly bounded by $\lambda_n^{-1} C$. Moreover, by Proposition 3.9, we have that $\|\sigma_{n+1}^k D_{(x,y)} \Psi_{n+1}^k\|$ is also uniformly bounded. Therefore,

$$\|\sigma_n^k D_{(x,y)}(E_{n+1} \circ \Psi_{n+1}^k)\| = \|\tilde{\sigma}_n^k D_{\Psi_{n+1}^k(x,y)} E_{n+1}\| \cdot \|\sigma_{n+1}^k D_{(x,y)} \Psi_{n+1}^k\|$$

$$= K \sup_{(x,y)} \|D_{(x,y)} E_{n+1}\|$$

$$< K \epsilon^{2^n}$$

for some universal constant $K > 0$. $\square$

By Propositions 3.9 and 3.11, we have the following theorem.

Theorem 3.12 *There exists a domain $U \subset \mathbb{C}^2$ that contains $\Omega \cup A_k(\Omega) \cup \Gamma \cup B_k(\Gamma)$ for all k sufficiently large such that the family $\{\sigma_n^k \Phi_n^k\}_{k=n}^{\infty}$ is normal on U.*

3.3 The Boundary of the Siegel Disk Is Not Smooth

Let $[t_l, t_r] \subset \mathbb{R}$ be a closed interval, and let $C : [t_l, t_r] \to \mathbb{C}$ be a smooth curve. For any subset $N \subset \mathbb{C}$ intersecting the curve C, we define the *angular deviation of C on N* as

$$\Delta_{\arg}(C, N) := \sup_{t,s \in C^{-1}(N)} |\arg(C'(t)) - \arg(C'(s))|, \tag{32}$$

where the function arg: $\mathbb{C} \to \mathbb{R}/\mathbb{Z}$ is defined as

$$\arg(re^{2\pi\theta i}) := \theta. \tag{33}$$

Lemma 3.13 *Let $\theta \in \mathbb{R}/\mathbb{Z}$, and let $C_\theta : [0, 1] \to \mathbb{C}$ be a smooth curve such that $C_\theta(0) = 0$ and $C_\theta(1) = e^{2\pi\theta i}$. Then for some $t \in [0, 1]$, we have*

$$\arg(C_\theta'(t)) = \theta.$$

Lemma 3.14 *Let*

$$q_2(x) := x^2 \quad \text{and} \quad A_r^R := \{z \in \mathbb{C} \,|\, r < |z| < R\}. \tag{34}$$

Suppose $C : [t_l, t_r] \to \mathbb{D}_R$ is a smooth curve such that $|C(t_l)| = |C(t_r)| = R$, and $|C(t_0)| < r$ for some $t_0 \in [t_l, t_r]$. Then for every $\delta > 0$, there exists $M > 0$ such that if $\mathrm{mod}(A_r^R) > M$, then either $\Delta_{\arg}(C, \mathbb{D}_R)$ or $\Delta_{\arg}(q_2 \circ C, \mathbb{D}_{R^2})$ is greater than $1/6 - \delta$.

Proof Without loss of generality, assume that $R = 1$, and $C(t_r) = 1$. We prove the case when $r = 0$, so that $C(t_0) = 0$. The general case follows by continuity.

Suppose that $\Delta_{\arg}(C, \mathbb{D}_R) < 1/6$. Then by Lemma 3.13, we have

$$1/3 < \arg(C(t_l)) < 2/3.$$

This implies that

$$-1/3 < 2\arg(C(t_l)) < 1/3.$$

Hence, by Lemma 3.13, we have $\Delta_{\arg}(q_2 \circ C, \mathbb{D}_{R^2}) > 1/6.$ $\qquad\square$

Corollary 3.15 *Let $W \subset \mathbb{C}$ be a simply connected neighbourhood of 0, let $C :$ $[t_l, t_r] \to \overline{W}$ and $E : [t_l, t_r] \to \mathbb{C}$ be smooth curves, and let $f : W \to \mathbb{C}$ be a holomorphic function with a unique simple critical point at $c \in \mathbb{D}_r$ for $r < 1$. Consider the smooth curve*

$$\tilde{C} := f \circ C + E.$$

Suppose $C(t_l), C(t_r) \in \partial W$, and $|C(t_0)| < r$ for some $t_0 \in [t_l, t_r]$. Then for every $\delta > 0$, there exists $\rho > 0$ and $M > 0$ such that if $\|E\| < \rho$ and $\mathrm{mod}(W \setminus \mathbb{D}_r) > M$, then either $\Delta_{\arg}(C, W)$ or $\Delta_{\arg}(\tilde{C}, f(W))$ is greater than $1/6 - \delta$.

Let $U \subset Z \subset \mathbb{C}$ be a simply-connected domain containing the origin. For all k sufficiently large, the unique critical point c_k of η_k is contained in U. Let $V_k := \eta_k(U)$. Then there exists conformal maps $u_k : (\mathbb{D}, 0) \to (U, c_k)$ and $v_k : (\mathbb{D}, 0) \to (V_k, \eta_k(c_k))$ such that the following diagram commutes:

$$
\begin{array}{ccc}
\mathbb{D} & \xrightarrow{\;u_k\;} & U \\
{\scriptstyle q_2}\downarrow & & \downarrow{\scriptstyle \eta_k} \\
\mathbb{D} & \xrightarrow{\;v_k\;} & V_k
\end{array}
$$

By Corollary 3.8, we have the following result:

Proposition 3.16 *The maps $u_k : (\mathbb{D}, 0) \to (U, c_k)$ and $v_k : (\mathbb{D}, 0) \to (V_k, \eta_k(c_k))$ converge to conformal maps $u_* : (\mathbb{D}, 0) \to (U, 0)$ and $v_* : (\mathbb{D}, 0) \to (\eta_*(U), \eta_*(0))$. Moreover, the following diagram commutes:*

$$
\begin{array}{ccc}
\mathbb{D} & \xrightarrow{\;u_*\;} & U \\
{\scriptstyle q_2}\downarrow & & \downarrow{\scriptstyle \eta_*} \\
\mathbb{D} & \xrightarrow{\;v_*\;} & \eta_*(U)
\end{array}
$$

Proof of Non-smoothness By Theorem 3.12, the sequence $\{\sigma_0^k \Phi_0^k\}_{k=0}^\infty$ has a converging subsequence. By replacing the sequence by this subsequence if necessary, assume that $\{\sigma_0^k \Phi_0^k\}_{k=0}^\infty$ converges. Consider the following commutative diagrams:

$$
\begin{array}{ccc}
\mathbb{D} \xrightarrow{\;u_k\;} U & \qquad & \Omega \xrightarrow{\;\Phi_0^k\;} \Omega \\
\Big\downarrow{\scriptstyle q_2} \qquad \Big\downarrow{\scriptstyle \eta_k} & \text{and} & \Big\downarrow{\scriptstyle A_k} \qquad \Big\downarrow{\scriptstyle A_0}\, . \\
\mathbb{D} \xrightarrow{\;v_k\;} V_k & \qquad & A_k(\Omega) \xrightarrow{\;\Phi_0^k\;} A_0(\Omega)
\end{array}
$$

Since $\{\sigma_0^k \Phi_0^k\}_{k=0}^\infty$ converges, we can choose $R > 0$ sufficiently small so that if

$$
X_k := u_k(\mathbb{D}_R) \subset U, \quad \text{and} \quad Y_k := v_k(\mathbb{D}_{R^2}) \subset V_k,
$$

then for any smooth curves $C_1 \subset \Omega := U \times U$ and $C_2 \subset A_k(\Omega)$ intersecting $X_k \times X_k$ and $Y_k \times Y_k$ respectively, we have

$$
\kappa \Delta_{\arg}(C_1, X_k \times X_k) < \Delta_{\arg}(\Phi_0^k \circ C_1, \Phi_0^k(X_k \times X_k))
$$

and

$$
\kappa \Delta_{\arg}(C_2, Y_k \times Y_k) < \Delta_{\arg}(\Phi_0^k \circ C_2, \Phi_0^k(Y_k \times Y_k))
$$

for some uniform constant $\kappa > 0$.

Write

$$
A_k(x, y) = \begin{bmatrix} a_k(x, y) \\ h_k(x, y) \end{bmatrix} = \iota(\eta_k)(x, y) + E_k(x, y) = \begin{bmatrix} \eta_k(x) + e_x(x, y) \\ \eta_k(x) + e_y(x, y) \end{bmatrix}.
$$

By (28), we have

$$
\|E_k\| \to 0 \quad \text{as} \quad k \to \infty.
$$

Consider the renormalization arc of Σ_n:

$$
\gamma_n(t) = \begin{bmatrix} \gamma_n^x(t) \\ \gamma_n^y(t) \end{bmatrix}.
$$

Recall that we have

$$
\gamma_n^x, \gamma_n^y \to \gamma_* \quad \text{as} \quad n \to \infty,
$$

where γ_* is the renormalization arc of the 1D renormalization fixed point ζ_*.

Let $0 < \delta < 1/6$. Choose $r > 0$ is sufficiently small so that the annulus $X_k \setminus \mathbb{D}_r$ satisfies the condition of Corollary 3.15. Next, choose K sufficiently large so that for all $k > K$, we have

$$
|c_k|, |\gamma_k^x(0)| < r \quad \text{and} \quad \|E_k\| < \rho,
$$

where $\rho > 0$ is given in Corollary 3.15.

Now, suppose towards a contradiction that the renormalization arc γ_0 of Σ_0, and hence the renormalization arc γ_k of Σ_k for all $k \geq 0$, are smooth. By the above estimates, we can conclude:

$$
\begin{aligned}
\Delta_{\arg}(\gamma_0, \Phi_0^k(X_k \times X_k)) &= \Delta_{\arg}(\Phi_0^k \circ \gamma_k, \Phi_0^k(X_k \times X_k)) \\
&> \kappa \Delta_{\arg}(\gamma_k, X_k \times X_k) \\
&> \kappa \Delta_{\arg}(\gamma_k^x, X_k)
\end{aligned}
$$

and

$$
\begin{aligned}
\Delta_{\arg}(\gamma_0, \Phi_0^k(Y_k \times Y_k)) &= \Delta_{\arg}(\Phi_0^k \circ \gamma_k, \Phi_0^k(Y_k \times Y_k)) \\
&> \kappa \Delta_{\arg}(\gamma_k, Y_k \times Y_k) \\
&= \kappa \Delta_{\arg}(A_k \circ \gamma_k, Y_k \times Y_k) \\
&> \kappa \Delta_{\arg}(a_k \circ \gamma_k, Y_k) \\
&= \kappa \Delta_{\arg}(\eta_k \circ \gamma_k^x + e_x(\gamma_k), Y_k).
\end{aligned}
$$

By Lemma 3.15, either $\Delta_{\arg}(\gamma_k^x, X_k)$ or $\Delta_{\arg}(a_k \circ \gamma_k, Y_k)$ is greater than $1/6 - \delta > 0$. Hence,

$$
\max\{\Delta_{\arg}(\gamma_0, \Phi_0^k(X_k \times X_k)), \Delta_{\arg}(\gamma_0, \Phi_0^k(Y_k \times Y_k))\} > l
$$

for some uniform constant $l > 0$. Since $\Phi_0^k(X_k \times X_k)$ and $\Phi_0^k(Y_k \times Y_k)$ both converge to a point in γ_0 as $k \to \infty$, this is a contradiction. $\qquad \square$

References

1. A. Avila, X. Buff, A. Chéritat, Siegel disks with smooth boundaries. Acta Math. **193**(1), 130 (2004)
2. E. Bedford, K. Kim, No smooth Julia sets for polynomial diffeomorphisms of $\mathbb{C}^2$ with positive entropy. J. Geom. Anal. (to appear)
3. E. Bedford, J. Smillie, Polynomial diffeomorphisms of $\mathbb{C}^2$: currents, equilibrium measure and hyperbolicity. Invent. Math. **103**(1), 69–99 (1991)
4. E. Bedford, J. Smillie, Polynomial diffeomorphisms of $\mathbb{C}^2$ II: stable manifolds and recurrence. J. Am. Math. Soc. **4**(4), 657–679 (1991)
5. A. Burbanks, Renormalization for Siegel disks. Ph.D. Thesis, Loughborough University, 1997
6. A. de Carvalho, M. Lyubich, M. Martens, Renormalization in the Hénon family, I: universality but non-rigidity. J. Stat. Phys. **121**(5/6), 611–669 (2006)
7. A. Douady, Disques de Siegel et anneaux de Herman, vol. 1986/87. Séminaire Bourbaki, Astérisque **152–153** (1987), 4, 151–172
8. E. de Faria, W. de Melo, Rigidity of critical circle mappings I. J. Eur. Math. Soc. **1**, 339–392 (1999)
9. D. Gaidashev, M. Yampolsky, Cylinder renormalization of Siegel disks. Exp. Math. **16**, 215–226 (2007)
10. D. Gaidashev, M. Yampolsky, Golden mean Siegel disk universality and renormalization. arXiv:1604.00717

11. D. Gaidashev, M. Yampolsky, Renormalization of almost commuting pairs. arXiv:1604.00719
12. D. Gaidashev, R. Radu, M. Yampolsky, Renormalization and Siegel disks for complex Henon maps. ArXiv:1604.07069
13. J.H. Hubbard, R.W. Oberste-Vorth, Hénon mappings in the complex domain I: the global topology of dynamical space. Pub. Math. IHES **79**, 5–46 (1994)
14. C. McMullen, Self-similarity of Siegel disks and Hausdorff dimension of Julia sets. Acta Math. **180**, 247–292 (1998)
15. S. Morosawa, Y. Nishimura, M. Taniguchi, T. Ueda, *Holomorphic Dynamics*, vol. 66. Cambridge Studies in Advanced Mathematics (Cambridge University Press, Cambridge, 2000)
16. G. Osipenko, *Center Manifolds, Encyclopedia of Complexity and Systems* (Science, 2009), pp. 936–951
17. F. Przytycki, M. Urbański, A. Zdunik, Harmonic, Gibbs and Hausdorff measures on repellers for holomorphic maps. II. Studia Math. **97**(3), 189–225 (1991)
18. M. Shub, *Global Stability of Dynamical Systems* (Springer, Heidelberg, 1987)
19. A. Stirnemann, Existence of the Siegel disc renormalization fixed point. Nonlinearity **7**(3), 959–974 (1994)
20. S. Warschawski, On differentiability at the boundary in conformal mapping. Proc. Am. Math. Soc. **12**, 614–620 (1961)
21. M. Yampolsky, Siegel disks and renormalization fixed points. Fields Inst. Commun. **53** (2008)

Appendix A
Some Wellington's Photos From Childhood to Academic Life

See Figs. A.1, A.2, A.3, A.4, A.5, A.6, A.7, A.8.

© Springer Nature Switzerland AG 2019

M. J. Pacifico and P. Guarino (eds.), *New Trends in One-Dimensional Dynamics*, Springer Proceedings in Mathematics & Statistics 285, https://doi.org/10.1007/978-3-030-16833-9

Fig. A.1 1., 3. and 5. : As a child, 2. with his parents and wife Gilza; 4. with his parents and sisters; 6. with his parents; 7. with his mother, grandmother and sister; 8. with his mother and sisters; 9. with his sisters; 10. with his sister

Fig. A.2 Sailing his boat Doisdu and dancing with his wife Gilza

Fig. A.3 1, 2, 3 and 4.: with Steve Smale, Jacob Palis, Mauricio Peixoto and formal students Maria Jos Pacifico, Antonio Gaspar Ruas, Edson Vargas, Daniel Smania, Alejandro Kocsard, Artur Avila and Pablo Guarino

Fig. A.4 with: 1. M. J. Pacifico, his brother Cleber de Melo, L. J. Díaz, D. Smania; 2. M. Peixoto, 3. J. Palis; 4. A. Kocsard, M. J. Pacifico, S. Van Strien; 5. A. Verjovsky; 6. Y. Lequain, A. M. Doering, M. J. Carneiro, J-M. Gambaudo; 8. S. Smale, S. Newhouse, M. J. Pacifico

Fig. A.5 with: 1. M. J. Pacifico, Gilza; 2. E. de Faria, Gilza, Alcilea Augusto, M. Peixoto, Benar Svaiter; 3. A. Katok, L. F. da Rocha, E. Zehnder, C. I. Doering; 4. F. Takens; 5. Y. Lequain, A. M. Doering; 6. J. Palis, Gilza; 7. J. Palis, S. Smale

Fig. A.6 with: 1. Sabrina, Gilza de Melo, Artur Avila; 2. Marcelo Viana, Konstantin Khanin, Artur Avila; 3. Dennis Sullivan; 4. Jean-Christophe Yoccoz, Yakov Sinai; 5. Sheldon Newhouse, Alberto Pinto, Sebastian Van Strien; 6. Gilza de Melo; 7. Suely Lima, Gilza de Melo, Konstantin Khanin, Marcelo Viana, Maria José Pacifico, Artur Avila; 8. Mikhail Lyubich; 9. Edson de Faria, Daniel Smania, Jorge Rocha, Roberto Markarian, José Alves, Gilza de Melo, J-M. Gambaudo; 10. Gilza de Melo

Fig. A.7 with: 1. Artur Avila; 2. Gilza de Melo; 3. Celebrating the 70 years; 4. Blowing the 70's candles; 5. Opening table of the meeting: Maria José Pacifico, Marcelo Viana, Steve Smale, Mauricio Peixoto, Welington de Melo and Jacob Palis

Fig. A.8 Group picture of the meeting "New trends on 1-dimensional dynamics" celebrating the 70 anniversary of Wellington de Melo